AF766131

High Performance Polymers - Polyimides Based - From Chemistry to Applications

High Performance Polymers - Polyimides Based - From Chemistry to Applications

Editor

Alano Riley

scitus academics

High Performance Polymers - Polyimides Based - From Chemistry to Applications
Edited by **Alano Riley**

ISBN: 978-1-68117-469-3
Library of Congress Control Number: 2016942917

© 2017 by
SCITUS Academics LLC,
www.scitusacademics.com
Box No. 4766, 616 Corporate Way,
Suite 2, Valley Cottage,
NY 10989

This book contains information obtained from highly regarded resources. Copyright for individual articles remains with the authors as indicated. All chapters are distributed under the terms of the Creative Commons Attribution License, which permits unrestricted use, distribution, and reproduction in any medium, provided the original author and source are credited.

Notice

Reasonable efforts have been made to publish reliable data and views articulated in the chapters are those of the individual contributors, and not necessarily those of the editors or publishers. Editors or publishers are not responsible for the accuracy of the information in the published chapters or consequences of their use. The publisher believes no responsibility for any damage or grievance to the persons or property arising out of the use of any materials, instructions, methods or thoughts in the book. The editors and the publisher have attempted to trace the copyright holders of all material reproduced in this publication and apologize to copyright holders if permission has not been obtained. If any copyright holder has not been acknowledged, please write to us so we may rectify.

Printed in United States of America on Acid Free Paper

Preface

Polymers are substances whose molecules have high molar masses and consist of many repeating units. Natural and synthetic polymers are available. Protein, cellulose, starch and latex are among the natural polymers. Synthetic polymers are commercially produced on a very large scale and also have a wide range of applications and properties. High performance plastics differ primarily from standard engineering and plastics because of their temperature stability, but also because of their chemical stability and physical properties, quantity of production and price. There are many synonyms for high-performance plastics, including: high-temperature plastics, high-performance polymers, high-performance thermoplastics or possibly high-tech plastics. We use high-performance polymers in almost every aspect of our lives from relatively obvious areas such as automotive, aerospace, many parts and the defense of the energy industry to areas where you might not immediately think of such applications as electronics, medical, sports and food. Without these high-performance plastics, composites and elastomers, it would simply not be technically feasible or possibly economically viable. In most applications, the problems endured by these materials in terms of chemical environment, mechanical stress, stress, pressure or temperature must therefore be carefully selected to balance the required properties, durability and performance with costs.

Focusing on synthesis, physio-chemical properties, ultimate properties, mechanisms, applications and processing of such high-end materials required in advanced technologies from chemistry to applications. This guide should be of enormous value to material scientists, polymer chemists, electronic component manufacturers and process engineers.

Table of Contents

1

CHAPTER

High Dielectric Performance of Polyamide 66/Poly(Vinylidene Fluoride) Flexible Blends Induced by Interfacial Copolymer for Capacitors

Rui Li, Zixuan Chen and Jianzhong Pei*

Highway School, Chang'an University, Xi'an 710064, China

ABSTRACT

The copolymer VAMA was synthesized from vinyl acetic and maleic anhydride. A new all-polymeric blend with a high dielectric constant (ε) has been developed by blending polyvinylidene fluoride (PVDF) with vinyl acetic-maleic anhydride modified polyamide (PA66-*g*-VM). The blend shows high dielectric constants ($\varepsilon_{blend} = 20$) and excellent mechanical properties. The SEM investigations suggest that the enhanced dielectric behavior originates from significant interfacial interactions between polymers. The XRD demonstrates that the compatibilizer affects the crystalline behavior of each component. Furthermore, the stable dielectric constants of the all-polymeric blends can be tuned by adjusting the content of the compatibilizer. The created high-ε all-polymeric blends represent a novel type of material that is technologically simple, easy to process, and of a relatively high dielectric constant, with application for flexible electronics.

Keywords: compatilbilizer; blend; dielectric property; interaction; all-polymer

1. INTRODUCTION

With the continuous development of science and technology, requirements become higher to the energy efficiency of materials, and it is hard for single polymer to meet requirements [1,2]. The polarization mechanism of polymer tells us that the structure, crystal form, and crystalline of polymer have great influence on the dielectric constants of polymer. Thus, two or more polymers blended to prepare polymer blending material make it possible to develop polymer film energy storage material. Polymer-based blend is combined with the respective advantages of two polymer materials. It has the advantages of processing convenience, fine flexibility, and the ability to take any shape [3,4]. It has extensive application prospects to blend different polymer matrixes in electronic systems. Especially, its applicability in preparing electrical appliance materials which can be used in modern electronic industry, such as the preparation of embedded capacitors, *etc.* [5,6,7].

According to the property index requirements of capacitor material, the paper selected two typical polar polymers which are polyvinylidene fluoride (PVDF) and polyamide (PA) to blend. In the previous study, the mechanical-electrical properties and improvement method were reported about PA11/PVDF blend system [8,9,10]. However, in the actual application, the expensive price of PA11 limited applications in many fields. As a nylon material which is frequently used, PA66 and its inherent excellent properties were focused on for a long time. However, there are still few studies about PA66/PVDF blend energy storage materials. As we know, the melting point of PA66 is 220–240 °C, but the processing temperature of PVDF is 160–180 °C. The melting point of PA66 was lowered by grafting process, thus the mechanical melt blending problem of PVDF and PA66 can be solved. In this chapter, the copolymer VM from acetic acid vinyl ester and maleic anhydride were synthesized. Next, the grafted copolymer VM-*g*-PA66 from PA66 and VM were prepared by melt blending graft. Grafting reaction can lower the melting point of PA66, and the blends of both can be obtained from VM-*g*-PA66 and PVDF by melt blending through torque rheometer, studying the mechanical properties and morphology of blends. Thus, this paper is going to test microstructure characterization on the electrical properties so as to discuss the relationship of polyamide structure and electrical properties from the changes of molecular chain structure and the position of amide groups to further improve the electrical properties of material.

2. EXPERIMENTAL

2.1. Materials

PVDF (T_m = 175 °C, ρ = 1.75 g/cm^3) and PA66 (T_m = 225 °C, ρ = 1.15 g/cm^3) powders were supplied by the Shanghai 3F Corporation (Shanghai, China) and the Sinopec Baling Branch Corporation (Changsha, China), respectively. The related products are commercially available.

2.2. Synthesis of Copolymer of Vinyl Acetate and Maleic Anhydride

Methylbenzene (200 mL) was added to a three-mouth flask equipped with a stirrer, thermometer, and reflux condensing tube, and then 0.10 mol powdered maleic anhydride was added and dissolved into methylbenzene slowly after stirring and heating. After the solution turns to be transparent, 0.10 mol acetic acid vinyl ester and 1.4% initiator azodiisobutyronitrile were added. The temperature was controlled at 80 °C. After the copolymer subsided, the reactant was obtained by cooling. After filtering and washing with dichloroethane five times, the unreacted maleic anhydride and initiator were removed. After vacuum drying under the circumstance of 60 °C, white powdered copolymer was obtained. This paper abbreviates the copolymerization of VAc and MAH to VM from now on.

2.3. Preparation of VM-g-PA66 and VM-g-PA66/PVDF

PA66 was dried for 24 h at 100 °C in drying oven. VM copolymer was placed for 8 h with vacuum drying at 60 °C. The PA66 was added to torque rheometer with 220 °C. The prepared and dried VM copolymer was added after PA66 was melted completely. After mixing the blends for a while, the material can be taken out and cooled to have standby application. See the grafting reaction process of PA66 and VM in Figure 1. PA66, VM-*g*-PA66 and PVDF were mixed for 20 min based on different volume ratios in an XSS-300 torque rheometer (Shanghai Kechuang Rubber Plastic Machinery Co., Ltd, Shanghai, China), and then a press vulcanizer was used to press the evenly pressed VM-*g*-PA66/PVDF blends into thin a blended sheet which was 2 mm thick. In addition, a micro tablet machine was used to press sheets into around 1 millimeter of thickness. Finally, the samples were polished and silvered with silver paint.

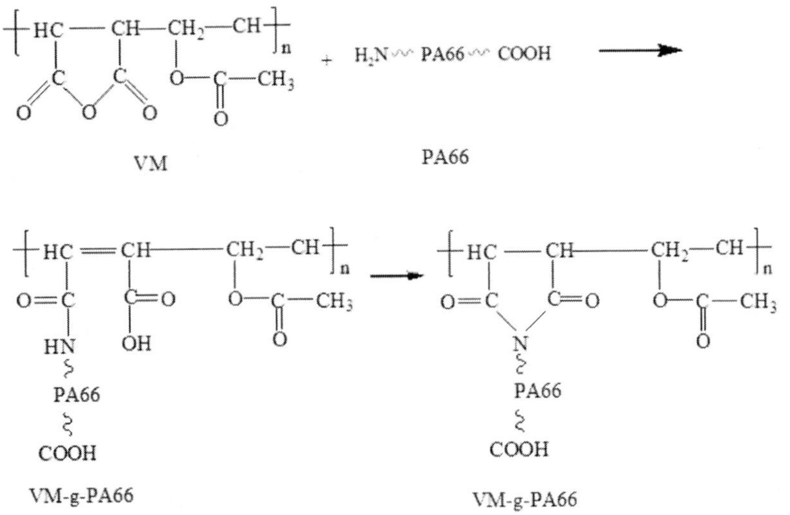

Figure 1. Reaction scheme of PA66 and VM copolymerization.

2.4. Instrumental Analysis

The dielectric properties of the samples were performed by HIOKI3532-50 LCR, the frequency was from 10^2 to 5×10^6 Hz at room temperature. The microstructures of the blends were characterized by using X-ray diffraction (XRD, Japanese Rigaku Company, Tokyo, Japan), differential scanning calorimetry (DSC, PerkinElmer Apparatus Co., Ltd, Berlin, German), Scanning Electron Microscopy (SEM, Japanese Electronics Corporation, Tokyo, Japan), and Fourier Transform Infrared Spectroscopy (FTIR, German Bruker Company, Billerica, MA, USA). Dynamic mechanical properties were measured using a dynamic mechanical thermal analyzer (DMTA, Shenzhen Xinsansi Electronics Corporation, Shenzhen, China) at 1 Hz and at a heating rate of 3 °C/min from −100 to 150 °C.

3. RESULTS AND DISCUSSIONS

3.1. Structure Analysis

The infrared spectrum of acetic acid vinyl ester and maleic anhydride copolymer is tested by Fourier transformation infrared spectrometric analyzer. It can be seen from the Figure 2 that the absorption peak of wave number is 1780 cm^{-1} for carboxylic acid C=O stretching vibration and 1231 cm^{-1} for C–O stretching vibration; they are the characteristic structures of acid anhydride. The wave number is 1780 cm^{-1} for carboxyl ester C=O stretching vibration and 1374 cm^{-1} for methyl groups C–H flexural vibration. The intensity of the absorption band in wave number 1374 cm^{-1} is greater than in 1433 cm^{-1}, which is an obvious characteristic of vinyl acetate structure [11,12]. Thereby, it proves the reaction of copolymerization between acetic acid vinyl ester and maleic anhydride. See the reaction equation of VM in Figure 2.

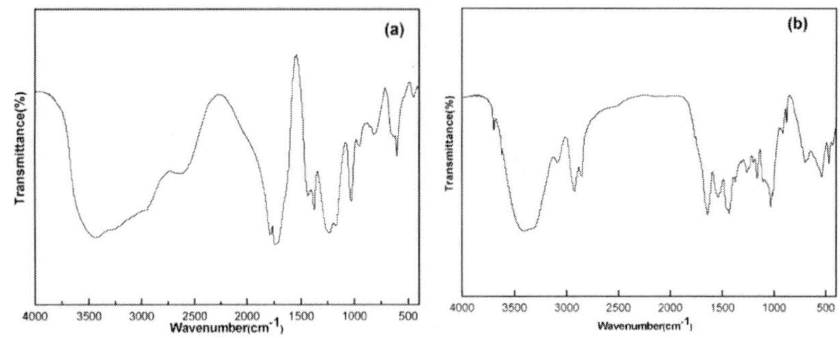

Figure 2. FT-IR spectrum of copolymer (**a**) VM; and (**b**) VM-*g*-PA66.

See the infrared spectroscopic analysis of grafted copolymer in Figure 2b. The peaks of 1650 and 1540 cm^{-1} are corresponding to the absorption peak of C=O and the absorption peak of N–H and C–N of amide. There was a large and wide peak envelope in the range of 2700–3600 cm^{-1}. This is the characteristic peak of the carboxyl group. However, 1433 and 1374 cm^{-1} are the characteristic absorption peaks of vinyl acetate, which proves that PA66 and VM can react to grafted copolymer by melt blending.

Figure 3a is the FT-IR spectrum of PA66/PVDF blends. It can be seen from the Figure that PA66 has a significant absorption peak at 3301 cm^{-1}, which is corresponding to the stretching vibration of N–H. The strong amide I band at 1640 cm^{-1} is produced by the C=O stretching vibration of amide group. The characteristic absorption peaks of amide at 1555 and 1405 cm^{-1} were produced by C–N stretching vibration and N–H flexural vibration. For PA66 and PVDF, FT-IR spectrum of PA66/PVDF blend film has no new peak or significant deviation. It means there is no bulking agent, PA and PVDF cannot have chemical reaction by melt blending.

Figure 3b is the infrared spectrum of PA66/PVDF and VM-*g*-PA66/PVDF. When the infrared spectral curve of VM-*g*-PA66/PVDF was compared with that of VM, it can be found that the two acid anhydride peaks disappear at 1780 and 1740 cm^{-1}. When the infrared spectral curve of VM-*g*-PA66/PVDF is compared with it of VM, it can be found that VM-*g*-PA66/PVDF characteristic peak of cyclic amide disappears at 1414 cm^{-1}, which shows the anhydride of VM molecule has chemical reaction with –NH of PA66 molecule.

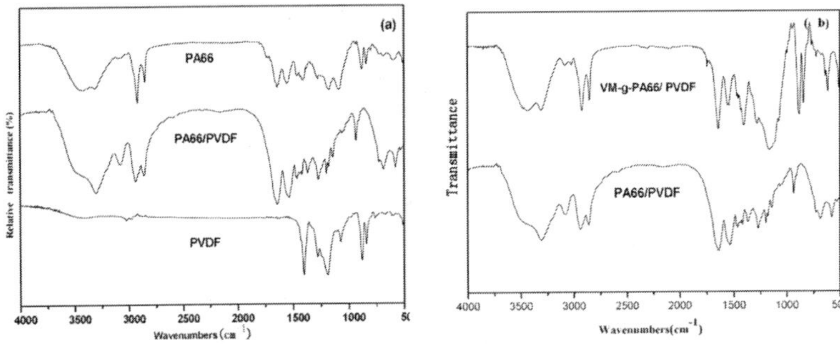

Figure 3. FT-IR spectrum of PA66/PVDF blends: (**a**) without VM; (**b**) with VM.

Figure 4 is DSC curve of PA66/PVDF blends. It can be seen from the Figure that the melting peak position of PA66 is at 223.2 °C. The melting peak of PA66/PVDF blends were coexisting after blending PA66 and PVDF, it shows PA66 and PVDF of PA66/PVDF blends can also be crystallized.

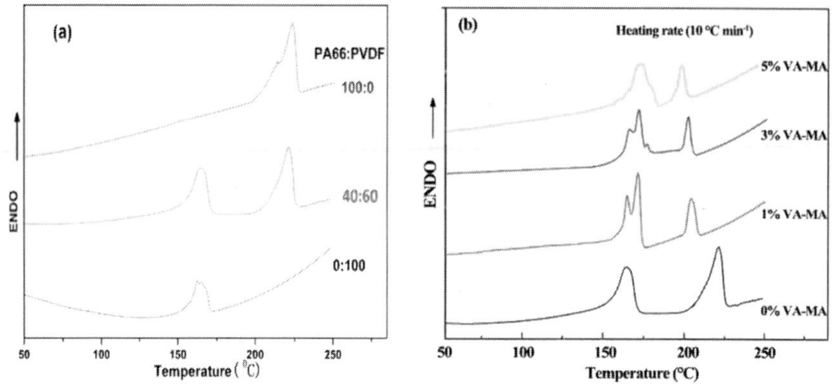

Figure 4. DSC thermograms of PA66/PVDF blends at 10 °C·min^{-1}: (**a**) without VM; (**b**) with VM.

Figure 5 is XRD pattern of PA66 and PVDF blends. It can be seen from the Figure that there are the α crystal-form diffraction peak of pure PA66 monomer at 2θ = 20.3° and β crystal-form diffraction peak at 2θ = 22.8°. It can be seen by blending PVDF and PA66 that the α crystal-form diffraction peak of PA66 monomer at 2θ = 20.3° was weakened with the accession of PVDF, but the β crystal-form diffraction peak at 2θ = 22.8° was enhanced, which shows that PVDF can inhibit the formation of α crystal form of PA66 and promote the β crystal of PA66. Besides, there are α crystals formed of PVDF in PVDF/PA66 blends, which shows that the accession of PA66 cannot promote to form β crystal or inhibit the formation of α crystal form in PVDF.

Figure 6 is the SEM image of PA66/PVDF blends. Figure 6a is the SEM image of PA66/PVDF. It can be seen from Figure 6a that PA66/PVDF blends present a unique structure of ordered arrangement which is evenly distributed. It can be seen clearly that the particles of PA66 are evenly distributed in PVDF. VM-*g*-PA66 is as the dispersion phase of the main ingredient which is dispersed in flakes, where the continuous phase gives priority to PVDF. With regard to the amplification of the continuous phase, it can also be found that there are small VM-*g*-PA66 particles. All of these are demonstrated that PA/PVDF blends are a two-phase system. Figure 6b is the SEM image of VM-*g*-PA66/PVDF blends after adding the bulking agent of VM. It can be seen from the Figure that PA66/PVDF blend film presents a two-phase separation structure in PA66/PVDF blends. PA66 as the dispersion phase of the main ingredient is dispersed in flakes in the continuous phase where it gives priority to PVDF, and the interface adhesion is not good.Figure 6b displays that the accession of 3 wt % bulking agent VM makes the interface of dispersion phase and continuous phase blurred with the melt blending of PA66 and PVDF. This shows that the accession of bulking agent VA-MA can improve interface adhesion between the PVDF phase and PA66 phase.

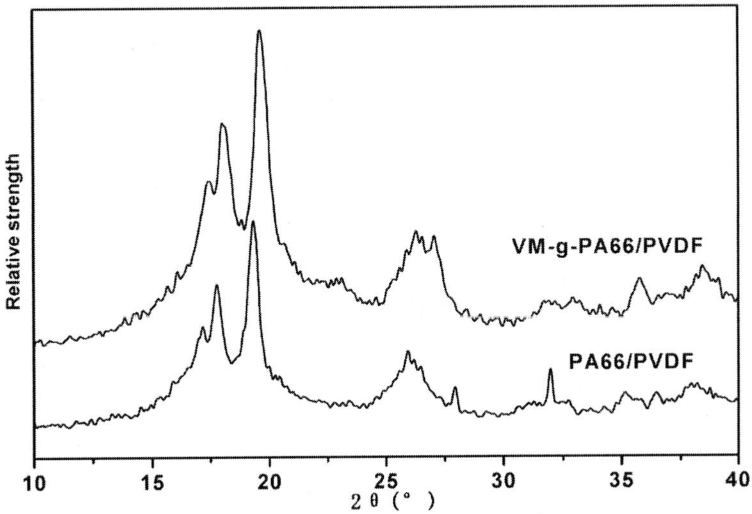

Figure 5. XRD pattern of PA66/PVDF blends.

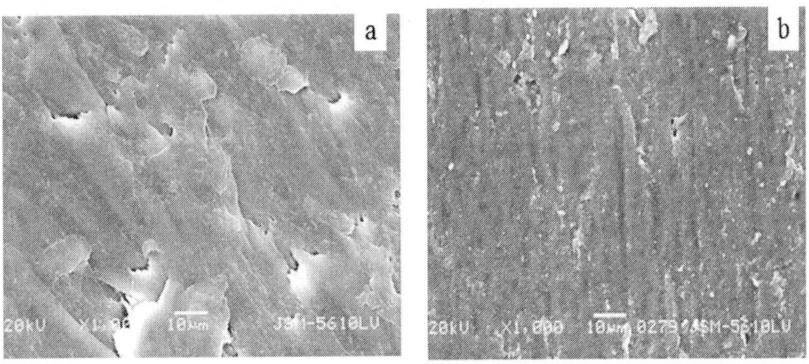

Figure 6. SEM micrographs for the PA66/PVDF and VMg-PA66/PVDF blends. (**a**) PA66/PVDF(40/60); (**b**) 3 wt %VM-*g*-PA66/PVDF(40/60).

3.2. Dielectric Properties

Figure 7 is the relationship of dielectric constant ε of PA66/PVDF blend system to frequency. In PA66/PVDF blend system. It can be seen from the Figure that the dielectric constant ε of PA/PVDF blend system has no significant change with the increase of electric field frequency, which presents excellent frequency stability. After PA and PVDF are melted and blended according to different volume ratios, the dielectric constants of all the volume ratios of PA66/PVDF blends decrease with the increase of frequency. It can also be seen from the Figure that the dielectric loss of PA66/PVDF blends in different volume ratios increases with the increase of frequency after PA66 and PVDF are melted and blended in different volume ratios, but it is slower than the speed of pure PVDF. In the low frequency region ($f < 10^5$ Hz), the dielectric loss tanδ of PA66/PVDF is less than 0.10, and there is no significant difference among the volume ratios. In the high frequency region (such as 10^6 Hz), the dielectric loss of PA66/PVDF blends decreases continuously with the increase of PA66 volume content in the blend system. This is because the dielectric loss of PA66 is lower than PVDF in high frequency region, and polar amide groups can promote the orientation of CF_2 in high frequency, decreasing thermal loss of the blends. This shows that the interaction of PA66 and PVDF requires a better compatibility with the melt blending of PA66 and PVDF.

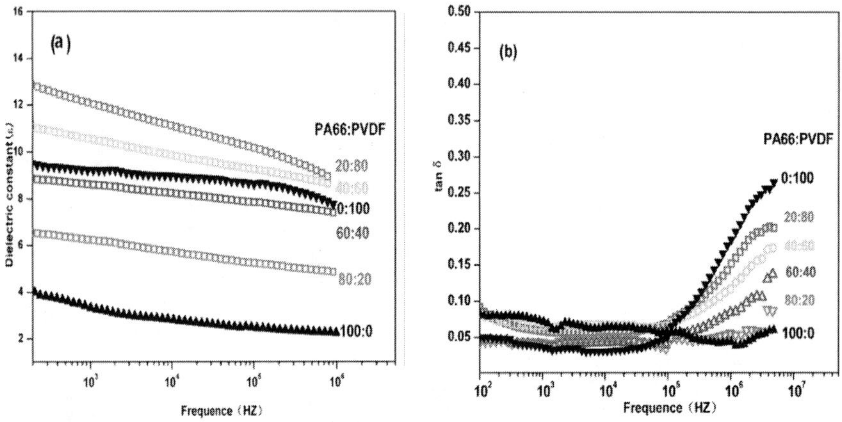

Figure 7. Frequency dependence of (**a**) ε and (**b**) tanδ for PA66/PVDF blends.

Figure 8 is the melt blending of PA66 and PVDF after VM is added, by which it can not only improve the dielectric constant of PA66/PVDF blends to some extent. Meanwhile, it can reduce the dielectric loss of blends. VA-MA makes polar dipoles easier to be oriented in the PA66/PVDF blend system, thus lowering the dielectric loss of blends.

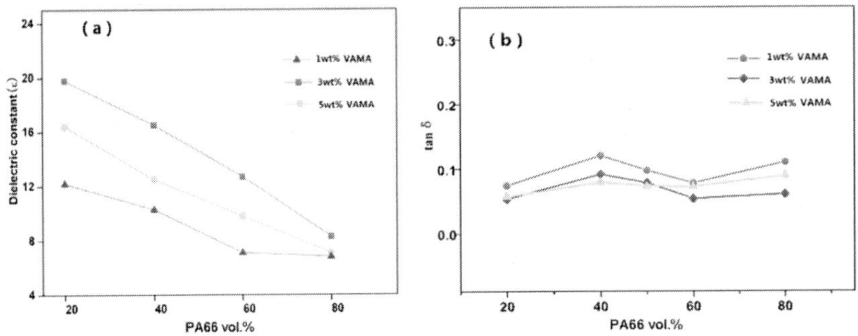

Figure 8. Variation of (**a**) ε and tanδ (**b**) *versus* VM content of VM-*g*-PA66/PVDF blends.

PA66/PVDF blends present excellent dielectric properties. However, combining the result of thermal property analysis on three PA/PVDF blends in the last section, it can be found that the melting point (T_m = 223.2 °C) of PA66 is high. During the melt blending with PVDF (T_m = 163.4 °C), it requires a very high processing temperature. In the temperature rising process, it may destroy the crystal form of the polymer and cause inconvenience for preparation. Therefore, VM is selected to provide graft modification on PA66 to study the influence it has on PVDF blends' other properties.

At room temperature, when the frequency is 10^3 Hz, the dielectric properties of VM-*g*-PA66/PVDF blends are obtained for analysis after adding different mass fractions of VM. The quantities of VM added are 1 wt %, 3 wt %, and 5 wt %. The relation curve of the dielectric constants of VM-*g*-PA66/PVDF blends to the content of VM is shown in Figure 8. Besides, the imaginary line means VM-*g*-PA66/PVDF blends of VM which are not added. Obviously, the influence of bulking agent VM-*g*-PA66/PVDF on the dielectric constant of blends was 3 wt % > 1 wt % > 5 wt %. When PA66 and PVDF are melted and blended, it can improve the flexibility of molecules with the accession of VM, and the dielectric constant of VM-*g*-PA66/PVDF can be improved. However, excessive VM can improve the grafting ratio of PA66, and the crystallinity of PA66 is declined, which is not good for the collection and delivery of system charge.

3.3. Mechanical Properties

Figure 9 is the influence curve of PA66 volume content on VM-*g*-PA66/PVDF blend stretching strength and breaking elongation. It can be seen clearly that stretching strength presents irregular variations with the increase of PA66 content, which decreases gradually first. When VM-*g*-PA66 was about 60 vol %, the minimum stretching strength was 16.13 MPa, and then it increased sharply. Meanwhile, breaking elongation is also declining and then rising with the increase of PA66 content. Coincidently, when the volume content of PA66 is 60 vol %, both of the values are the minimum. The elongation of blends is over 20%, and stretching strength was over 18 MPa, which shows that the melting blends of VM-*g*-PA66 and PVDF were suitable to be used for flexible dielectric material.

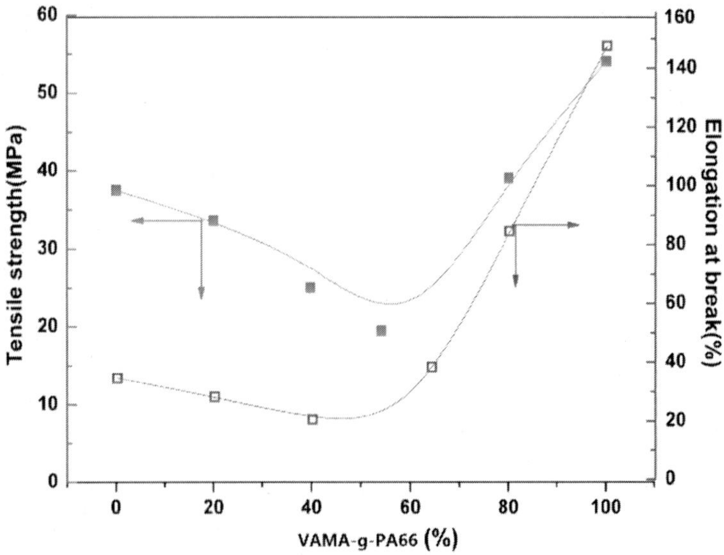

Figure 9. Variations of the tensile strength and the tensile elongation of VM-*g*-PA66/PVDF blends *versus* VM-*g*-PA66 volume fraction.

4. CONCLUSIONS

This paper demonstrated a simple, efficient, and repeatable route to produce discontinuous all-polymer blends composed of VM-*g*-PA66 and PVDF. The alternating copolymer VM prepared can be grafted on the molecular chain of PA66. VM-*g*-PA66 can influence α crystal orientation and crystal transformation of PVDF. By adding VM at 1 wt %, we can improve the dielectric constant ε of VM-*g*-PA66/PVDF blends without impact on dielectric loss. The all-polymer blends having a relatively high-ε, low dielectric loss, and excellent storage modulus as induced by interfacial copolymer VM. When the quantity of VM added was 3%, the dielectric constant ε of VM-*g*-PA66/PVDF blends declined sharply. The dielectric constant ε of PA66/PVDF blends can be up to 25 after capacity expansion. The elongation of blends was over 20%, stretching strength was over 16.13 MPa. The melted blends of PA66 and PVDF can be used as flexible dielectric material.

ACKNOWLEDGEMENTS

The research was supported by the National Natural Science Foundation of China (Grant Nos. 51378073 and 51408048), the Key Program for International Science and Technology Cooperation Projects of Shaanxi province (Grant Nos. 2012KW-09 and 2014KW10-03), the China Postdoctoral Science Foundation (Grant No.2015M570805), the Fundamental Research Funds for the Central Universities of Chang'an university (Grant Nos.310821152005 and 310821153502) and the Research Fund for the Doctoral Program of Higher Education of China (Grant No. 20120205120010).

AUTHOR CONTRIBUTIONS

Jianzhong Pei and Rui Li conceived and designed the experiments; Rui Li performed the experiments; Rui Li and Zixuan Chen analyzed the data; Jianzhong Pei contributed reagents/materials/analysis tools; Rui Li and Zixuan Chen wrote the paper.

REFERENCES

1. Zhang, Q.M.; Li, H.F.; Poh, M. An all organic composite actuator material with a high dielectric constant. *Nature* **2002**, *419*, 284–287.

2. Li, R.; Pei, J. High dielectric performance of polyamide 11/poly(vinylidene fluoride) blend films induced by interfacial glycidyl methacrylate. *Polym. Sci. Ser. A* **2015**, *57*, 792–798.

3. Zhang, Q.M. Braille display pin actuators. *WW-EAP Newslett.* **2006**, *8*, 10–11.

4. Dang, Z.M.; Zhang, Y.H.; Tjong, S.C. Dependence of dielectric behavior on the physical property of fillers in the polymer matrix composites. *Synth. Metals* **2004**, *146*, 79–84.

5. Dang, Z.M.; Lin, Y.H.; Nan, C.W. Novel ferroelectric polymer_matrix composites with a high dielectric constant at the percolation threshold. *Adv. Mater.* **2003**, *15*, 1625–1628.

6. Bayer, I.S.; Biswas, A.; Megaridis, C.M. Biocompatible polyvinylidene fluoride cyanoacrylate composite coatings with tunable hydrophobicity and bonding strength. *Appl. Phys. Lett.***2008**, *93*, 173902.

7. Raffaele, M.; Janne, R.; Glenn, H. Templating organic semi-conductors via self-assembly of polymer colloids. *Science* **2003**, *299*, 1872–1874.

8. Li, R.; Xiong, C.X.; Kuang, D.L.; Dong, L.J.; Lei, Y.A. Polyamide 11/Poly(vinylidene fluoride) blends as novel flexible materials for capacitors. *Macromol. Rapid. Commun.* **2008**, *29*, 1449–1454.

9. Cheng, H.; Jiang, Y.L. Colossal dielectric and eletromechanical responses in self-assembled polymeric nanocomposites. *Appl. Phys. Lett.* **2005**, *87*, 182901.

10. Lehmann, W. Giant lateral electrostriction in ferroelectric liquid-crystalline elastomer.*Nature* **2001**, *410*, 447–450.

11. Kuang, D.L.; Li, R.; Pei, J.Z.H. Polyamide 11/poly(vinylidene fluoride)/vinyl acetate maleic anhydride copolymer as novel blends flexible materials for capacitors. *Polymers* **2014**, *6*, 2146–2156.

12. Tjong, S.C.; Bao, S.P.; Liang, G.D. Polypropylene/montmorillonite nanocomposites toughened with SEBS-*g*-MA: Structure-property relationship.*J. Polym. Sci. B Polym. Phys.***2005**, *43*, 3112–3126.

2

CHAPTER

Recent Progress of High Performance Polymer OLED and OPV Materials for Organic Printed Electronics

Chizu Sekine, Yoshiaki Tsubata, Takeshi Yamada, Makoto Kitano and Shuji Doi

Sumitomo Chemical Co. LTD, 6 Kitahara Tsukuba Ibaraki, Japan

ABSTRACT

The development of organic printed electronics has been expanding to a variety of applications and is expected to bring innovations to our future life. Along with this trend, high performance organic materials with cost-efficient fabrication processes and specific features such as thin, light weight, bendable, and low power consumption are required. A variety of organic materials have been investigated in the development of this field. The basic guidelines for material design and the recent progress of polymer-based organic light-emitting diodes (OLEDs) and organic photovoltaic cells (OPVs) are reported.

Keywords: polymer OLED, OPV, conjugated polymer, efficiency, lifetime, degradation, PCE, printing technology

1. INTRODUCTION

The development of organic printed electronics has been expanding to a variety of applications such as organic light-emitting diodes (OLEDs), organic photovoltaic cells (OPVs), electronic papers, wearable electronics and various sensors. It is expected that flexible, thin, and light-weight devices with cost-efficient productivity should bring innovation to our lives, which makes the research and development (R&D) in this area so competitive and global.

As self-emitting devices, OLEDs have features of fine image-quality, are ultra thin, and are light weight. They have become familiar to general consumers as they are equipped in smart phones such as Samsung's Galaxy series. Currently, an evaporation process is adopted for the mass production of OLEDs. Although this process is practical when used for the mass production of small and middle size OLED displays, it is still considered that some difficuties exist in the fabrication of large TV displays from technical and cost points of view.

We have been developing organic materials applicable to printed electronics based on conjugated polymer technology, because we expect the printing process to be essential for cost-competitive mass production of large sized TVs or flexible devices [1, 2].

Based on the R&D of OLEDs, we have also developed OPV materials whose working mechanism is the reverse to that of OLEDs. Recently, new applications or quality assurance projects have been announced by some companies. OPVs are still in the middle of development for use in materials, devices, modules, and fabrication processes. As with OLEDs, a printing process should be the key technology in the cost-efficient fabrication of large sized OPV modules, especially when it is combined with a roll-to-roll process. For these reasons, knowledge gained in the development of OLEDs can be effectively applied to OPV development.

In this article, material technologies in polymer OLEDs and OPVs are reviewed, from their development history to recent progress, including our experiences. A lot of materials used for organic printed electronics applications have been reported, and systematic reviews are frequently updated [3]. Such reports are very useful for grasping the characteristics of each conjugated polymer structure. On the other hand, organic electronics materials exhibit their performances as electric devices, hence their performance depends strongly on the combinations of materials, layer structures, process conditions in their fabrication, etc Theses circumstances make analysis and speculation complicated. Furthermore, device durability is also an important factor from the practical application perspective. The durability is governed by not only the chemical or physical stability of materials, but also by factors other than material properties. Therefore, parallel activities of basic analytical research and experimental screening are needed to promote material development. Such intrinsic circumstances in this field may lead to a situation where a few years (or more) are needed to find new materials and/or working mechanisms to bring them to completion for practical usage. We hope that these peculiarities of material research, strongly tied with device architecture and device fabrication processes, will be understood.

2. POLYMER OLED

2.1. Development history and characteristics of polymer OLED materials

It was big news when Sony launched their eleven inch OLED-TV 'XEL-11' in 2007. Currently, Samsung's Galaxy series, equipped with OLED displays, are very popular. Although small or middle size displays have so far been the main products for OLEDs, the recent trends in OLED applications are shifting to large sized TVs or flexible/wearable display fields. As this trend develops, soluble light emitting materials have received lots of attention from a viewpoint of low-cost mass production and/or their applicability to flexible plastic substrates.

In 1989, a group from Cambridge University observed electroluminescence (EL) in a conjugated polymer [4]. Almost at the same time, we observed a similar phenomenon [5]. This was only a weak emission of 0.1% external quantum efficiency (EQE), and the device had a very short lifetime of a few minutes. After this finding, Sumitomo Chemical Co. Ltd, Covion in Germany, Dow Chemical in the US, and Cambridge Display Technology (CDT) in the UK started the energetic development of polymer OLED materials, as well as the optimization of device structures. As the results of over 20 years of R&D, high EQE of 5 ~ 10%, and a long operating lifetime of over several tens of thousands hours have been achieved. Polymer OLEDs show not only red, green, and blue (RGB) emissions but also white color emissions, which are essential to various applications.

OLED emissive materials are categorized into a small molecule group and a polymer group as shown in Figure 1. The polymer group is further classified into non-conjugated or conjugated subgroups. Dendritic compounds can be used as intermediate materials for the small molecule and polymer groups [6]. Currently, evaporation is practically the main process for OLEDs and small molecule are used in the panels. As detailed later, the trend of fabrication process is shifting from dry process (evaporation) to wet process. Along this trend, the development of solubilized small molecules was started with a sense of urgency. The introduction or replacement of substituent groups by small molecules is the main modification to acheive solubility in ink solvents. Figure 2 shows an example of how dithienylbenzothiadiazole derivatives were modified to become soluble materials [7]. As explained in the next section, small molecule OLEDs contain multilayer device architecture. Because of the requirement to form stacked layers, further modifications in chemical structures are needed, in addition to solubility. Although a lot of patents regarding this technology have been filed, solubilized small molecule OLEDs seem to still be under development for overcoming such requirements.

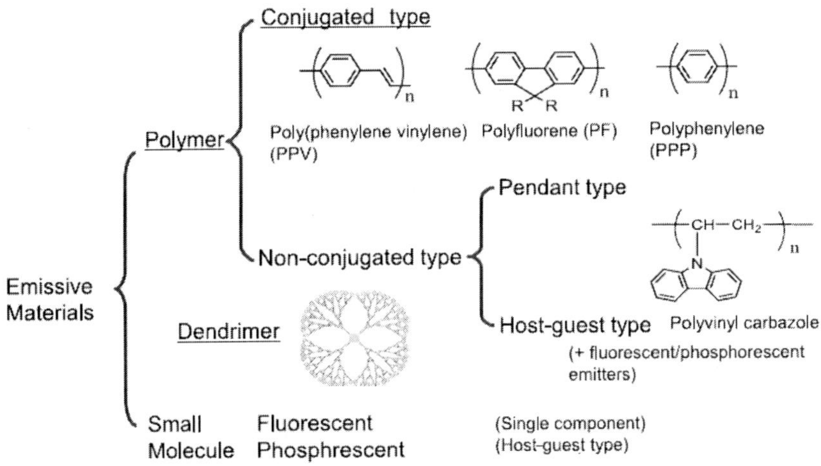

Figure 1. Schematic classification of polymer and dendrimer emissive materials.

Polymer OLED materials are readily soluble by ink solvents, and this results in appropriate materials for wet processes (printing processes) like inkjet, die-coating, etc which are considered to be appropriate for the fabrication of large sized OLED devices. Conjugated polymers, having sp2 carbons in the main chain, exhibit high electron and hole mobility originating from the delocalized π-electrons system. Many conjugated polymers with this characteristic have been reported as emissive materials in the visible region. Poly-phenylenvinylene (PPV) [4, 8], polyfluorene (PF) [9–12], and poly-p-phenylene (PPP) [13] are typical examples of conjugated polymer OLED materials.

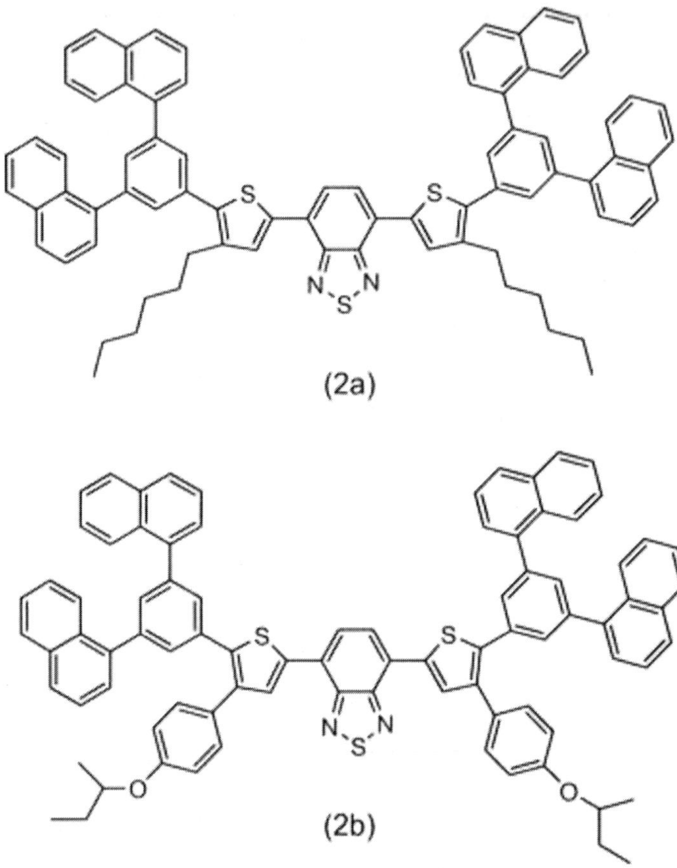

Figure 2. Solubilized emitters of dithienyl benzothiadiazole derivatives [7].

Two types of non-conjugated polymer OLEDs, the pendant type and the host-guest blend type, have been developed to maximize device performance. In the former, functional units are attached to the non-conjugated backbone (polymer) and, in the latter, functional guests are doped into the host polymer. Of these non-conjugated types, polyvinylcarbazole (PVCz) is one of the typical polymers, and its optical and electrical properties were well-analyzed in the early stage of OLED developments [14, 15] where PVCz was used extensively as a host material. A fluorescent [16] or phosphorescent [17] host-guest polymer OLED containing PVCz in an emitting layer has also been reported. Furthermore, PVCz has been developed as a pendant type phosphorescent polymer where an iridium complex is incorporated in its side chain [18,19]. As the fluorescent pendant version, triarylamine pendant type polymers have been studied, which offer the potential benefit of ease of synthesis compared to the conjugated type polymers [20]. Their OLED performance was evaluated as part of a mixture with the conjugated polymer in order to complement mutually insufficient functions. Owing to its ease of synthesis, non-conjugated type polymers have been used for basic studies of OLEDs rather than forpractical development because the independent controllability of emissive and carrier transporting functions can be relatively easily achieved by selecting suitable functional units.

Dendritic compounds are also useful for the printing process, particularly in combination with polymers. Dendrimers containing an iridium complex as the phosphorescent emitting core are known in red [21], green [22], and blue [23]. Polymer OLEDs utilizing conjugated polymers and red dendrimers already achieve a high enough performance for their practical usage [24].

The conjugated polymer system works well without a complex multilayer device structure, because the required functions are introduced into polymer chains by co-polymerization technology. A conjugated polymer consists of functional backbone units, electron affinity units, hole affinity units, or emissive units, as shown in Figure 3. Emission color, which is one of the most important properties, is usually controlled by introducing an emissive unit to the polymer chains. Therefore, backbone units are generally selected from wide band gap structures. From this standpoint, polyfluorene [25–29] or polycarbazole [30], whose band gap is wide enough even for blue emission, are commonly used for a backbone structure. As a typical design of a polymer, fluorene-acene co-polymers are shown in table 1 [25]. Each acene unit was expected to emit RGB color; anthracene for blue, naphthalene for green, and pentacene for red. Devices with PDOFA (fluorene: anthracene = 9:1) with CIE color point of (0.21, 0.19), PDOFN (fluorene: naphthalene = 9:1) with (0.37, 0.45) and PDOFP (fluorene: pentacene = 9:1) with (0.64, 0.3) showed blue, green, and red EL emissions, respectively, as designed. In a similar scheme, a variety of unit structures were co-polymerized with fluorene units to obtain polymers with the desired colors [31, 32].

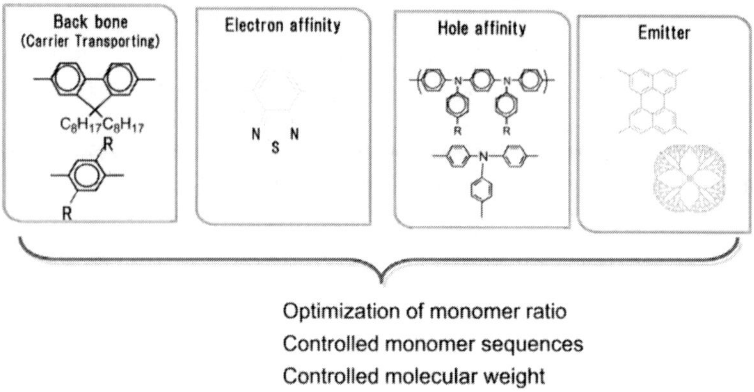

Figure 3. Copolymerization structure of functional units of a typical light-emitting polymer.

Table 1. EL emission color of fluorene-acene copolymers [25].

Polymer	Structure	Peak of EL emission (nm)	CIE (*x*, *y*)
PDOFA	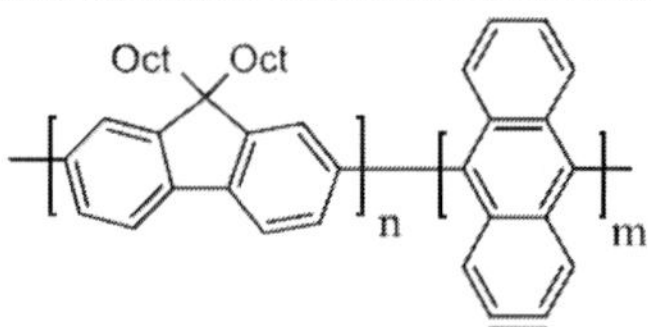	435	(0.21, 0.19) Blue
PDOFN	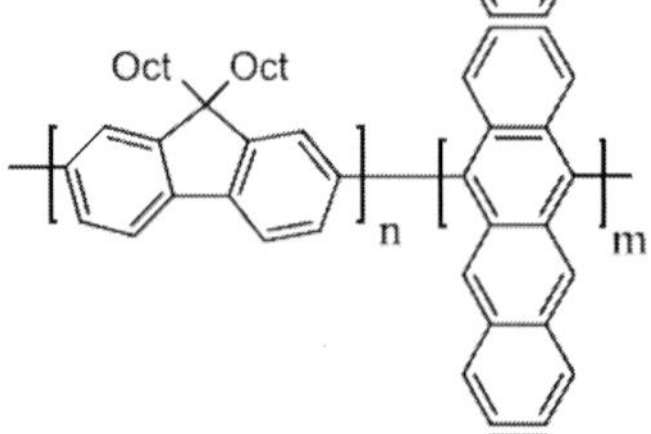	525	(0.37, 0.56) Green
PDOFP	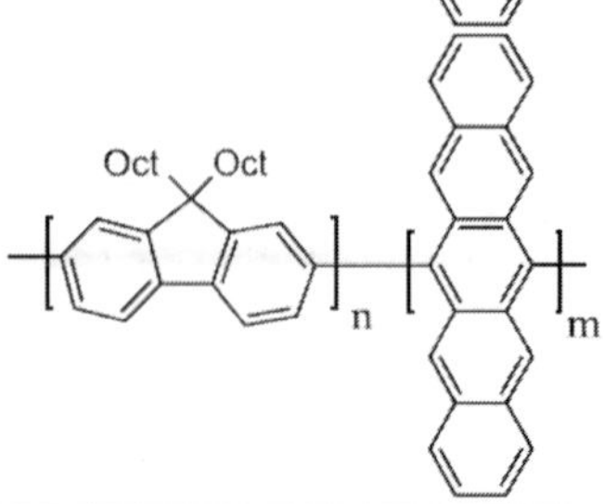	625	(0.64, 0.30)Red

2.2. Features of polymer OLED

Polymer OLEDs, as well as small molecule OLEDs, have characteristics of (a) high contrast ratio coming from a self-emissive device (luminance-on/-off), (b) wide viewing angle, (c) vivid colors, (d) thin devices, (e) high speed image switching, and (f) low power consumption. The noteworthy feature of polymer OLEDs is the applicability of a cost efficient fabrication process in mass production. As shown in table 2, conventional small molecule OLEDs, consisting of complex device structures (multilayer structure), are fabricated by a vacuum evaporation process. On the other hand, polymer OLEDs have a simple device architecture, which can be fabricated with a printing device like an inkjet printer.

Table 2. Comparison of characteristics between small molecule type and polymer type OLED.

	Small Molecule	Polymer
	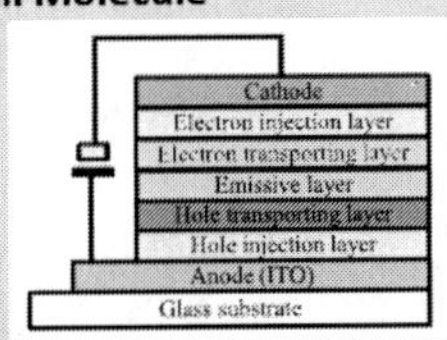	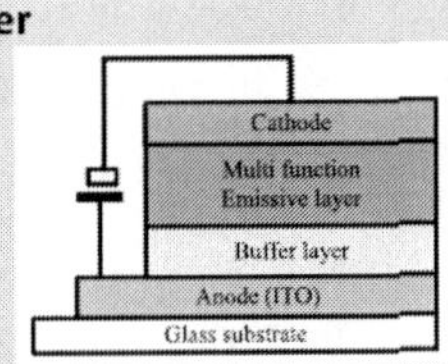
Process	Dry process (Vacuum evaporation) 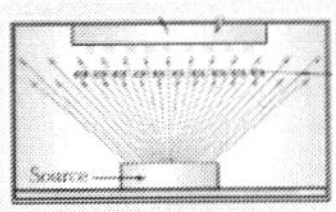	Wet process 
Patterning	Shadow mask	Printing (IJ etc.)
Structure	Complex layer structure (5-6) →Complex process	Simple layer structure (2-3) →Simple process, scalable
Material	Separated function	Integrated function
Issue	Layer structure complexity Difficulty in mask patterning	Performance (esp. LT) Patterning technology

A typical polymer OLED consists of an anode, a hole injection layer, an emissive layer, and a cathode. The introduction of an interlayer between the hole injection layer and the emissive layer has brought a big improvement in device performance [33]. The cross-linking properties of the interlayer makes it possible to stack three layers of hole injection, interlayer, and emissive layers by a printing process.

2.3. Polymer OLED device performance

2.3.1. Emission efficiency

The OLED emission efficiency can be expressed as [34]

$$\varphi = \gamma \cdot \eta_{\mathrm{e-h}} \cdot \varphi_{\mathrm{ph}} \cdot (1 - Q) \tag{1}$$

where φ is EL quantum efficiency, γ is carrier balance of electrons and holes, $\eta_{\mathrm{e-h}}$ is recombination rate, φ_{ph} is photo luminescence efficiency, and Q is quenching factor by cathode.

According to equation (1), higher photoluminescence efficiency and recombination rate, well-gcontrolled injection of electrons and holes, and suppression of cathode quenching leads to the improvement of the OLED efficiency. The basic understanding and primitive design policy for high efficiency were reviewed in a previous article [35] in detail. We would like to introduce our typical approach to polymer design with amine units to obtain blue polymers with high efficiency.

After screening amine unit candidates, promising candidates are tested as co-polymers with fluorene backbone units of various co-polymerization ratios. In these polymers, amine units work as hole transporters as well as blue emitters. Figure 4 indicates that the amine content can control the hole mobility in the range of 10^{-7} to 10^{-3} $cm^2\ Vs^{-1}$, and efficiency reaches its maximum at the lowest mobility [36]. It is considered that a small amount of amines work as hole traps in the polymer host, resulting in the increase of electron and hole recombination.

In the history of small molecule OLED development, the introduction of emissive dopants into double-layer devices was one of the breakthroughs for increasing emission efficiency [37]. As already pointed out, an interlayer is one of the remarkable technologies for a dramatic increase in the emission efficiency of polymer OLEDs [38]. The interlayer is a layer introduced between the hole injection layer and the emission layer, and is considered to work as an electron and exciton blocking layer in addition to its hole transporting function. It was demonstrated that a thin interlayer (~10 nm) with a conjugated polymer, poly[9,9-dioctyl fluorine-co-N-(4-butylphenyl) diphenylamine] (TFB), between poly(3,4-ethylenedioxythiophene) and poly(styrene sulfonic acid) (PEDT:PSS, hole injection layer) and an emissive polymer poly(9,9-dioctyl fluorine-co-benzothiadiazole) (F8BT) prevented exciton quenching by the PEDT:PSS, resulting in an improvement in the device efficiency (see Figure 5).

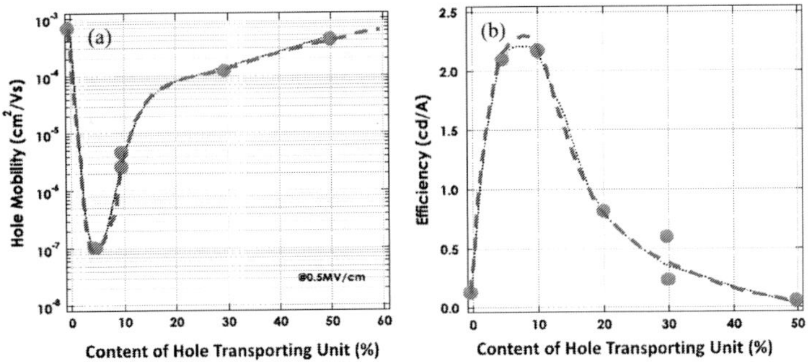

Figure 4. The dependence of (a) hole mobility and (b) the current efficiency of EL devices on the content of hole transporting units in light emitting polymers [36].

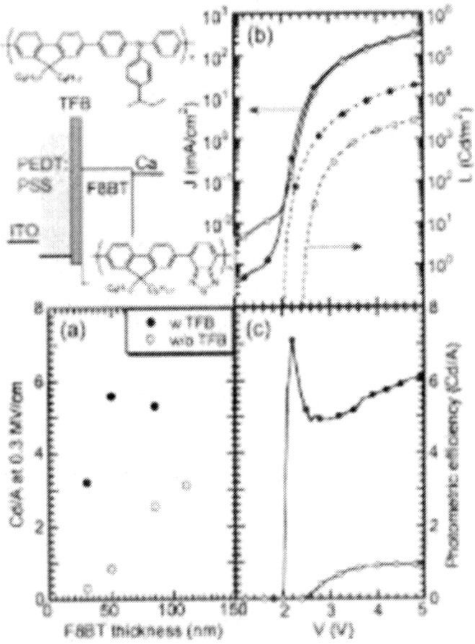

Figure 5. (a) EL efficiency at an internal electric field (0.3 MV cm^{-1}) of ITO/PEDT:PSS (65 nm) / F 8 BT (x nm) / Ca (5 nm) / Al (400 nm) LEDs with/without the TFB interlayer as a function of F8BT film thickness, (b) Current–voltage–luminance and (c) EL efficiency-voltage characteristics of LEDs ($x = 50$ nm). Schematic of the device structure and the chemical structures of TFB and F8BT are shown. F8BT is a highly luminescent semiconducting polymer that emits green–yellow light, and has a large ionization potential (5.8 eV) and a large electron affinity (2.95 eV). (Reprinted with permission from [38], ©2005 American Institute of Physics.)

After this finding, a lot of investigations and material screenings were performed so that lower hole mobility by trapping could make carrier recombination confined to near the interface of the interlayer and the emission layer. In summary, the introduction of the interlayer is considered to play two roles: (1) separation of the emission zone and the hole injection layer; and (2) accumulation of electrons at the interface of the interlayer/emission layer due to its electron blocking property. For the maximum use of the interlayer function, it is preferable for the electron mobility of the interlayer to be lower than that of emission layer.

One of the fundamental limits to the efficiency of a fluorescent OLED is the ratio of emissive singlet excitons to non-emissive triplet excitons formed in the electroluminescent process. Although, from simple spin statistics, only 25% of the excitons formed are singlets, there are a number of suggested mechanisms for exhibiting greater efficiency than the statistically expected value. In small molecule OLEDs, it was reported that triplet-triplet annihilation (TTA) increased the emission efficiency by converting triplet excitons into singlets [39, 40]. It was also shown that approximately 20% of the device efficiency originated from the production of singlet excitons by TTA in the fluorescent conjugated polymer OLED using F8-PFB and an interlayer of F8-TFB [41]. A similar phenomenon was observed in a polymer OLED with a spirofluorene derivative [42]. This mechanism yielded a very high emission efficiency of 10% EQE with $y = 0.13$ in a blue-emitting device [43].

These technologies brought about remarkable progress in the performance of polymer OLEDs and became close to that of small molecule type OLEDs with a multilayer device structure.

2.3.2. Device operating lifetime

Device operating lifetime means the decay of luminance by constant current driving. We measured lifetime and the change of PL intensity in a device to understand the degradation phenomena. The fabricated device had a structure of ITO/HIL/IL/LEP/Alkali metal compound/Al, where ITO stands for indium tin oxide, HIL is hole injection layer (50 nm thickness), IL is interlayer (20 nm thickness) and LEP is light-emitting polymer (80 nm thickness). The polymer OLED was prepared using the following procedure. HIL ink was passed through a 0.45 μm filter before being deposited on ITO by spin coating in air, and then dried at 200 °C for 10 min on a hotplate. The IL was then deposited onto the film by spin coating a polymer solution in mix-xylene on top of the HIL. It was annealed at 180 °C for 60 min in a glove box. After that, the LEP solution in mix-xylene was coated by spin coating on the top of the IL, and then dried at 130 °C for 10 min in a glove box. The device fabrication was completed by depositing a thin layer of alkali metal compound (2 nm) and Al (100 nm) at a pressure of around 10^{-5} Pa. Finally, the device was encapsulated using a UV-cured glue. The EL, EL spectra and PL spectra were measured by an OLED TEST SYSTEM (Tokyo System Kaihatsu). The device lifetime was measured by an OLED lifetime system PEL-100T (EHC CO. Ltd). We observed a decrease in PL as EL intensity decreased in the polymer OLED (see Figure6). From this observation, PL decay is considered as the main cause of EL decay. However, at the same time we should also consider possibilities of the change in carrier balance and/or some quenching sites due to diffusion of electrode materials as other degradation pathways.

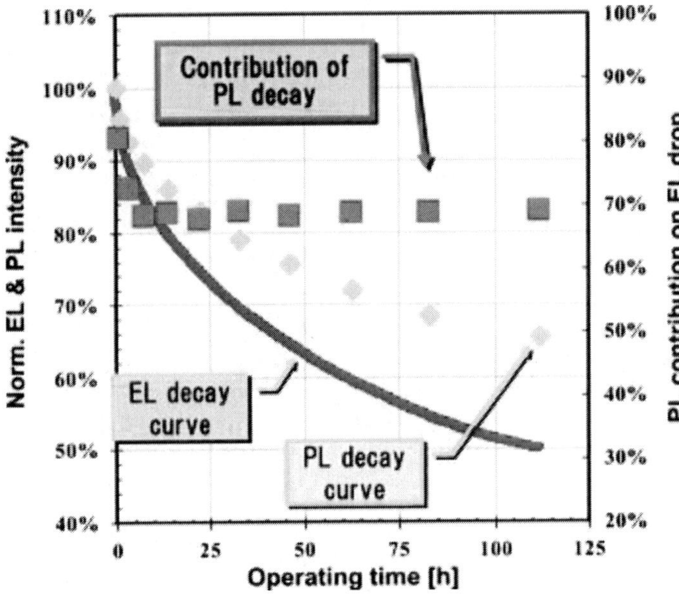

Figure 6. Degradation curves of EL and PL peak intensities of EL device containing a blue polymer during constant-current operation. Device structure: ITO/ HIL/ IL(20 nm)/ LEP(60 nm)/ cathode.

In the investigation of the PL decay mechanism, we have found that PL decay was not observed in the driving of mono-polar devices (hole-major or electron-major device) as shown in Figure7. These devices were fabricated with a similar procedure to the polymer OLED fabrication described above. The device structures of the hole-major and electron-major devices are ITO/HIL/IL/LEP/Au and Al/LEP/Alkali metal compound l/Al, respectively. We also performed degradation analysis of the host-guest polymer OLED device by reverse engineering techniques. The polymer OLED device was fabricated with a host polymer and guest emitter compounds. After degradation, the device was decomposed for chemical analysis of the host and guest materials, and the following results have been obtained:

- no change in the photoluminescence efficiency of the guest emitter;
- formation of an insoluble layer of polymer.

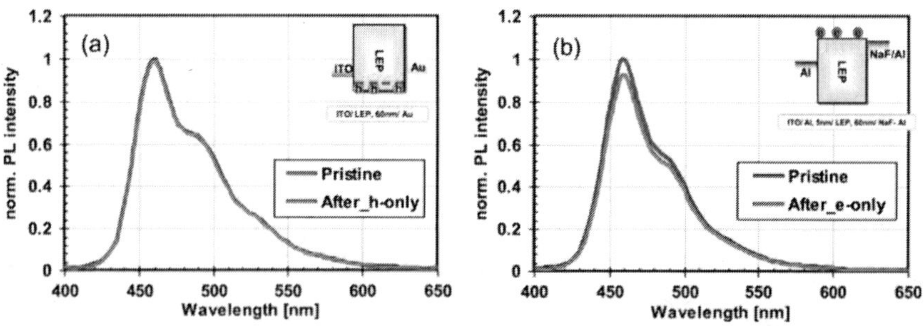

Figure 7. PL spectra of OLEDs before and after they were operated at 25 mA cm^{-2} for 130 h: (a) hole-driven device and (b) electron-driven device.

These observations indicate that (1) the PL decay of emissive materials is related to the excited state by the recombination of carriers, (2) PL decay does not originate from the degradation of the emissive unit, but from the creation of quenching sites in the host polymer.

Based on these findings, we designed more robust polymer materials by optimizing substituents to backbone and/or using more rigid units to reach double digit improvements of device lifetime.

2.4. Future trend of R&D in polymer OLED materials

2.4.1. Improvement of device performance

In practical applications, the further improvement of operating lifetime is a critical issue. Regarding the lifetime, sticking image lifetime (LT95) is important in addition to half-life time in luminance (LT50).

Self-emissive displays like OLEDs tend to get into a sticking image problem, which is the reduction of luminance in a specific area caused by continuous emission. To solve this issue, both material improvements to suppress the

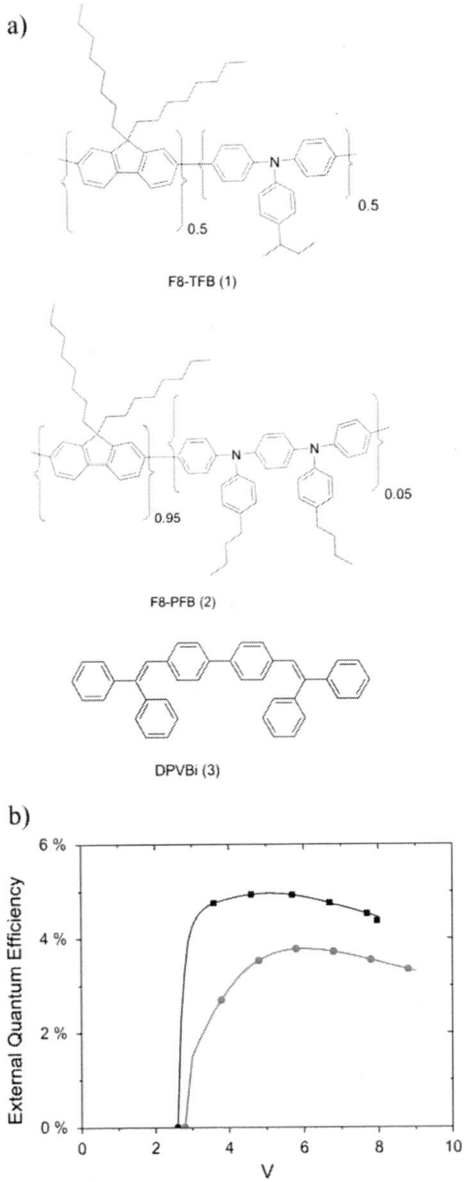

Figure 8. (a) Chemical structures of the materials under study. (b) External quantum efficiency of a prototypical PLED device comprising an F8-PFB LEP and a TFB interlayer, without (▪) and with (•) the addition of the triplet quenching additive (3). (Reprinted with permission from [41], ©2011 American Institute of Physics.)

image sticking phenomenon and the development of operating methods are necessary. In the case of small molecule OLEDs, new materials showing LT95 of more than 500 h have been reported [44]. On the other hand, the development of polymer materials for a longer LT95 is still ongoing towards a practical level, however some investigations have been giving hints on how to solve this issue [41]. It was observed that removal of the triplet exciton from the backbone polymer could significantly improve the stability of the device, particularly at the early stage of driving, i.e. introducing a triplet quenching additive leads to the improvement of device lifetime (see Figures 8 and 9). Careful adjustment is expected to satisfy both the need for high efficiency and a stable lifetime. Such comprehensive efforts, including material developments and detailed analysis, can bring polymer OLEDs to the goal of device performance required for practical usage.

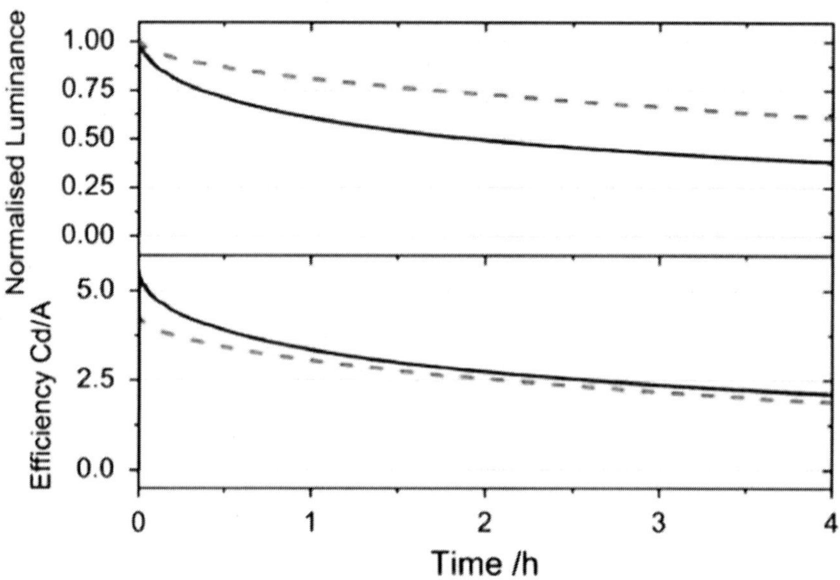

Figure 9. The decay of the device luminance (upper panel) with time and the change in efficiency of the device with time (lower panel) for a standard device (solid line) and a device containing the triplet quenching additive (dashed line). The devices were both driven from a starting luminance of 5000 cd m^{-2}. (Reprinted with permission from [41], ©2011 American Institute of Physics.)

Another trend is low power consumption as a general environmental countermeasure. Hole injection materials and electrodes are newly developed for this purpose as well as emissive materials. These kinds of materials used in small molecule OLED devices are also being tested in polymer OLEDs.

2.4.2. Next direction of R&D for polymer emissive materials

As already mentioned, polymer emissive materials have the unique feature that various required functions can be introduced into a single polymer by designing functional units and co-polymerization techniques. This feature takes advantage of the applicability of printing processes for large area device fabrication or color patterning. In terms of the aforementioned points, many kinds of emissive polymers have been investigated [45]. For higher efficiency, phosphorescent materials with a combination of small molecular phosphorescent compounds and carrier transporting polymers have been proposed as polymer phosphorescent OLED materials. In that study, non-conjugated polymers with carrier transporting units attached to their side chains were commonly used. This type of polymer was considered to be easier for controlling singlet or triplet energy levels and other physical parameters. Basic research for understanding the effect of carrier transporting units on phosphorescent units using non-conjugated polymers other than PVCz has continued [46, 47]. The combination of polymers and phosphorescent emitters is very important to obtain high efficiency in further investigations. Recently new conjugated polymers which can be used as blue phosphorescent materials were reported [48]. They have polyphenylene chains connected by meta-linkage in the main chains, and exhibit 4.69 cd A^{-1} efficiency originating from a typical light-blue iridium complex, Firpic. As conjugated polymers with delocalized electrons, they possess excellent carrier transport properties, and the design policy of soft-controlled conjugation is studied for better and improved materials.

As another approach, research on the phase separation in polymer blends is ongoing to uncover clues about high efficiency. In some trials, adding polystyrene to conjugated polymers has shown an increase in efficiency [49], and the investigation of a relation between EL properties and phase separation in conjugated polymer mixtures was reported [50].

We would like to introduce new topical high efficiency materials. There have been two approaches of materials systems to obtain high efficiency; TTA fluorescent materials using an up-conversion mechanism, and phosphorescent materials utilizing emissive triplet levels. As an up-conversion mechanism other than TTA, a thermally activated delayed fluorescence (TADF) process was proposed for converting non-emissive triplet excited states into emissive singlet excited states [51]. The energetic development of TADF materials was started to give us other options to avoid potential issues such as the use of the rare metal iridium, difficulties in developing pure blue phosphorescent materials, and so on. As in the energy diagram shown in Figure 10, the feature of TADF materials is considered as a small energy difference (ΔE_{ST}) between the singlet (S_1) and the triplet (T_1) energy levels [52, 53]. TADF materials, which thermally activate the reverse intersystem-crossing (ISC) from the triplet excited state to the singlet excited state, lead to an increase of fluorescence intensity, therefore high EL efficiency can be achieved in OLEDs with TADF materials. Furthermore, the roll-off characteristics of the EL efficiency can be reduced if the rate constant of the reverse ISC is significantly larger than the phosphorescent decay rate. Adachi's group started a search for efficient TADF materials from tin (IV) fluoride-porphyrin complexes. TADF materials with various colors were reported [53, 54], and remarkable materials of pure blue with 10% EQE [55], deep blue with 14.5% EQE [56] and very high EQE of 16.5% [57] were obtained. This material design is expected to be introduced into polymer OLED materials.

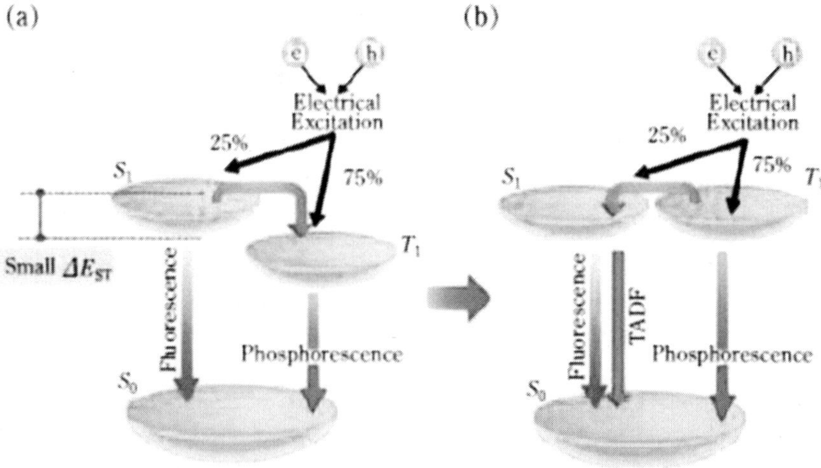

Figure 10. Process of formation of singlet and triplet excitons in the electroluminescence process (from Figure 2 [53]).

2.4.3. Progress of soluble emissive materials

Small molecule OLEDs have a longer development history, so it takes a lead in practical usage compared with polymers. A vacuum evaporation method is used to form layers of OLED and many technical improvements have been developed in this area.

Recently, soluble small molecule materials have received much attention as many material companies have reported or press released their progress. Introducing soluble groups to existing emissive structures are common molecular designs [58]. As an OLED performance of soluble phosphorescent small molecules, LT50 of about 130 000 h in red, 150 000 h in green, and 8000 h in blue [59] have been observed, and are estimated to be close to evaporated type OLEDs.

Soluble small molecules are becoming competitive technologies versus polymers. It is considered that further progress in these soluble materials should accelerate the expansion and growth of the organic printed electronics market, including OLEDs.

3. ORGANIC PHOTOVOLTAIC CELL (OPV)

3.1. Features of OPV and its development history

It is well known that OPVs have features of thinness, light-weight and unbreakable flexibility. A high temperature fabrication process is not necessary for OPVs, and that makes it possible to use plastic films as substrates. OPVs also have flexibilities in the design of cells. Continuous manufacture using roll-to-roll process etc can reduce production costs, resulting in the creation of new application markets utilizing a variety of OPVs.

The R&D of OPVs was accelerated after Tang *et al* reported a 1% power conversion efficiency (PCE) by p-n heterojunction stacking type cell. Afterwards, Hiramoto [60, 61], Yoshino [62], and Heeger *et al* [63] independently proposed a bulk heterojunction (BHJ) type, which triggered further improvements in efficiency. In the case of BHJ cells, it is understood that the high PCE characteristics come from the efficient charge separations of photogenerated excitons at the inherent wide interface area of the p-n junction. Remarkable improvements in efficiency were achieved in around 2000 by using poly(3-hexyl thiophene) (P3HT) and [6,6]-phenyl-C_{61}-Butyric Acid([C_{60}]-PCBM) [64] for p-type and n-type semiconductors respectively. Along with this trend, an efficiency of 4 ~ 5% was achieved in 2005 [65, 66]. In subsequent developments, newly proposed materials and tandem structures contributed to very rapid progress in efficiency, achieving higher PCEs of over 10%.

3.2. Improvement of OPV efficiency

3.2.1. Working mechanism of OPV and guiding principles for higher efficiency

The output characteristics of photovoltaic cells (PVs) are shown in Figure 11. The maximum area of an inscribed square in the current-voltage curve is the maximum power generation of the PV. The maximum efficiency η is expressed as

$$\eta = J_{sc} \times V_{oc} \times FF/\text{incident light energy} \tag{2}$$

where J_{sc} is the current density at a voltage of 0 V (short circuit current density), V_{oc} is voltage at the current density of 0 mA cm^{-2} (open circuit voltage), and *FF* is the fill factor—the area of an inscribed square divided by ($J_{sc} \times V_{oc}$).

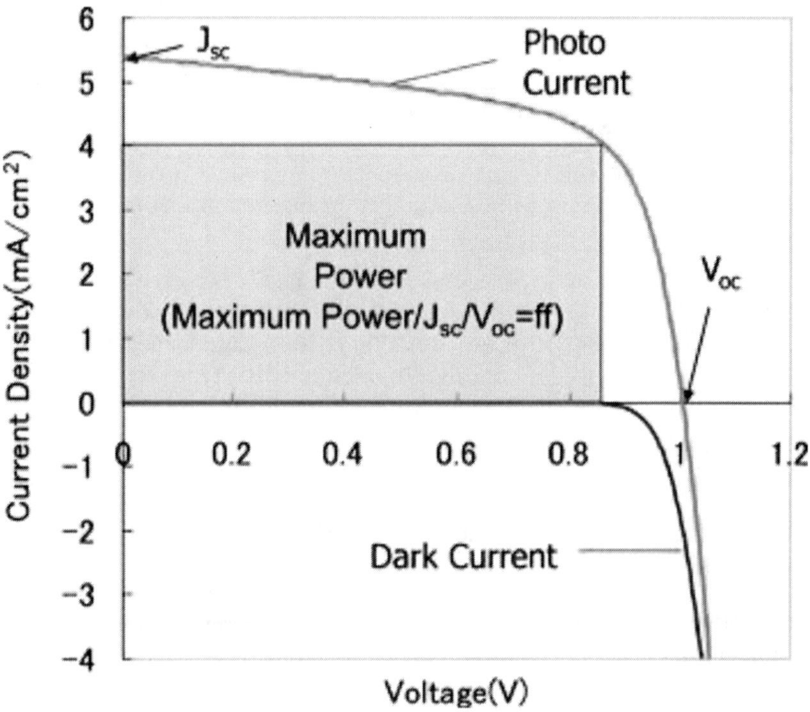

Figure 11. The current–voltage curve and the parameters of solar cell.

Generally the efficiency was measured by using a quasi-sunlight spectrum with 100 mW cm^{-2}as a light source.

The mechanism of the charge separation process is understood, as shown in Figure 12. The process is considered to consist of the following four steps:

a. a photogenerated exciton is created in a photoelectric conversion layer;

b. the exciton moves to the interface of the p/n-junction;

c. the exciton separates into a hole and an electron when they reach the interface;

d. the hole and the electron are transferred to the cathode and anode respectively, and collected there.

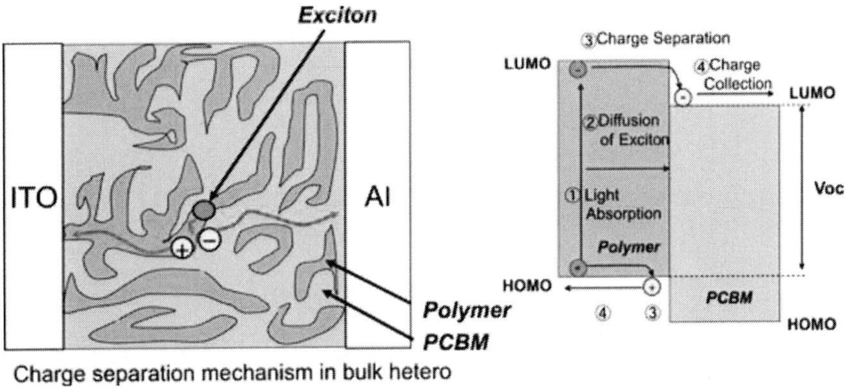

Figure 12. Device structure of bulk hetero type and mechanism of charge separation.

Recently, analytical techniques for the photogenerated charge separation process by femto-to-nanosecond laser have made an advance in exploring the detailed mechanism [67, 68] and the improvement of materials.

For higher conversion rates, increasing J_{SC}, V_{OC}, and *FF* is necessary. As a guideline for increasing J_{SC}, absorbance, the rate of the photogenerated charge separation process, and suppression of the recombination of hole and electron can be listed.

Widening the range of absorption wavelengths is proposed, to increase absorbance. So far P3HT is a commonly used p-type semiconductor material. However P3HT absorbs wavelengths of up to about 650 nm so incident light with long wavelengths is not used effectively for power generation. Investigation of long wavelength absorbance polymers has been actively promoted. For this purpose, polymers were designed by placing electron donor and acceptor units alternately in a polymer chain. Such a configuration causes intra-molecule charge transfer to absorb longer wavelength irradiation. In Figure 13, examples of newly developed polymers with long wavelength absorbance are shown. As table 3 shows, polymer 4 exhibited over 7% PCE in conjunction with appropriate electrode and buffer layers.

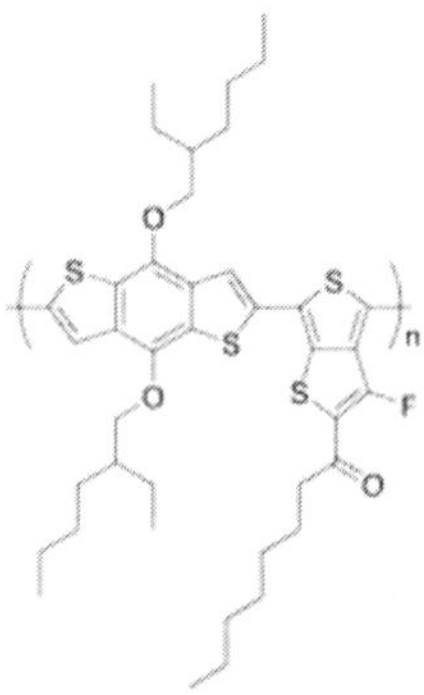

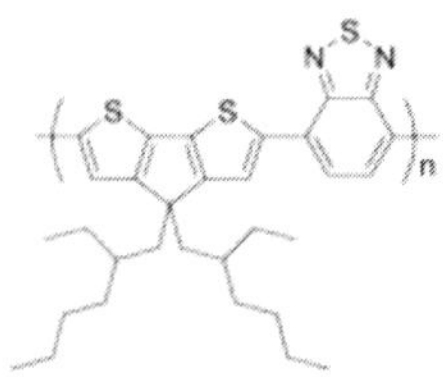

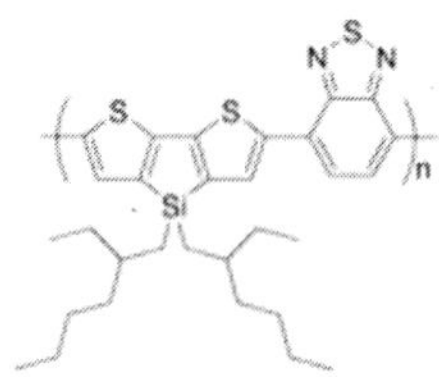

1 2 3

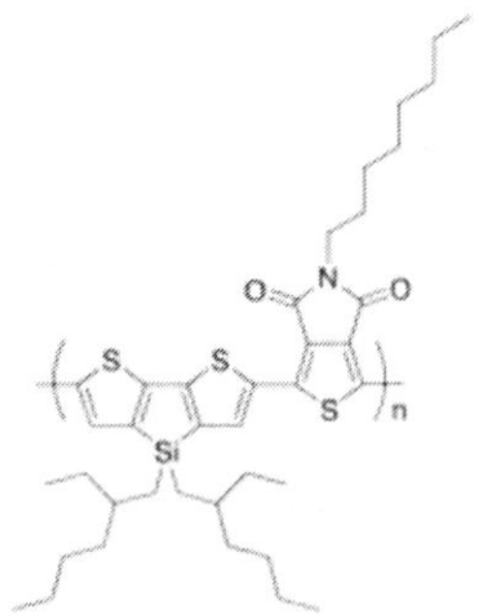

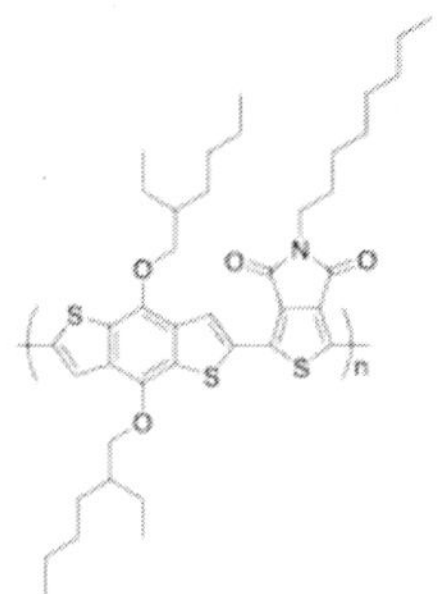

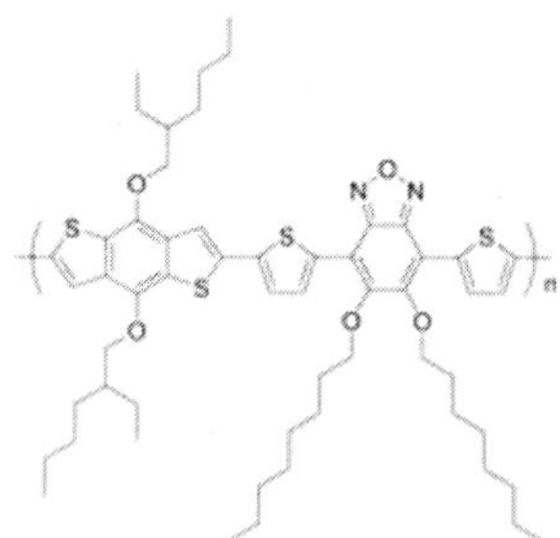

4 5 6

Figure 13. Examples of highly efficient p-type polymers.

Table 3. Examples of OPV performances of long-wavelength absorption polymers.

Polymer	J_{sc} (mA cm^{-2})	V_{oc} (V)	*FF*	PCE (%)	Ref
1	13.4	0.76	66.4	6.77	[69]
2	16.2	0.62	55	5.52	[70]
3	12.7	0.68	55	5.1	[71]
4	12.2	0.88	68	7.3	[72]
5	9.81	0.85	66	5.5	[73]
6	10.4	0.86	66.4	5.7	[74]

The control of photogenerated charge separation or the recombination of hole and electron is strongly related to phase separations as discussed in a later section.

V_{oc} is related to the energy gap between the highest-occupied molecular orbital (HOMO) of a p-type semiconductor, and the lowest unoccupied molecular orbital (LUMO) of an n-type semiconductor. For the purpose of increasing V_{oc}, the development of p-type semiconductor materials with a deep HOMO level and n-type semiconductor materials with a shallow LUMO level are required. The density-functional calculation is adopted extensively to pre-estimate these energy levels for molecular design.

It is considered that the *FF* is strongly correlated with the resistance of a PV cell. Utilizing high carrier mobility materials in the active layer, decreasing the resistance at the interface of each layer, and increasing the parallel resistance of PV equivalent circuits by removing defects in layers are all effective in increasing the *FF* value. OPV is expected to achieve high conversion efficiency by appropriate material designs.

3.2.2. Development of p-type polymers

We have explored p-type polymers for OPV materials based on our core technology of conjugated polymers. We developed fluorene polymers mainly in the early stage, and over a 6% PCE was obtained in conjunction with appropriate device process techniques.

Afterwards, we started to develop polymers with longer absorption wavelengths for the further improvement of efficiency. Our material design policy is shown in Figure 14. In addition to the basic policy of band-gap reduction by strong interaction between electron donor units and electron acceptor units, planar donor units were introduced, giving an expansion of conjugation and high hole mobility, which are important for the carrier transfer after photogenerated charge separation. The molecular design and calculated absorption wavelength supported our efficient screening of units to obtain new polymers with very long absorption wavelength of over 900 nm.

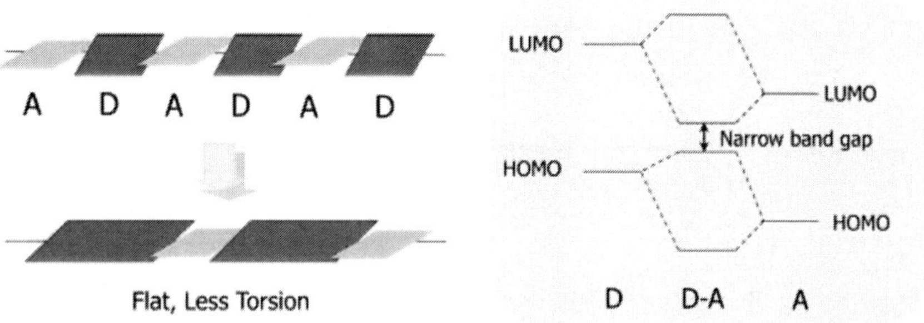

Figure 14. Concept of molecular design for narrow band gap polymers.

Besides compressing the bandgap to obtain higher PCE, new materials with high FFs were also proposed [75]. At high internal electric fields or at near short-circuit conditions, enhanced bulk charge mobility can improve transport and reduce space charge build-up. However, at low internal fields or near V_{oc}, long carrier lifetimes determine charge extraction and thus the steepness of the current-voltage curve. Accordingly, both carrier mobility and lifetime are target parameters to enhance, so as to increase the FFs. To optimize the *FF*, it was considered that highly ordered polymers with close $\pi-\pi$ stacking could maximize mobility. Based on these considerations, the two units of thieno [3,4–c] pyrrole-4,6-dione (TPD) and 2,2'-bithiophene-3,3'-dicarboximide (BTI), shown in Figure 15, were selected as effective building blocks for high mobility polymers. By introducing terthiophene (3T) with alkylation to promote ordering via side-chain interdigitation, it was assumed that combining BTI or TPD as the in-chain acceptor with 3T as the in-chain donor could imbue poly[5-(2-hexyldodecyl)-1,3-thieno [3,4–c] pyrrole-4,6-dione-alt-5,5-(2,5-bis (3-dodecylthiophen-2-yl)-thiophene)] (PTPD3T) and poly[N-(2-hexyldodecyl)-2,2'-bithiophene-3, 3'-dicarboximide-alt-5,5-(2,5-bis(3-decylthiophen-2-yl)-thiophene)] (PBTI3T). According to the evaluation of these polymers in the combination of [6,6]-phenyl-C_{71}-butyricacid methyl ester ($PC_{71}BM$) with an inverted bulk-hetero junction solar cell, they exhibited a high *FF* of ~78% (see table 4).

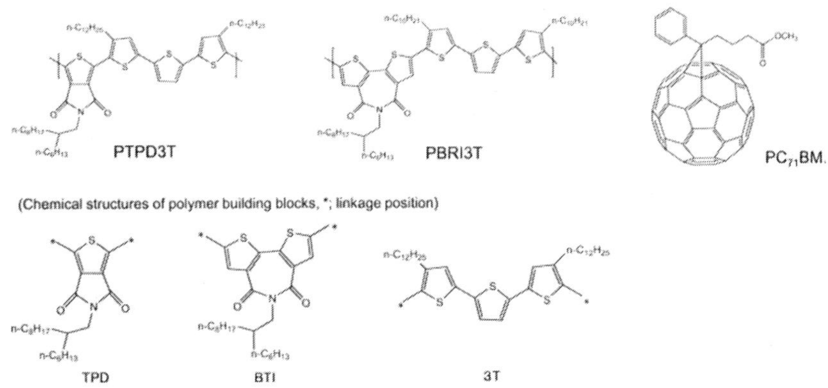

Figure 15. Chemical structures of polymer donors (PTPD3T, PBTI3T) and fullerene acceptor ($PC_{71}BM$) [75].

Table 4. Optimized photovoltaic response data for PTPD3T:$PC_{71}BM$ and PBTI3T:$PC_{71}BM$ inverted BHJ [75].

Polymer	V_{oc} (V)	J_{cs} ($mAcm^{-2}$)	FF (%)	PCE (%)
PTPD3T	0.786	12.3	78.7	7.72
PBTI3T	0.859	12.8	76.3	8.42

3.2.3. Controlling micro phase separation of active layer

Increasing charge separation and charge transfer is useful for obtaining higher conversion efficiency. The average drift path of an exciton is about 10 nm so the micro phase separation size of p-type and n-type semiconductors should be controlled with a similar size of 10 nm.

For the investigation of controlling phase separation, quantification of the phase separation structure is needed. We observed the structure by processing transmission electron microscopy (TEM) images. We introduced a new parameter of interface length per unit area (unit; 1 μm^{-1}) as an indicator to estimate the degree of phase separation, and to analyze the relation between conversion efficiency and phase separation.

As shown in Figure 16, a clear correlation was observed, and an appropriate 'interface length per unit area' was determined.

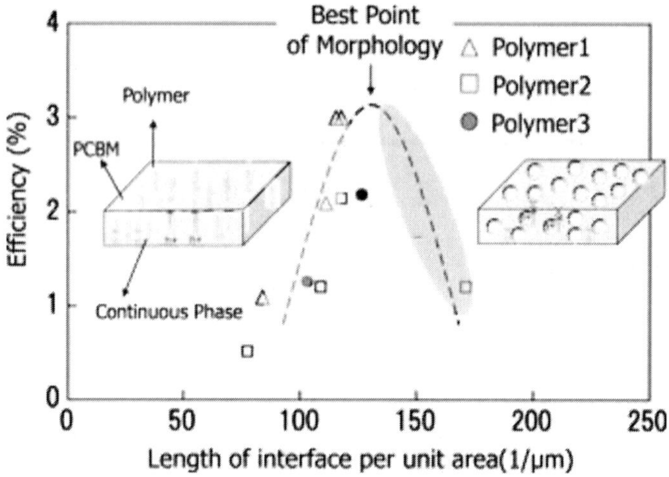

Figure 16. Relation between morphology and efficiency.

In the TEM tomography images showing microscopic phase-separated structures, an ideal formation of continuous domains in the direction of layer thickness was confirmed.

The phase separation structure was formed by an inkjet, or bar-coat, printing and drying process. It is known that such a structure size varies with solvents of inks [76, 77]. We took note of the solubility parameter (SP) value in selecting solvents and thus controlling phase separation size.

When a solvent with an SP value similar to that of fullerene derivatives, the size of the structure became smaller and the interface length became longer. This trend is understood, in that the p-type polymer is precipitated first to form fine networks, and afterwards the n-type fullerene derivative fills in the network to form small size phase separation structures. We achieved over 8% PCE by making use of this phase separation techniques.

3.2.4. Design and optimization of cell structures

As the impact of cell configuration on the efficiency is not small, the design and optimization of layer structures or thicknesses are critical factors in addition to the active materials or the control of phase separation. A thicker active layer gives a higher J_{sc} by increasing the absorbance of light. However this thicker layer results in decreasing charge collection by lowering the internal electric field. It also results in lowering the *FF* by increasing the resistance of the active layer. Furthermore, resistance at interfaces or external circuits has a harmful effect on efficiency. The optimization of cell structures should be tuned in consideration of these various factors.

A tandem device is one of the effective structures for obtaining high efficiency. Stacked active layers make it possible to absorb a wide wavelength range of incident light. In a tandem cell the active layers are connected in series, therefore V_{oc} is higher than that in a single cell. On the other hand, J_{sc} becomes lower than in a single cell due to reduced absorbance in each stacking structure.

We optimized the layer thicknesses of such tandem cells, and the material selection of the interconnecting layer to obtain 10.6% PCE with Prof. Yang of UCLA [78].

3.3. Next direction of R&D for polymer OPV materials

Given that the OPV industry is well on its way to market entry, R&D in practical aspects will be more active. The investigation of high efficiency polymers has been widely examined so far [79–81], and will be continued. Not only the simple polymer design for higher performance but also the orientation of polymer chains or morphology will be focused on as research subjects. As mentioned in the previous section, the impact on device performance is strongly tied with morphology because the power generation process is affected by charge separation, carrier mobility, etc In succession to the previously reported studies [82, 83], further active research is expected.

We would like to point out that specifically designed polymers for OPV applications, (*i.e.* transparent polymers, specific colored polymers, etc) will be explored competitively to meet strong requests from industrial design. This is one of the next trends in OPV material developments.

From a standpoint of quality assurance of products, one of the most important issues is an understanding of degradation. OPV will not always be used in the same circumstances as Si type solar systems. A variety of durability test conditions and evaluation methods will be explored and established so OPV products can be evaluated appropriately. At the same time, degradation analysis will also be investigated in order to design new robust OPV materials.

The R&D of OPV modules and the fabrication process are also important in order to achieve practical products. A big gap in performance exists between large size modules and small size test cells. New materials should contribute to solving this problem by detailed gap analysis and adequate material design for large size panel fabrications.

4. CONCLUSIONS AND PERSPECTIVE

In this article, the material technology in polymer OLEDs and OPVs was reviewed from their development histories to recent progress, including our experiences.

In the section on OLEDs, the features of OLEDs with polymers (PLEDs) were explained in comparison with OLEDs with small molecules. The main strengths of PLEDs are suitability to the printing process for device fabrication, and their simple device structure. Conjugated polymers can be well desinged to meet the required performances by co-polymerization technology. A variety of units have been developed, and polyfluorene is widely and practically used from standpoints of wide band gap and high conductivity of charge carriers.

Regarding polymer OLED performances, higher emission efficiency and longer operating device lifetime were picked up as the main subjects for development. We introduced an example of polymer design whereby controlling the hole mobility with amine unit contents in the polymer chain optimizes the emission efficiency. The impact of introducing an interlayer on the efficiency is an epoch-making technology in PLED, and TTA that is commonly used in fluorescent materials as a up-conversion technology was also introduced. By using these techniques, over 10% of EQE in a blue fluorescent PLED was obtained. Concerning device operating lifetimes, forming excited states was considered to be a key cause of the degradation process from the comparison of photoluminescence changes by operation of bipolar and monopolar devices.

In the last section on OLED, future trends of polymer OLED materials were indicated. Improvements of sticking image lifetime by the development of new materials is expected, and such investigation are predicted to become more active. Combinations of polymer and phosphorescent emitters, especially blue, was listed as one of these new materials. TADF, that is another up-coversion type material, was also introduced as a topical high efficiency material. As very high efficiency has already been obtained with small molecule type materials, it is expected that TADF will be applied to polymer materials as well.

In the OPV section, the main issue in the improvement of power PCE was explained, from the working mechanism of OPV, to guiding thoughts for higher efficiency. After the proposal of a BHJ type, remarkable improvements of PCE were obtained of over 4–5% with a combination of P3HT and [C_{60}]-PCBM. A lot of p-type polymers were developed to improve the absorption wavelength range, J_{sc}, V_{oc}, and *FF*. According to our material design policy, very long absorption wavelengths of over 900 nm were obtained. High *FF* materials were also designed by the concept of highly ordered polymers with close π–π stacking to enhance carrier mobility and lifetime. As a result, high FFs of around 78% were obtained. For higher PCE, the control of micro phase separation and the design of cell structures should also be considered in additon to materials development.

In the last part of the OPV section, the next direction of R&D for polymer OPV materials was discussed. In addition to continuous material development, R&D for the establishment and evaluation of durabilility tests, and further investigation and development of phase separation techniques will also be active.

For both of the technologies of polymer OLED and OPV, it can be said that fundamental material performances have become close to practical levels. However, we should understand that further efforts are needed to finish the material development to be used for electrical device products. Printed electronics materials are strongly tied to device architectures and fabrication processes to demonstrate their performances, thus parallel developments in materials and device/process are expected to be effectively promoted.

ACKNOWLEDGEMENTS

We acknowledge that a part of the OPV development reported here was supported by NEDO.

REFERENCES

1. Gustafsson G, Cao Y, Treacy G M, Klavetter F, Colaner N and Heeger A J 1992 Nature **357** 477
2. Tekoglu S, Hernandez-Sosa G, Kluge E, Lemmer U and Mechau N 2013 Org. Electron. **14** 3493
3. Guo X, Baumgarten M and Müllen K 2013 Prog. Polym. Sci. **38** 1832
4. Burroughes J H, Bradley D D C, Brown A R, Marks R N, Mackay K, Friend R H, Burn P L and Holmes A B 1990 Nature **345** 539
5. Ohnishi T, Nakano T, Doi D and Noguchi T Tokukaihei 3-244630
6. Burn P L, Victor C, Lo S-C, Pillow J, Gerard N, Mark L J and Samuel I D W WO02/66552
7. Huang J et al 2009 Adv. Funct. Mater. **19** 2978

8. Greenham N C, Moratti S C, Bradley D D C, Friend R H and Holmes A B 1993 Nature **365** 628

9. Bernius M, Inbasekaran M, O'Brien J and Wu W 2000 Adv. Mater. **12** 1737

10. Bernius M, Inbasekaran M, Woo E, Wu W and Wujkowski L 2000 J. Mater. Sci.: Mater. Electron.**11** 111

11. Bernius M, Inbasekaran M, Woo E, Wu W and Wujkowski L 2000 Thin Solid Films **363** 55

12. Inbasekaran M, Woo E, Bernius M and Wujkowski L 2000 Synth. Met. **111-112** 397

13. Grem G, Leditzky G, Ullrich B and Leising B 1992 Adv. Mater. **4** 36

14. Partrige R H 1983 Polymer **24** 733

15. Partrige R H 1983 Polymer **24** 755

16. Mori Y, Endo H and Hayashi Y 1992 Oyo Butsiri **61** 1044

17. Takizawa S, Nishida J, Tsuzuki T, Tokito S and Yamashita Y 2007 Inorg. Chem. **46** 4308

18. Wang X-Y, Kimyonok A and Weck M 2006 Chem. Comm. **37** 3933

19. Tokito S, Suzuki M, Sato F, Kmamachi M and Shirane K 2003 Org. Electron. **4** 105

20. Gupta A, Watkins S E, Scully A D, Singh T B, Wilson G J, Rozanski L J and Evans R A 2011 Synth. Met. **161** 856

21. Frampton M J, Namdas E B, Lo S C, Burn P L and Samuel I D W 2004 J. Mater. Chem. **14** 2881

22. Lo S C, Anthopoulos T D, Namdas E B, Burn P L and Samuel I D W 2005 Adv. Mater. **17** 1945

23. Lo S C, Harding R E, Shipley C P, Stevenson S G, Burn P L and Samuel I D W 2009 J. Am. Chem. Soc. **131** 16681

24. Pollow J, Liu Z, Sekine C, Mikami S and Mayumi M 2005 SID'05 Digest 1071

25. Tokito S, Weinfurtner K H, Fujikawa H and Taga Y 2000 Gekkan Display **6** 26

26. Ego C, Grimsdale A C, Uckert F, Yu G, Srdanov G, Inbasekaran M and Woo E 2002 Adv. Mater.**14** 809

27. Miteva T, Meisel A, Knoll W, Northofer H G, Scherf U, Müller D C, Meertholz K, Yasuda A and Neher D 2001 Adv. Mater. **13** 565

28. Lee J H and Hwang D H Chem. Comm. 2836 203

29. Vak D, Chun C, Lee C L, Kim J J and Kim D Y 2004 Mater. Chem. **14** 1342

30. Ohmiri Y 2007 Oyo Butsuri **76** 522

31. Wu W et al 2004 Microelectron. J. **35** 343

32. Hou Q, Zhou Q, Zhang Y, Yang W, Yang R and Cao Y 2004 Macromolecules **37** 6299

33. Morteani A C, Dhoot A S, Kim J-S, Silva C, Greenham N C, Murphy C, Moons E, Ciná S, Burroughes J H and Friend R H 2003 Adv. Mater. **15** 1708

34. Tokito S, Adachi C and Murata H 2004 Organic EL Displays, 21, Ohm-sha

35. Tsutsui T and Takata N 2013 Jan J. Appl. Phys. 110001

36. Yamada T, Tsubata Y, Sekine C and Ohnishi T 2008 **39** 404

37. Tang C W, VanSlyke S A and Chen C H 1989 J. Appl. Phys. **65** 3610

38. Kim J S, Friend R H, Grizzi I and Burroughes J H 2005 Appl. Phys. Lett. **87** 023506

39. Kondakov D Y, Pawlik T D, Hatwar T K and Spindler J P 2009 J. Appl. Phys. **106** 124510

40. Kuma H 2008 The 7th Yuki EL Toronkai S9-1

41. King S M, Cass M, Pintani M, Coward C, Dias F B, Monkman A P and Roberts M 2011 J. Appl. Phys. **109** 074502

42. Monkman A P 2013 ISRN Mater. Sci. **19** 670130

43. Roberts M, King S, Cass M, Pintani M, Coward C, Akino N, Nakajima H and Anryu M 2011 **42**820

44. Liu J and Pei Q 2010 Current Organic Chemistry **14** 2133

45. Lee C, Lee K and Kim J 2000 Appl. Phys. Lett. **77** 2280

46. Motomura 2009 The 9th Yuki EL Toronkai S4-2

47. Motomura 2010 The 10th Yuki EL Toronkai S2-2

48. Liu J and Pei Q 2010 Macromolecules **43** 9608

49. He G, Li Y, Liu J and Yang Y 2002 Appl. Phys. Lett. **80** 4247

50. Kim K, Doherty W, Salaneck W, Murphy C, Friend R and Kim J 2010 Nano Lett. **10** 385

51. Endo A, Ogasawara M, Takahashi A, Yokoyama D, Kato Y and Adachi C 2009 Adv. Mater. **21**4802

52. Uoyama H, Goushi K, Shizu K, Nomura H and Adachi C 2012 Nature **492** 235

53. Adachi C 2013 Oyo Butsuri **82** 458

54. Tanaka H, Shizu K, Nakanotani H and Adachi C 2013 Chem. Mater. **25** 3766

55. Zhang Q, Li J, Shizu K, Huang S, Harata S, Miyazaki H and Adachi C 2012 J. Am. Chem. Soc.**134** 14706

56. Wu S, Aonuma M, Zhang Q, Huang S, Nakagawa T, Kuwabara K and Adachi C 2014 J. Mater. Chem. C **2** 421

57. Nasu K, Nakagawa T, Nomura H, Lin C J, Cheng C H, Tseng M R, Yasuda T and Adachi C 2013Chem. Commun. **49** 10385

58. Kim K, Doherty W, Salaneck W, Murphy C, Friend R and Kim J 2010 Nano Lett. **10** 385

59. Bohm E et al 2011 SID-ME Spring Meeting

60. Hiramoto M, Fujiwara H and Yokoyama M 1991 Appl. Phys. Lett. **58** 1062

61. Hiramoto M, Fujiwara H and Yokoyama M 1992 J. Appl. Phys. **72** 3781

62. Yoshino K, Morita S, Kawai T, Araki H, Yin X H and Zakhidov A A 1993 Synth. Met. **56** 2991

63. Yu G, Gao J, Hummelen J C, Wudl F and Heeger A J 1995 Science **270** 1789

64. Hummelen J C, Knight B W, LePeq F, Wudl F, Yao J and Wilkins C L 1995 J. Org. Chem. **60** 532

65. Li G, Shrotriya V, Huang J, Yao Y, Moriarty T, Emery K and Yang Y 2005 Nat. Mater. **4** 864

66. Reyes-Reyes M, Kim K and Carroll D L 2005 Appl. Phys. Lett. **87** 83506

67. Yamamoto S, Ohkita H, Benten H and Ito S 2012 J. Phys. Chem. C **116** 14804

68. Ohkita H, Kosaka J, Guo J, Benten H and Ito S 2011 J. Photon. Energy **1** 011118

69. Chen H-Y, Hou J, Zhang S, Liang Y, Yang G, Yang Y, Yu L, Wu Y and Li G 2009 Nat. Photonics **3**649

70. Peet J, Kim J Y, Coates N E, Ma W L, Moses D, Heeger A J and Bazan G C 2007 Nat. Mater. **6**497

71. Hou J, Chen H-Y, Zhang S, Li G and Yang Y 2008 J. Am. Chem. Soc. **130** 16144

72. Chu T-Y, Lu J, Beaupré S, Zhang Y, Pouliot J-R, Wakim S, Zhou J, Leclerc M, Li Z, Ding J and Tao Y 2011 J. Am. Chem. Soc. **133** 4250

73. Zou Y, Najari A, Berrouard P, Beaupré S, Aïch B R, Tao Y and Leclerc M 2010 J. Am. Chem. Soc.**132** 5330

74. Jiang J-M, Yang P-A, Chen H-C and Wei K-H 2011 Chem. Commun. **47** 8877

75. Guo X et al 2013 Nat. Photonics **7** 825

76. Zhou E, Cong J, Tajima K, Yang C and Hashimoto K 2012 J. Phys. Chem. C **116** 2608

77. Zhou E, Cong J, Tajima K and Hashimoto K 2010 Chem. Materials **22** 4890

78. Sumitomo Chemical 2012 http://www.sumitomo-chem.co.jp/newsreleases/2012/

79. Osaka I, Kahara T, Takemura N, Koganezawa T and Takimiya K 2013 J. Am. Chem. Soc. **135**8834

80. Chang W H, Gao J, Dou L, Chen C C, Liu Y and Yang Y 2013 Adv. Energy Mater. **1**

81. Li Y et al 2013 Macromolecules **46** 5497

82. Osaka I, Saito M, Koganezawa T and Takimiya K 2014 Adv. Mater. **26** 331

83. Bartelt J A et al 2013 Adv. Energy Mater. **3** 364

3

CHAPTER

BPDA-PDA Polyimide: Synthesis, Characterizations, Aging and Semiconductor Device Passivation

S. Diaham, M.-L. Locatelli and R. Khazaka

[1] University of Toulouse – UPS – INPT – LAPLACE Laboratory– CNRS, Toulouse, France

1. INTRODUCTION

Polyimides (PIs) are advanced polymeric materials well-known for their excellent thermal, electrical, mechanical and chemical properties [1]. PIs are particularly attractive in the microelectronics industry due to their high thermal stability (*Td*), high glass transition temperature (*Tg*), low dielectric constant, high resistivity, high breakdown field, inertness to solvent, radiation resistance, easy processability, etc [2,3]. Recently, the emergence of novel wide bandgap semiconductor (SiC, GaN or Diamond) devices aiming to operate between 200 °C and 400 °C make PIs as one of the most potential organic materials for the surface secondary passivation [4]. In such a high temperature range of operation with large thermal cycling constraints imposed by both the devices and the ambient temperature, the thermo-mechanical properties of passivation PIs appear as fundamental to ensure long lifetime of the materials and reliable behaviour of the devices. For instance, Table I shows some electrical, thermal and mechanical properties of wide band gap semiconductors compared to those of silicon (Si).

Table 1. Main physical properties of wide band gap semiconductor materials at 300 K.

	E_g (eV)	ε	E_{br} (MV/cm)	λ_{th} (W/m/K)	CTE (ppm/°C)
Si	1.1	11.8	0.2	150	3
GaN	2.3-3.3	9-11	1.3-3.3	80-130	5.6
SiC	2.2-3.3	10	1.2-2.4	450	4
Diamond	5.4	5.5	5-10	2000	1

E_g: forbidden energy band gap; ε: dielectric constant; E_{br}: breakdown field; λ_{th}: thermal conductivity; CTE: coefficient of thermal expansion.

It is possible to observe that the semiconductor materials own a low coefficient of thermal expansion (CTE) below 6 ppm/°C. Thus, among the criteria for the PI passivation material choice, the CTE, *Tg*and *Td* temperatures are of prior importance. Table II shows some electrical, thermal and mechanical properties of the main aromatic PIs.

The classical poly(4,4'-oxydiphenylene pyromellitimide) (PMDA-ODA) appears as not well adapted for a severe thermal cycling operation due to the strong mismatch between its CTE (30-40 ppm/°C) and the one of semiconductor materials (<6 ppm/°C). This mismatch induces strong mechanical stresses in PMDA/ODA films coated on Si wafer (29 MPa). The poly(4-4'-oxydiphenylene biphenyltetracarboximide) (BPDA-ODA), the poly(4-4'-oxydiphenylene benzophenonetetracarboximide) (BTDA-ODA), the poly(*p*-phenylene benzophenonetetracarboximide) (BTDA-PDA) and the poly(*p*-phenylene oxydiphthalimide) (ODPA-PDA) have a chemical packing which leads to similar issues on Si wafer due to large CTEs mismatching (≥28 ppm/°C). Moreover, PMDA-ODA, BPDA-ODA, BTDA-ODA and ODPA-PDA exhibit the lowest degradation temperature values due to the degradation of the C–O–C ether bond present in the PI monomer unit [1,15,16].

On the contrary, PIs synthesized from pyromellitic dianhydride (PMDA) or 3,3',4,4'-biphenyltetracarboxilic dianhydride (BPDA) with *p*-phenylene diamine (PDA) in order to form PMDA-PDA and BPDA-PDA, respectively, present a higher thermal stability (*Td*=610 °C and 595 °C, respectively) than PIs owning C–O–C ether bonds. Moreover, both appear as better candidates for the wide band gap semiconductor passivation due to their low CTEs (2 ppm/°C and 3-7 ppm/°C, respectively). For instance, PMDA-PDA and BPDA-PDA show internal stresses of -10 MPa (compression) and 5 MPa when they are coated on Si wafers (3 ppm/°C). Thus, BPDA-PDA seems to be, as given by the main thermo-mechanical properties, the most compatible PI for SiC and GaN semiconductor passivation while PMDA-PDA should be preferred for diamond passivation.

Table 2. Thermal, electrical and mechanical properties of the main aromatic PIs.

	T_g (°C)	T_d (°C)	ε	E' (GPa)	TS (MPa)	Film stress on Si wafer (MPa)	CTE (ppm/°C)
PMDA-ODA	>380	586	3.4	3.0	170	29	30-40
PMDA-PDA	>450	610	3.0	12.2	296	-10 (compression)	2
BPDA-ODA	325	531	3.3	3.0	230	34	28
BPDA-PDA	>330	595	3.1	10.2	597	5	3-7
BTDA-ODA	280	554	3.0	3.0	150	40	40
BTDA-PDA	>370		2.98	7.1	248	30	35
ODPA-PDA	>370		3.0	8.1	263	42	35

T_g: glass transition temperature; T_d: degradation temperature defined at 10% wt. loss; ε: dielectric constant; E': Young's modulus; TS: tensile strength; CTE: coefficient of thermal expansion between 50 °C and 300 °C. Film stress is given for film below 20 μm of thickness [5-14].

In this chapter, a particular attention is focused on the electrical properties of unaged BPDA-PDA and their evolution during a thermal aging on Si wafers in both oxidative and inert atmospheres. A comparative aging study with higher CTE's PIs (PMDA-ODA and BPDA-ODA) is carried out in order to highlight the longer lifetime of BPDA-PDA. Prior to this, a paragraph dealing with the optimization of the thermal imidization of BPDA-PDA is reported through a simultaneous analysis of the infrared spectra and the electrical properties evolutions as a function of the imidization curing temperature. Finally, an application of BPDA-PDA to the passivation of SiC semiconductor devices will be presented through the PI on-wafer etching process and the electrical characterization of bipolar diodes at high temperature and high voltage.

2. SYNTHESIS AND OPTIMIZATION OF THE IMIDIZATION OF BPDA-PDA POLYIMIDE

The final physical properties of PIs and their integrity during aging depend strongly on the control and on the optimization of the imidization reaction (i.e. the curing process) [17,18]. This process step appears as crucial for industrial applications. Unfortunately, it is quite difficult to predict *a priori* the imidization temperature optimum which leads to the best electrical properties. Literature presents a large range of imidization temperature from 200 °C to 425 °C without always indicating if it corresponds to an optimum [6,18-23]. Moreover, these works present mainly the optimization of the imidization reaction from a chemical point of view only based on *qualitative* infrared measurements. Even if the dielectric properties are strongly linked to the PIs chemical structure, it would be more adequate to optimize the imidization reaction taking into account both the chemical structure and the dielectric properties simultaneously. Indeed, dielectric characterizations can be more sensitive than infrared measurements regarding the determination of the imidization temperature optimum. In the section 2, the results are extracted from [24].

2.1. Material, sample preparation and curing process

BPDA-PDA PI was purchased as a polyamic acid (PAA) solution. It was obtained through the two-steps synthesis method from its precursor monomers [25]. The PAA solution was obtained by dissolving the precursor monomers in an organic polar solvent N-methyl-2-pyrrolidone (NMP). Two different vicosity types of the PAA solution were used for controlling the thickness. To convert PAA into PI, the solution was heated up to remove NMP and to induce the imidization through the evaporation of water molecules. Figure 1 shows the synthesis steps of BPDA-PDA.

The PAA solution was spin-coated on both square stainless steel substrates (16 cm^2) and highly doped 2" Si N^{++} wafers ($<3\times10^{-3}$ Ω cm). PAA was first spread at 500 rpm for 10 seconds followed by a spin-cast at different rotation speeds between 2000 rpm and 4000 rpm for 30 seconds. Two successive curing steps followed the coatings. After a soft-bake (SB) at a low temperature (*TSB*) of 150 °C for 3 minutes on a hot-plate in air, coatings were hard-cured (HC) at a higher temperature (*THC*) in a regulated oven under nitrogen atmosphere and during a time *tHC*. In the following, the *THC* temperature represents the imidization temperature. The final film thicknesses have been measured using a KLA Tencor Alpha-Step IQ profilometer in a range from 1.5 to 20 μm depending on the spin-coating parameters, the viscosity of PAA and *THC*. In order to perform the electrical measurements, an upper gold metallization was evaporated after imidization onto the PI film surface under vacuum (10^{-4} Pa). This metal layer was then patterned using successively a photolithography step through a selective mask and a humid gold etching to form different circular electrodes from 300 μm to 5 mm in diameter.

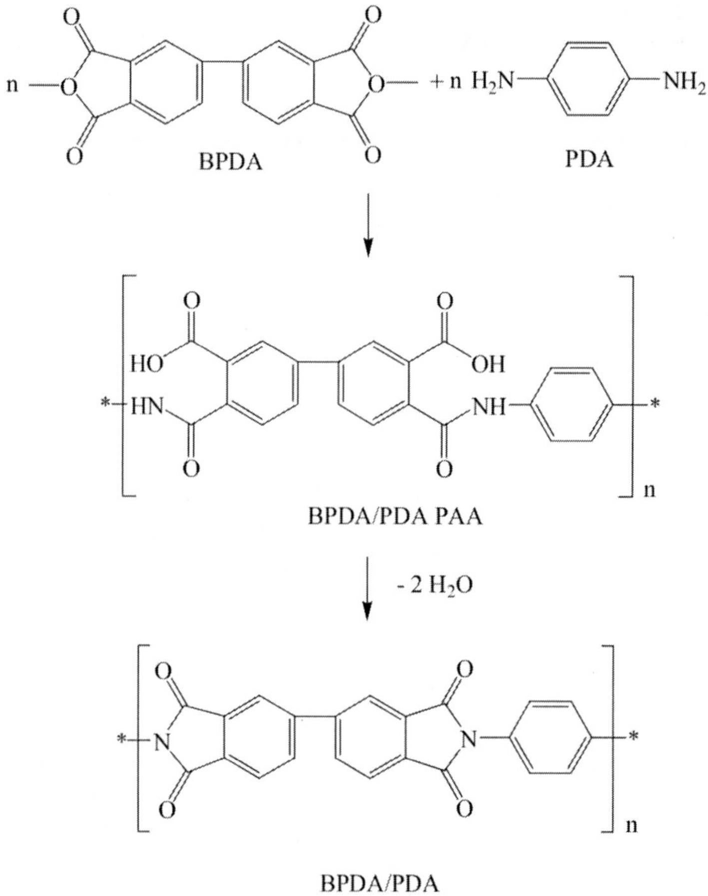

Figure 1. Synthesis steps of the BPDA-PDA polyimide.

The imidization cure is necessary to drive off solvent (boiling point of 202 °C for NMP), and to achieve the conversion of the PAA into PI by the formation of the imide rings. PAA coatings were hard-cured at*THC* in the range from 175 to 450 °C under nitrogen for a time *tHC* of 60 minutes. The heating and cooling rates for all the samples were 2.5 and 4 °C min^{-1}, respectively.

2.2. Optimization of the Imidization Reaction

2.2.1. Fourier transform infrared spectroscopy (FTIR)

In order to detect the chemical bond changes during the imidization of PAA into PI, assignments of the absorption bands in FTIR spectra are necessary to identify the amide and imide peaks. The characteristic IR absorption peaks were assigned thanks to previous works [17,26-36]. Usually, PAA spectra are compound of the N–H stretch bonds at 2900–3200 cm^{-1}, the C=O carbonyl stretch from carboxylic acid at 1710–1720 cm^{-1}, the symmetric carboxylate stretch bonds at 1330–1415 cm^{-1}, the C=O carbonyl stretch of the amide I mode around 1665 cm^{-1}, the 1540–1565 cm^{-1} amide II mode and the 1240–1270 cm^{-1} band due to the C–O–C ether aromatic stretch (if present in the monomer).

After the conversion reaction, the absence of the absorption bands near 1550 cm^{-1} (amide II) and 1665 cm^{-1} (amide I) indicates that PAA has been converted into PI. Simultaneously, this is confirmed by the occurrence of the C=O stretch (imide I) peaks at 1770–1780 cm^{-1} (symmetric) and 1720–1740 cm^{-1}(asymmetric), the typical C–N stretch (imide II) peak around 1380 cm^{-1}, the C–H bend (imide III) and C=O bend (imide IV) absorption bands respectively in the ranges of 1070–1140 cm^{-1} and of 720–740 cm^{-1}. The presence in PI films of a large absorption band between 2900 and 3100 cm^{-1} is associated to the C–H stretch bonds. Finally, the measurements may highlight the occurrence of a shoulder on the asymmetric C=O stretch bonds at 1710 cm^{-1} which corresponds to the out-of-plane optical response of the imide I conformation [37].

Figure 2 shows FTIR spectra of both PAA coatings after the SB at 150 °C and PI films after a HC at 250 °C. Spectra have been normalized to the classical C=C absorption band appearing at 1518 cm^{-1}. The spectrum performed after the SB shows the typical absorption bands of PAA coatings. The large absorption band observed between 2300 and 3400 cm^{-1} corresponds to the N–H stretch vibration modes, the C–H stretch bonds and the O–H stretch bonds present in both the PAA and NMP solvent. The FTIR spectrum of PI films already shows the typical completion of the imidization reaction with the presence of the four absorption bands from the imide rings. They occur at 1775/1734 cm^{-1} (imide I), 1371 cm^{-1} (imide II), 1124/1080 cm^{-1} (imide III) and 737 cm^{-1} (imide IV). Moreover, it is possible to observe the large absorption band induced by the C–H stretch vibration modes between 2600 and 3100 cm^{-1}. At 1415 cm^{-1}, a shoulder appears near the C–N stretch peak. This absorption band could be attributed to symmetric stretch of carboxylate ion COO^-. The carboxylic acid groups present in PAA appear through the O–H stretching bonds at 3400 cm^{-1} but free carboxylic acid groups can be deprotonated by the weak amine base [36]. Consequently, COO^- carboxylate ions are usually also present in PIs exhibiting two peaks around 1606 and 1415 cm^{-1}. This could explain the release of mobile H^+ protons (from COOH) responsible of electrical conduction in PI [38].

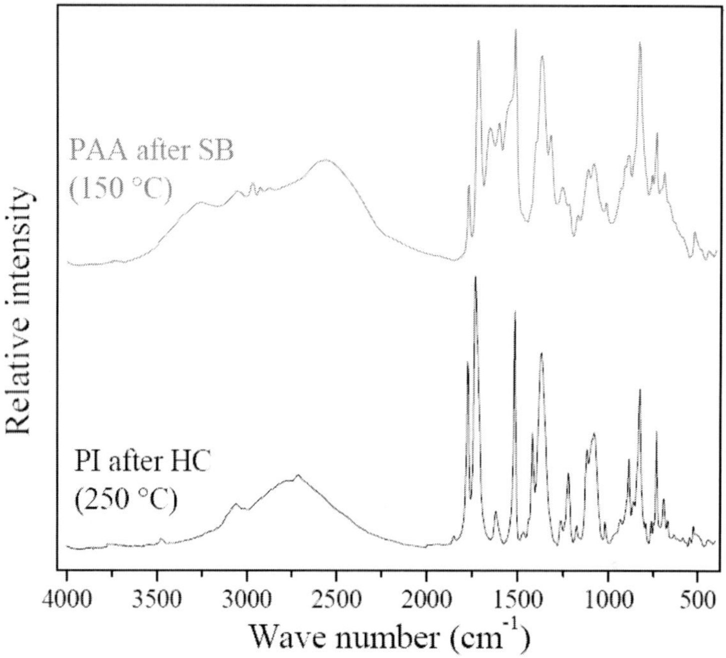

Figure 2. FTIR spectra of BPDA-PDA PAA coatings and PI films (thickness: 1.5 μm). Taken from [24].

FTIR measurements have been performed for different imidization temperatures *THC* to determine the completion of the imidization reaction of the PI films. This analysis rests on the absorption peak magnitude changes in the functional groups or in the characteristic linkages during the reaction. Figure 3a shows the changes in FTIR spectra of BPDA-PDA for different imidization temperatures. When *THC* increases, a general increase in all the absorption peaks is observed. This suggests that residual PAA monomers continue to be converted into PI. This evolution is stabilized after exposure to temperature above 350 °C.

To study the imidization kinetics of PI films, the peak of aromatic ring (C=C) stretching around 1500 cm^{-1} is chosen as a reference and the peak height method is adopted to calculate the amount of the appearing imide groups formed. The degree of imidization (DOI) is thus defined by comparing the intensity of an imide absorption peak normalized to the intensity of the C=C reference band and is given by [27]:

$$DOI_{T_{HC}} = \frac{(A/A^{*})_{T_{HC}}}{(A/A^{*})_{i}} \tag{1}$$

where A^* is the peak height of the C=C reference band at 1518 cm^{-1} and A is the imide peak height (i.e. 1775, 1732, 1420, 1371, 1080 and 737 cm^{-1}). Subscripts i and THC indicate the reaction at the initial and a given imidization temperature, respectively.

Figure 3b shows the extent of imidization of the main bonds of BPDA-PDA versus the imidization temperature. Most of the imidization reaction takes place rapidly with a conversion rate as high as 70-85% at 250 °C and still continues slightly up to 400 °C as shown through the increase in the magnitude of the imide bands. However, it is difficult to detect the optimum imidization temperature (i.e. the highest magnitude) to not exceed in order to preserve PI from degradation. For instance at 450 °C, the imide II and IV absorption bands decrease of 20% and 10% respectively showing the initiation of a desimidization of the structure. Therefore, the use of complementary electrical measurements as a probe of the imidization advancement can allow obtaining a higher accuracy regarding the optimal temperature of the curing.

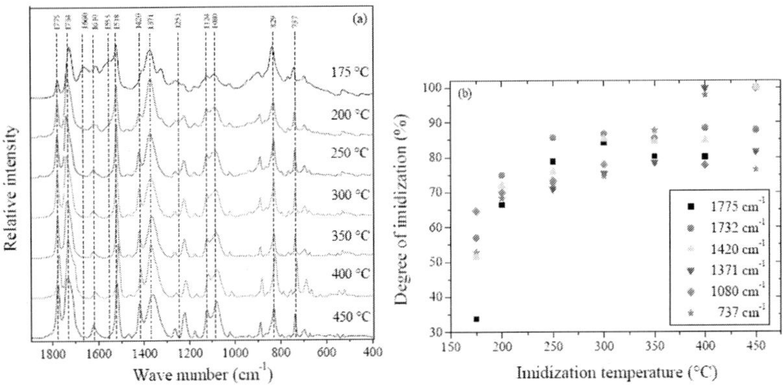

Figure 3. a) FTIR spectra of BPDA-PDA for different *THC* (thickness: 1.5 µm). (b) Degree of imidization for the main absorption bands versus the imidization temperature. Taken from [24].

2.2.2. Electrical properties

As for the DOI, the electrical properties strongly depend on the imidization temperature. Changes in the electrical conductivity, dielectric properties or in the dielectric breakdown field of the PI films can be used to determine precisely the optimal imidization temperature. Larger the DOI is, better the electrical properties are expected due to a lower impurities amount in the PI films.

Current-Field (*j-F*) measurements up to 1 MV/cm show that, as soon as the low field range (10 kV/cm), the minimum of the dc conductivity (σDC, evaluated from the *j-F* curves) is obtained for an imidization cure of 400 °C (as seen in Figure 4a). Whereas the σDC values are poorly dispersed between 10^{-17} and 10^{-16} Ω^{-1} cm^{-1} up to 200 kV/cm whatever *THC*, the measurements show a strong divergence at high fields from 400 kV/cm. In this field range, the dc conductivity of BPDA-PDA increases much more for imidization curings at 350 °C and 450 °C (as seen in Figure 4a). Thus, the highest insulation quality (i.e. the lowest σDC) has been obtained for the films imidized at 400 °C. For *THC*=400 °C, the charges density and/or their mobility seem to be strongly reduced with this optimal imidization temperature. Usually, the electrical conduction in PI films is related to the motion of H^+ protons coming from unreacted PAA [38]. This is in likelihood agreement with the evolution of the COO^- band intensity which reaches its lowest magnitude at 400 °C before it increases again at 450 °C (as seen in Figure 3b). This observation is emphasized by the changes in the dielectric breakdown field (*EBR*) as a function of the imidization temperature (as seen in Figure 4b). Indeed, after a continuous increase in *EBR* up to*THC*=400 °C, a sudden decrease in its magnitude when *THC* is raised up to 450 °C is observed. The degradation of the electrical properties above *THC*=400 °C appears as a consequence of desimidization (i.e. the decrease in the imide bands) of BPDA-PDA structure leading to the release of free mobile charges in the bulk.

3. THICKNESS INFLUENCE ON THE STRUCTURAL AND DIELECTRIC PROPERTIES OF BPDA-PDA POLYIMIDE

3.1. Influence on The Chemical Structure

The influence of the thickness of PI films on the chemical structure is rarely investigated. Figure 5shows FTIR spectra of BPDA-PDA imidized at 400 °C for different film thicknessses. As represented by the downward arrows, one can observe that the quantity and the intensity of the bands corresponding to the amide bonds increase when increasing the film thickness. Hence, for higher film thicknesses, the conversion rate of PAA into PI is strongly affected either due to a bad diffusion of the temperature within the medium of the coating bulk during the curing process (presence of unreacted PAA) or due to a higher difficulty to remove by-products such as solvent and water molecules inherent in the imidization reaction. Unfortunately, this issue cannot be solved by higher temperatures or longer curings because in this range the desimidization of PI starts. Consequently, all these remaining impurities can act as ionizable centers supplying free mobile charges.

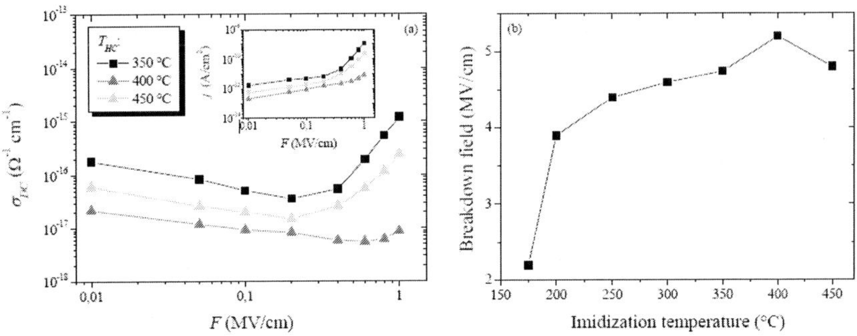

Figure 4. a) Volume conductivity versus electric field of BPDA-PDA for different *THC* (thickness: 1.5 µm). (b) Breakdown field versus the imidization temperature (mean of 20 tested samples). Measurements performed at room temperature. Taken from [24].

3.2. Influence on the Electrical Properties

Wheras such a phenomenon can be negligeable for low temperature applications (< 150 °C) because of the low mobility of free charges, this can be more influent at high temperature (>200 °C). Figure 6ashows the temperature dependence of the dc conductivity of BPDA-PDA between 200 °C and 350 °C for thicknesses from 1.5 µm up to 20 µm. At 200 °C, the dc conductivity is one order of magnitude higher for the thickest films compared to the one of the thinnest films. In comparison at 300 °C, the dc conductivity is two orders of magnitude higher for the thickest films than for the thinnest ones. The fact that the low field dc conductivity is thickness-dependent, particularly in the high temperature range, is directly related to the presence into BPDA-PDA of unreacted PAA impurities for which temperature supplies sufficient energy to the free charges to become mobile.

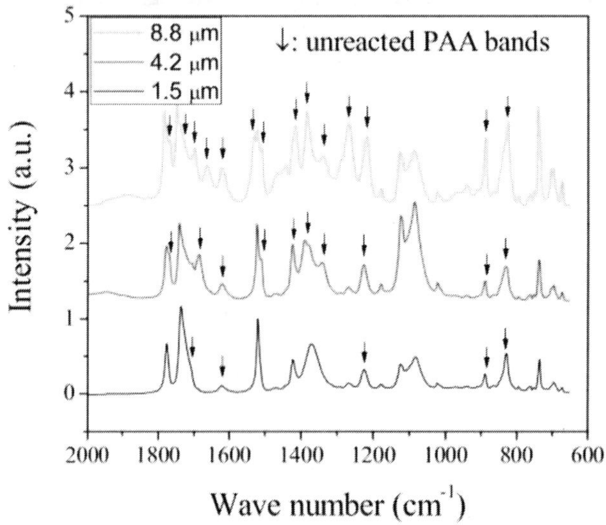

Figure 5. FTIR spectra of 400 °C-imidized BPDA-PDA for different film thicknesses.

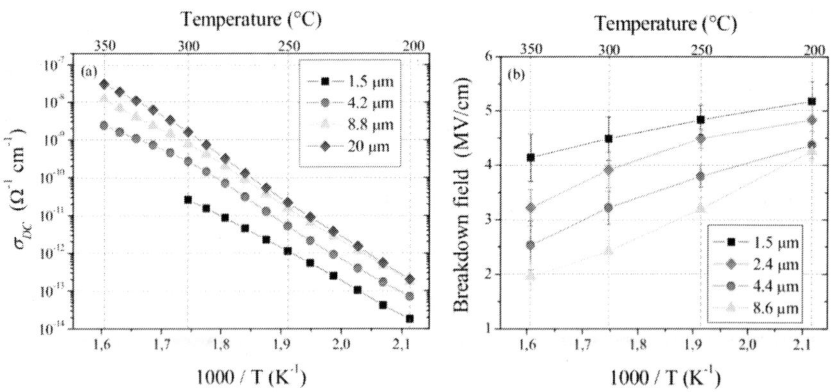

Figure 6. Temperature dependence of the dc conductivity (a) and breakdown field (b) of BPDA-PDA for different film thicknesses.

Figure 6b shows the temperature dependence of the dielectric strength of BPDA-PDA between 200 °C and 350 °C for thicknesses from 1.5 μm up to 8.6 μm. Same findings can be done in this high electric field region. The larger presence of PAA impurities in the thickest films leads to substantially decrease the dielectric breakdown field of 15% at 200 °C (compared to thinnest films) and of 55% at 350 °C. In these thick films, the earlier breakdown event could be explained by a prematured Joule effect occuring when a higher conduction current magnitude happens across the film during the voltage raising. Thus, the breakdown channel appears for lower applied electric fields.

4. THERMAL AGING OF BPDA-PDA POLYIMIDE

The effect of long time aging of polyimide at high temperature (>200 °C) and in oxidative environment on the mechanical properties [39], weight loss [40,41], and chemical properties [42,43], was widely investigated for thick polyimide matrix composites (1 mm thick) used in high temperature aerospace applications. It was found that while thermal degradation occurred throughout the material, the oxidative degradation occurs mainly within a thin surface layer where oxygen diffuses into the material. Few papers discussed the effect of thermal aging on the electrical properties of PIs and this is always for thick and freestanding films [44,45]. Consequently, an overall understanding of the thermo-oxidative aging mechanisms (for PI thickness <20 μm) and their effects on the electrical properties are still lacking.

All the measurements presented below were performed on PI films deposited on highly doped N^{++}cleaned silicon wafers (resistivity < 3×10^{-3} Ω cm) and aged at high temperature. The time at the beginning of the aging is noted as *t0*. At scheduled times, a part of the specimens were removed from the oven and subjected to non-destructive and destructive tests. The thicknesses noted in the text and labels refer to the initial ones measured at *t0*.

4.1. Thermal Stability of Pis

For PIs, it has been shown that the increase in the number of benzene rings contributes to an increase in the degradation temperature [1]. However, the degradation temperature can be also affected by the presence of low thermo-stable bonds in the macromolecular structure. For instance, even if BPDA-PDA and PMDA-ODA (Kapton-type) own the same number of benzene rings (i.e. three in elementary monomer backbone), the absence of the C—O—C ether group in the case of BPDA-PDA allows increasing *Td* (defined at 10% wt. loss) of 48 °C in nitrogen and 100 °C in air in comparison to *Td* of PMDA-ODA (see Figure 7). Moreover, if the diamine ODA is replaced by PDA in BPDA-based PIs, *Td*increases of 68 °C in nitrogen and 105 °C in air. Indeed, this is due to the lower thermal stability of the ether bonds inducing earlier degradations than the rest of the structure [1,44].

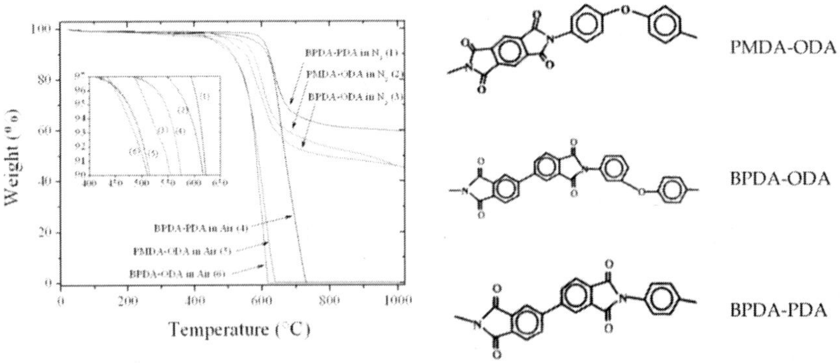

Figure 7. Dynamical TGA of PMDA-ODA, BPDA-ODA and BPDA-PDA in air and nitrogen.

4.2. Thermal Aging in Inert Atmosphere

Figure 8 shows the evolution of the FTIR spectrum of BPDA-PDA before and after an aging at 300 °C in nitrogen during 1000 h and the evolution of the film thickness and the related breakdown field during this aging. One can notice that at this temperature in inert atmosphere, no change in the vibration bonds is remarkable even after a long period of aging. This is in agreement with a good stability of both the film thickness and the high field dielectric properties.

Moreover, a similar observation were done for aging at higher temperature. Indeed, up to *Tg* at 360 °C in nitrogen, both a stability of the chemical structure and the breakdown field were observed during 1000 h. This concludes that BPDA-PDA does not evolved up to 360 °C in inert atmosphere.

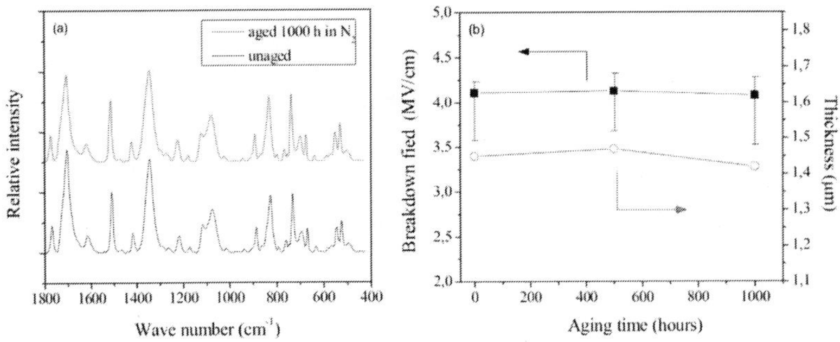

Figure 8. a) FTIR spectra before and after aging in N_2 at 300 °C and (b) breakdown field and thickness during aging for 1.5 µm-thick BPDA-PDA films. The breakdown field is measured at 300 °C.

4.3. Thermal Aging Below *Tg* In Oxidative Atmosphere

The aging effects up to 5000 h at 300 °C in air on different properties of BPDA-PDA were measured on three different initial thicknesses varying from 1.5 µm to 8 µm. For comparison, a same aging was performed on BPDA-ODA, with a glass transition *Tg* of330 °C, up to 1000 h.

The chemical structure variation was measured by FTIR and the spectra of the 4.2 µm-thick film is presented in Figure 9a. A quasi-stabilisation of almost all the peaks can be revealed during 5000 h specially the imide ones (see Figure 9b). However, an increase in the peak localized at 1212 cm^{-1}related to the asymmetric vibration of the C-O-C band can be observed. This can indicate the occurrence of additional oxidation of the unreacted polyamic acid, which was not completely imidized during the curing cycle.

On the contrary, for BPDA-ODA films (see Figure 9c and 9d), the same aging at 300 °C during 1000 h shows as soon as the first 200 h a strong decrease in all the main vibration bonds. Consequently, such an aging affects the chemical integrity of the chemical backbone and the physical properties would be modified.

If we look at the film thicknesses, the BPDA-PDA films do not show any thickness variation during 5000 h of aging, indicating that neither densification nor degradation occurred. On the other hand, the BPDA-ODA films loose more than 50% of their initial thickness after 1000 h of aging, reflecting the strong degradation in this case.

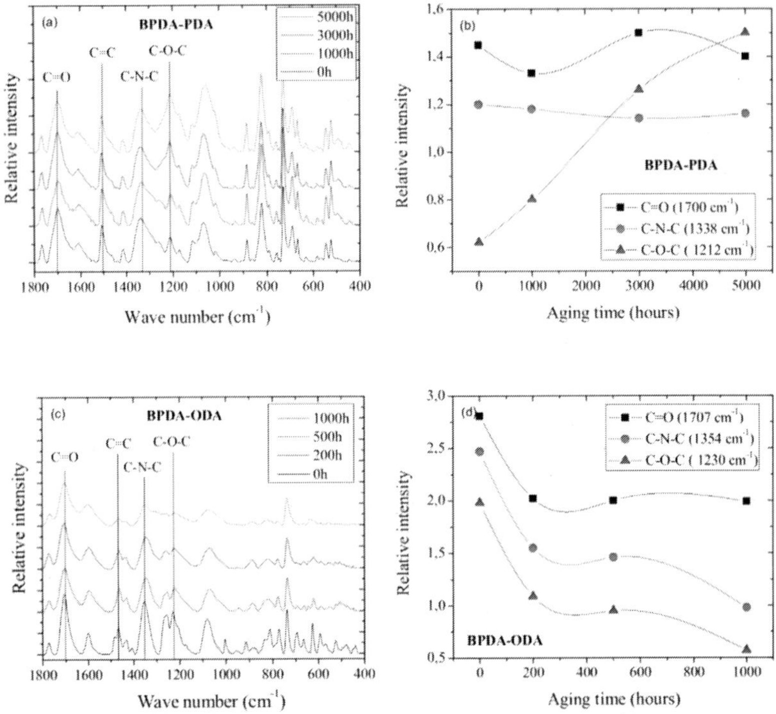

Figure 9. FTIR spectra during the aging in air at 300 °C for 4.2 µm-thick BPDA-PDA and 13.7 µm-thick BPDA-ODA films.

The effect of the aging under air atmosphere on the breakdown field and low field dielectric properties measured at the aging temperatures, for the BPDA-PDA and BPDA-ODA films, are now presented and discussed. The breakdown field, performed by polarizing positively the gold electrode, for different initial BPDA-PDA thicknesses and one BPDA-ODA thickness are presented in Figure 11. Whereas a stabilization of the breakdown field during the 5000 h aging is observed for the BPDA-PDA films, a continuous decrease is observed for the BPDA-ODA films. This invariance of the breakdown field of BPDA-PDA is in good agreement with the good stability of FTIR spectra during aging. On the contrary, the strong decrease in the breakdown field of BPDA-ODA after only 1000 h highlights the progressive and fast degradation of the imide bonds in this kind of PIs.

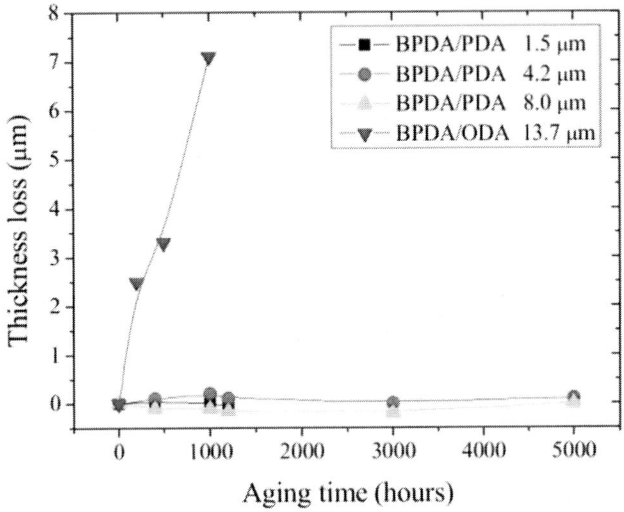

Figure 10. Thickness loss during the aging at 300 °C in air for three initial thicknesses of BPDA-PDA and a 13.7 µm initial thickness of BPDA-ODA.

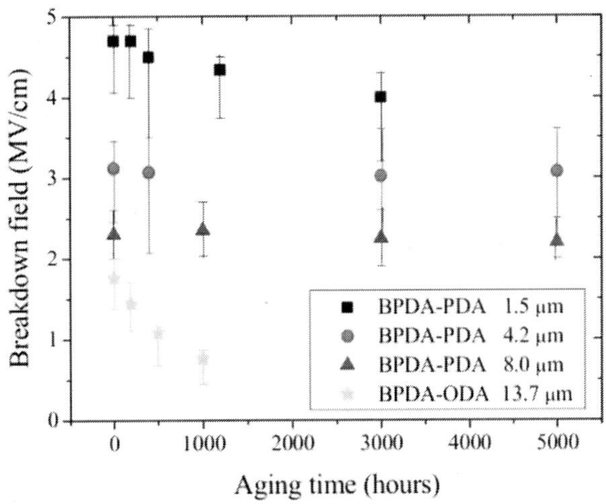

Figure 11. Breakdown field during the aging at 300 °C in air for three different thicknesses of BPDA-PDA and the 13.7 µm-thick films of BPDA-ODA. The breakdown field is measured at 300 °C.

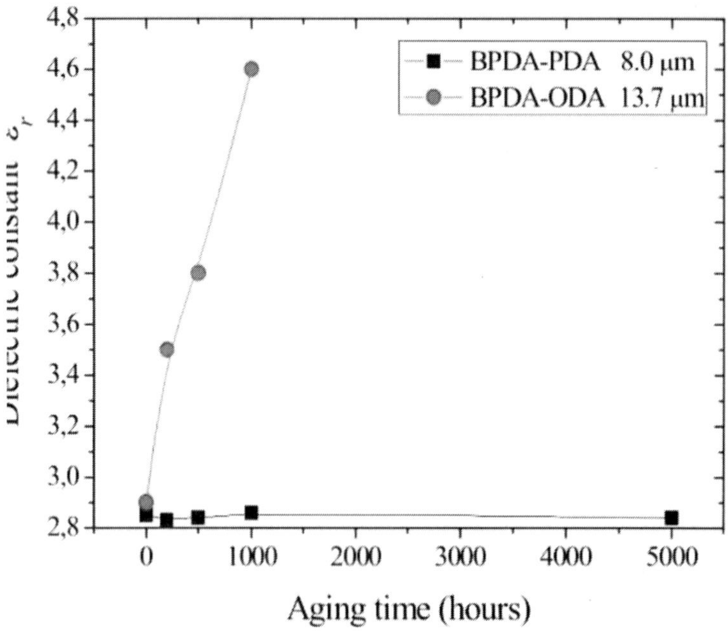

Figure 12. Dielectric constant measured at 300 °C and at 1 kHz during the aging for BPDA-PDA and BPDA-ODA films.

The dielectric constant εr measured at 300 °C and at a frequency of 1 kHz during the aging is illustrated in Figure 12. Here also, while an invariance in the εr values can be observed for the BPDA-PDA films, a huge increase in εr is observed during the aging for the BPDA-ODA films. This increase could be attributed to the chain breakage, already observed on the FTIR spectra, leading to the formation of polar groups in the dielectric bulk. Different authors observed such behavior during the aging of PIs films at high temperature in air [45, 46]. The degradation of BPDA-ODA films is also reflected on the DC conductivity values (not presented here) that increase from 2×10^{-12} to 4×10^{-11} $\Omega^{-1}cm^{-1}$ after 1000 h of aging.

4.4. Thermal Aging Above *Tg* in Oxidative Atmosphere

In order to check the stability of BPDA-PDA films at higher temperatures, aging has been performed at different temperatures above *Tg*. Figure 13 shows a comparison of the evolution of FTIR spectra of BPDA-PDA and PMDA-ODA during an accelerated aging at 400 °C during 200 minutes in air.

It can easily be observed that BPDA-PDA remain stable after aging while PMDA-ODA is strongly degraded and even if the latter owns a comparable thermal stability and a higher *Tg* (see Table II).

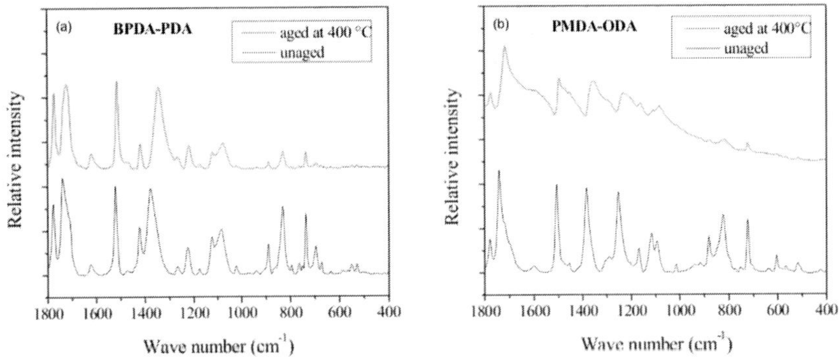

Figure 13. FTIR spectra before and after aging at 400 °C in air for 200 minutes for BPDA-PDA and PMDA-ODA films.

BPDA-PDA films 4.2 μm-thick were aged in air at 360 °C during 800 h. The FTIR spectrum does not show any variation in the imide bonds during this aging period (not shown here) indicating that the bulk of the material is not affected by the aging. Here also, the measured relative permittivity at 1 kHz remains constant with a value of 2.8 during all the aging period (not shown here), indicating that no additional polar groups are formed during the aging. In contrast the film thickness and the breakdown field measured at 300 °C, present a slight continuous decrease during the aging as illustrated in Figure 14. After 800 h a breakdown field decrease of about 50 % while a thickness reduction of 14% can be measured. During this aging, an increase in the surface roughness of BPDA-PDA and the formation of craters were observed (not shown here). They can cause local field intensification and assist the dielectric breakdown mechanism. Such a surface state degradation can explain the breakdown field decrease during the aging. So, it is believed that the observed degradations are related to the oxygen presence since no variation occurs during aging at the same temperature but in inert gas. Thus, BPDA-PDA is attacked at the outer layer face exposed to oxygen.

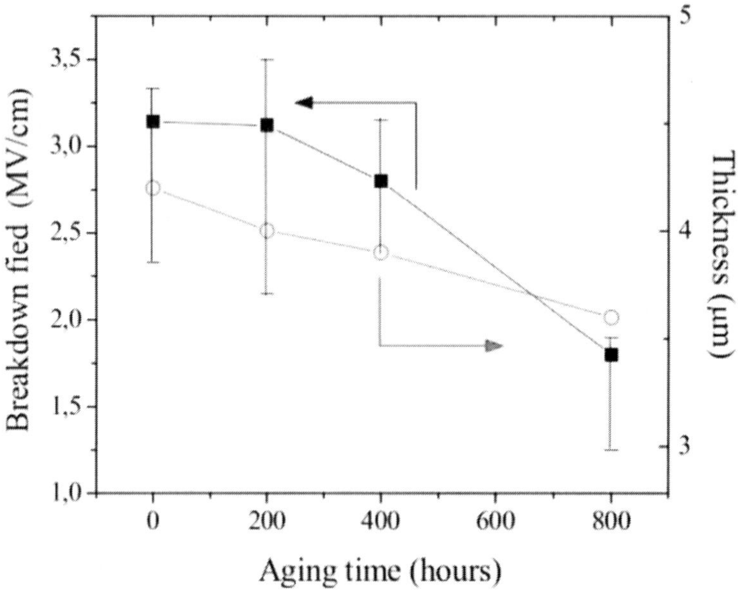

Figure 14. Breakdown field and thickness for 4.2 μm-thick BPDA-PDA films during aging at 360 °C in air. The breakdown field is measured at 300 °C.

All these results lead to show that BPDA-PDA is a reliable kind of PIs in order to passivate wide band gap semiconductor devices up to 360 °C during extremely long duration in inert atmosphere without any remarkable properties degradation.

5. SECONDARY PASSIVATION OF SIC BIPOLAR DIODES WITH BPDA-PDA POLYIMIDE

The secondary passivation is the last fabrication step, at the wafer level, of a semiconductor device. It aims to reinforce the die protection against mechanical aggressions, chemical contamination, and surface electric flashover under blocking voltage operation. Note that after the wafer dicing and die packaging, a complementary insulating environment (moulding case, silicone gel, ...), may be required over the device depending on the maximal voltage rating, in order to avoid electrical arcing in air during operation. Coming back at the wafer level, the component secondary passivation fabrication step consists in two main phases: the first one corresponds to the wafer PI coating, the second one corresponds to the PI film etching at the component metal electrode areas, in order to allow ulterior electrical contacting.

The feasibility of using BPDA-PDA polyimide for the passivation of high voltage silicon carbide bipolar diodes has been studied [47]. As BPDA-PDA material is not photosensitive and not removable using wet etching, its local etching requires applying a plasma process through a previously deposited metal mask. The following sections will first present the aimed diode structure and the used BPDA-PDA coating and ectching processes. Then, the component electrical characterizations will be reported. The results of this paragraph are extracted from [47].

5.1. Diodes Fabrication, Passivation Process and Plasma Etching

4H-SiC PiN diodes were realized on a 2"-SiC wafer in a 60 µm-thick N^- epilayer (ND=8.4×10^{14} cm^{-3}), with a Al-implanted P^+ emitter (NE=10^{19} cm^{-3}), and a Al-implanted junction terminaison extention (JTE) periphery (with $LJTE\geq125$ µm, an implanted-Al dose of 1.1×10^{13} cm^{-2}, assuming a 80% electrical activation [48]). The common post-implantation annealing was performed at 1650 °C for 30 minutes in Ar. The theoretical breakdown voltage of these PiN JTE diodes, derived from ionization integral calculations realized by numerical 2D-simulations, taking into account the Konstantinov's ionization coefficients [49], is equal to 7.8 kV for the 0.9×10^{13} cm^{-2} JTE doping dose assumed. The primary passivation layer is a 40 nm-thick thermal SiO_2 oxide grown by dry oxidation at 1150 °C for 2 h followed by a post oxidation annealing at 1150 °C for 1 h in N_2. Then, after the SiO_2 layer selective opening, a first Ni/Al thin metallization was deposited, patterned and annealed to realize the low resistive ohmic contact to the diode P^+ emitter areas (of 250 µm to 1 mm in diameter). Then a 2 µm-thick Al metallisation was evaporated and patterned as well to obtain thick anode plots. Note that some electrical tests at that stage were performed on the wafer outside the clean room.

Then, the second passivation layer has been realized with the spin-coated BPDA-PDA through a multi-layer (3 layers) coating process onto the wafer. In order to reinforce the adhesion with the primary passivation layer, an adhesion promottor (containing silane groups) was first spin-cast onto the wafer at 3000 rpm. A first layer of BPDA-PDA PAA solution was spin-coated at 500 rpm followed by a rotation speed at 3000 rpm. The wafer was consecutively baked at 100 °C for 1 minute and 175 °C for 3 minutes on a hot plate. A second and a third BPDA-PDA PAA layers were coated with the same process in order to increase the passivation layer thickness. The wafer was finally hard cured *(apart from the clean room)* in a regulated oven under nitrogen atmosphere at 200 °C for 15 minutes (to drive off NMP solvent) followed by a slight raising slope of 2 °C min^{-1} up to 400 °C for 1h. After the imidization, the wafer was cooled down with a low slope of -4 °C min^{-1}. The resulting BPDA-PDA thickness onto the wafer was 4 μm.

Before the etching, a mask metallic layer of 300 nm was evaporated into vacuum onto the BPDA-PDA passivation layer. The metallic layer was opened (metal etch) just above the SiC diode anodes using the photolithography technique. Plasma etching of the unmasked BPDA-PDA areas was performed into a plasma reactor containing a 100% O_2 atmosphere at pressure of 1 mTorr. The injected micro-wave electrical power was 600 W. An auto-polarization voltage of -84 V was applied between the plasma electrodes. The incident RF power was 70 W while the reflected power was negligeable (around 3 W). The wafer was placed onto the ground electrode of the reactor which was cooled with water to maintain the temperature of the wafer around 10 °C. The etching was performed during 20 minutes by short steps of 5 minutes long in order to avoid an increase in the temperature of the wafer and until the total removing of exposed BPDA-PDA areas. Figure 15 shows optical and SEM images of the upper electrode of a SiC diode after the plasma etching of BPDA-PDA.

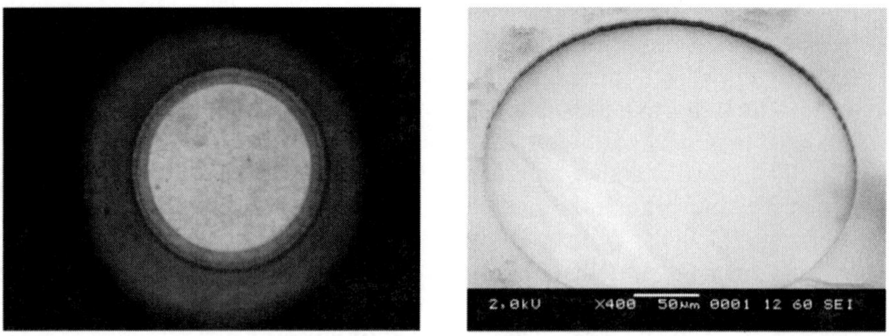

Figure 15. a) Optical microscope image and (b) SEM image of the topside view of a SiC diode after the etching of BPDA-PDA secondary passivation.

5.2. Electrical Characterizations of Bpda-Pda Passivated Sic Bipolar Diodes

The diode forward and reverse current *vs.* voltage I(V) characteristics were measured using a Keithley 2410 source meter, at low levels (less than 1 A/cm^2 forward current density and -600 V reverse voltage) to avoid stressing the devices. These preliminary measurements were performed at different temperatures (the wafer sample being attached to a heating chuck), and in air ambient at low pressure (10^{-3} mbar). Then the diodes exhibiting the lowest reverse currents (less than 10^{-6} A up to -600 V) were selected for high reverse voltage measurements, performed in a perfluoropolyether oil (PFPE Galden® fluid, with a high dielectric strength around 16 kV/mm) in order to avoid arcing in air ambient. Figure 16 presents an example of diode forward I(V) characteristics measured at different temperatures before and after the secondary passivation . It is observed that the latter did not affect the functionality of the component. On the contrary, an improvement of the diode ideality factors at low forward current levels resulted compared to the non coated diode ones, especially at higher temperatures. A positive effect could be noticed on the reverse I(V) curves up to -600 V applied voltage as well (not shown here).

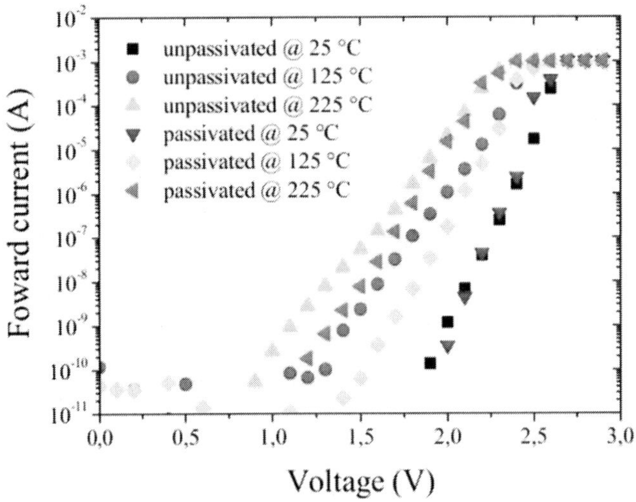

Figure 16. Forward current vs. voltage characteristics at three different temperatures before and after the BPDA-PDA passivation realization. The current was limited at 1 mA. The diode P$^+$-emitter diameter and JTE length are 400 µm and 500 µm, respectively. Characterizations were performed in vacuum.

The typical high voltage reverse characteristics measured at ambient temperature for the selected best diodes are shown on Figure 17a. The breakdown event always occurred suddenly, at a voltage value between 5 kV and 6 kV, and leaving a visible mark on the sample located relatively far from the tested component (at a distance longer than 1 mm) as presented on Figure 17b.

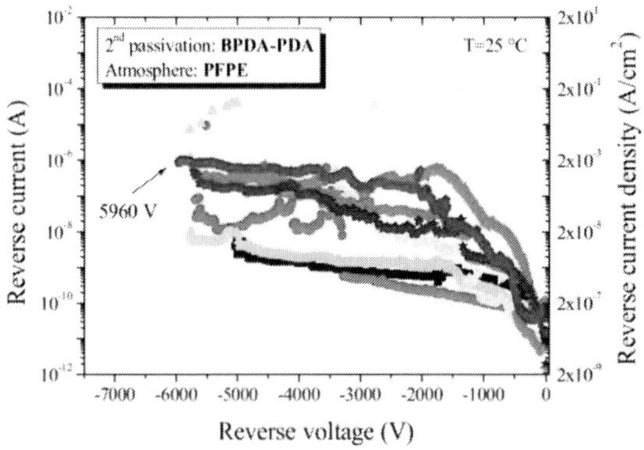

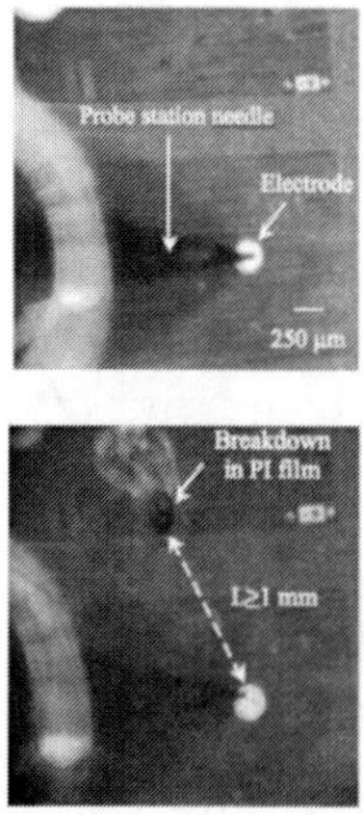

Figure 17. a) Typical reverse characteristics of low leakage diodes passivated with BPDA-PDA (measurements performed in PFPE environment at 25 °C). (b) Views of the same probed diode in PFPE oil, before (top image) and after (bottom image) high voltage measurements.

The tested diodes were not destroyed after the first high voltage measurement, exhibiting approximately the same breakdown voltage values when polarized again several times. Considering the reverse voltage values achieved and the distance observed between the resulting crater and the device, the second passivation layer is certainly at the origin of the failure (the PFPE environment being able to withstand around three times higher voltages), due to the presence of local defects. For another diode, probably situated in a better quality area, a maximal breakdown voltage was measured with a value of 7.3 kV, with a I(V) characteristics presenting a current knee before breakdown, as can be seen on Figure 18a. Such a breakdown voltage value is very close to the theoretical maximal value assuming an avalanche breakdown in the SiC tested structure. Such a SiC avalanche mechanism in this particular case was further supported by the observation of the post-breakdown degraded zone, situated at the diode itself, the latter being totally destroyed as exhibited on Figure 18b.

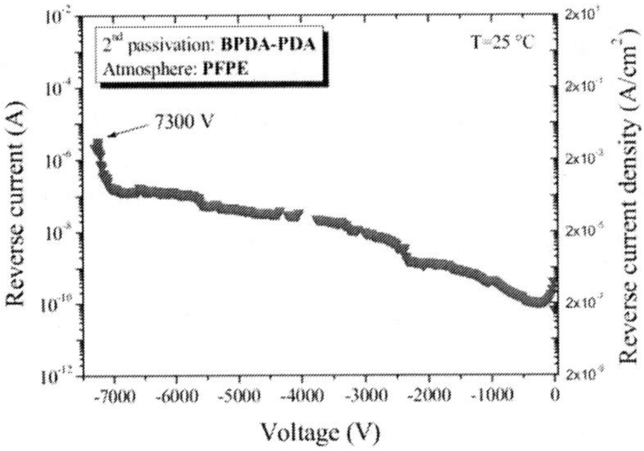

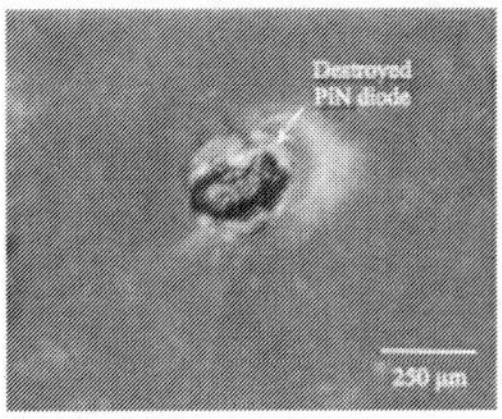

Figure 18. a) Reverse characteristic of a BPDA/PDA passivated diode with the maximal breakdown voltage obtained (performed in PFPE environment at 25 °C). (b) View of the destroyed diode in PFPE oil after the high voltage measurement.

So the feasibility and the potentiality of the BPDA-PDA films for high voltage SiC device secondary passivation could be experimentally demonstrated, positively affecting the current voltage characteristics and allowing high breakdown voltage typical values with a maximum value close to the theoretical limit to be reached. Moreover, a significant improvement in the BPDA-PDA protection efficiency should result from a fabrication entirely performed in clean room conditions and from the use of thicker PI layers.

4. CONCLUSION

This chapter deals with polyimide materials (PIs), having in mind the emerging high temperature semiconductor devices currently demanded for high temperature and high power electronics. Among several PIs already well known for their best thermal properties, very good dielectric characteristics, chemical and radiation resistance, and easy processability, this chapter focuses on BPDA-PDA polyimide, evaluating its superiority for semiconductor insulating coating in the temperature range up to 400 °C.

It is shown that the BPDA-PDA's CTE, which is the closest to the semiconductors (as SiC and GaN) ones, is not the only advantage of this material with regards to the targeted application. In fact, though exhibiting comparable *Tg* and *Td* to PMDA-ODA and PPDA-ODA ones in particular,

BPDA-PDA on silicon substrate demonstrates a higher properties stability under thermo-oxidative ambient aging than its counterparts, thanks to its chemical nature exempt of C-O-C ether bonds. Indeed, the presented results highlight the insulation long term reliability at 300 °C in air, and at 360 °C in nitrogen ambient, of BPDA-PDA films on semiconductor (no chemical and no electrical degradations having been evidenced up to 5000 h of aging).

The presence of impurities (source of free charges) within the PI films playing a major role in the degradation of their dielectric characteristics above 200 °C, the highest degree of imidization has to be looked for, as considered in this chapter. An imidization cure (400 °C-temperature, 1 h-duration) is found optimal for maximizing both the low field resistivity and the dielectric strength, in correlation with FTIR spectrometry analysis. Because of its impact on the intrinsic free charges density as well, the film thickness parameter is also taken into account. Its strong influence on the high temperature dielectric properties is underlined, which can not be solved by a higher temperature or longer curing (leading to desimidization). As an example, at 350 °C, the mean dielectric strength of a 8.6 μm-thick film is measured two fold lower than that of a 1.5 μm-thick layer; however it is remaining as high as 2 MV/cm, so comparable to SiC critical field.

Going up to the application, the chapter finally describes an experiment demonstrating the feasibility of the secondary passivation of 7.8 kV SiC bipolar diodes, using BPDA-PDA. The PI coating and etching processes are detailed, resulting in a 4 μm-thick PI layer. The electrical characterization results arise that the applied final fabrication step positively affected the high temperature forward I(V) curves of the diodes. In reverse bias, the typical breakdown voltage, of around 70% of the theoretical maximum value, could be attributed to the presence of local defects throughout the PI coating. So, such a first experiment already attests the potentiality of BPDA-PDA for high voltage secondary passivation, knowing that one can expect an even higher protection efficiency using clean room elaboration conditions, and thicker PI layers.

REFERENCES

1. C E Sroog, A L Endrey, S V Abramo, C E Berr, W M Edwards, K L Olivier, 1965J Polym, Sci. A Polym. Chem. 3: 1373.

2. G Hougham, G Tesero, J Shaw, 1994Macromolecules 2736423649

3. M K Gosh, K L Mittal, 1996Polyimides, Fundamentals and Applications. New-York: Mercel-Dekker.

4. Wayne Johnson RWang C, Liu Y, Scofield J D (2007IEEE Trans. Elec. Pack. Manuf. 30: 182–193

5. M Ree, S Swanson, W Volksen, 1994Polymer. 34: 1423–1430

6. M Ree, K Kim, S H Woo, H Chang, 1997J Appl, Phys. 81: 698–708

7. H S Chung, C K Lee, Y I Joe, H S Han, 1998J Kor, Inst. Chem. Eng. (Hwahak Konghak). 36: 329-335 (in Korean language).

8. H Chung, C Lee, H Han, 2001Polymer. 42: 319–328

9. K Cho, D Lee, M S Lee, C E Park, 1997Polymer. 38: 1615–1623

10. T W Poon, B D Silverman, R F Saraf, A R Rossi, P S Ho, 1992Phys. Rev. B. 46: 11456–11462

11. S Numata, S Oohara, K Fujisaki, J Imaizumi, N Kinjo, 1986J Appl, Polym. Sci. 31: 101–110

12. H C Liou, P S Ho, R Stierman, 1999Thin Solid Films. 339: 68–73

13. M Ree, T J Shin, S W Lee, 2001Kor. Polym. J. 9119

14. G Maier, 2001Prog. Polym. Sci. 2: 63–65

15. M Tsukiji, W Bitoh, J Enomoto, 1990Proc. Inter. Symp. Elec. Insul. 88.

16. S H Hsaio, Y J Chen, 2002Eur. Polym. J. 38: 815.

17. K P Pramoda, S Liu, T S Chung, 2002Macromol. Mater. Eng. 287: 931.

18. M Ree, C W Chu, M J Goldberg, 1994J Appl, Phys. 75: 1410.

19. T J Shin, M Ree, 2007J Phys, B Chem,

20. S A Lee, T Yamashita, K Horie, T Kozawa, 1997J Phys, B Chem,

21. J Sung, D Kim, C N Whang, M Oh-e, H Yokoyama, 2004J Phys, B Chem,

22. B J Factor, T P Russell, M F Toney, 1991Phys. Rev. Lett. 66: 1181.

23. T Sasaki, H Moriuchi, S Yano, R Yokota, 2005Polymer. 46: 6968.

24. S Diaham, M L Locatelli, T Lebey, D Malec, 2011Thin Solid Films. 519: 1851–1856

25. C E Sroog, 1991Prog. Polym. Sci. 16: 561.

26. K M Chen, T H Wang, J S King, A Hung, 1993J Appl, Polym. Sci. 48: 291.

27. T Hsu, C J Liu, Z L (1992J Appl, Polym. Sci. 46: 1821.

28. H Ishida, S T Wellinghoff, E Baer, J L Koenig, 1980Macromolecules. 13: 826.

29. B Thomson, Y Park, P C Painter, R W Snyder, 1989Macromolecules. 22: 4159.

30. R W Synder, B Thompson, B Bartges, D Czerniawski, P C Painter, 1989Macromolecules. 22: 4166.

31. C A Pryde, 1993J Polym, Sci. A: Polym. Chem. 31: 1045.

32. I Karamancheva, V Stefov, B Soptrajanov, G Danev, E Spasova, J Assa, 1999Vibr. Spectr. 19: 369.

33. E Spassova, (2003). Vacuum. 70: 551 EOF561 EOF.

34. H Deligöz, T Yalcinyuva, S Özgümüs, S Yildirim, 2006Eur. Polym. J. 42: 1370.

35. M B Saeed, M S Zhan, 2006Eur. Polym. J. 42: 1844.

36. M Anthamatten, S A Letts, K Day, R C Cook, A P Gies, T P Hamilton, W K Nonidez, 2004J Polym, Sci. A: Polym. Chem. 42: 5999.

37. G D Hietpas, D L Allara, 1998J Polym, Sci. B Polym. Phys. 36: 1247.

38. Y Ito, M Hikita, T Kimura, T Mizutani, 1990J Jpn, Appl. Phys. 29: 1128.

39. M B Ruggles-wrenn, J L Broeckert, 2009J Appl, Polym. Sci. 111: 228–236

40. G A Schoeppner, G P Tandon, E R Ripberger, 2007Composites: Part A. 38: 890–904

41. G P Tandon, K V Pochiraju, G A Schoeppner, 2008Mater. Sci. Eng. A. 498: 150–161

42. M Meador, A B Johnston, C J Cavano, P J Frimer, A A Macromolecules, 30: 3215–3223

43. M Meador, A B Johnston, C J Frimer, A A Gilinsky-sharon, P Macromolecules, 32: 5532–5538

44. M Tsukiji, W Bitoh, J Enomoto, 1990Proc. IEEE Int. Symp Elec. Insul. 8891

45. L Li, N Bowler, P R Hondred, M R Kessler, 2011J Phys, Chem. Solids. 72: 875–881

46. R A Dine-hart, D Parker, B V Wright, W W Br, J Polym, 3: 222–236

47. S Diaham, M L Locatelli, T Lebey, C Raynaud, M Lazar, H Vang, D Planson, 2009Mater. Sci. Forum. 615-617: 695-698.

48. M Lazar, C Raynaud, D Planson, J-P Chante, M-L Locatelli, L Ottaviani, P Godignon, 2003J Appl, Phys. 94: 2992–2999

49. A O Konstantinov, Q Wahah, N Nordell, U Lindefelt, 1997Appl. Phys. Lett. 71: 90–92

4

CHAPTER

Structural Performance of Polymer Fiber Reinforced Engineered Cementitious Composites Subjected to Static and Fatigue Flexural Loading

Mohamed A.A. Sherir, Khandaker M.A. Hossain * and Mohamed Lachemi

Department of Civil Engineering, Ryerson University, 350 Victoria Street, Toronto, ON M5B 2K3, Canada

ABSTRACT

This paper presents the influence of silica sand, local crushed sand and different supplementary cementing materials (SCMs) to Portland cement (C) ratio (SCM/C) on the flexural fatigue performance of engineered cementitious composites (ECCs). ECC is a micromechanically-based designed high-performance polymer fiber reinforced concrete with high ductility which exhibits strain-hardening and micro-cracking behavior in tension and flexure. The relative high cost remains an obstacle for wider commercial use of ECC. The replacement of cement by SCMs, and the use of local sand aggregates can lower cost and enhance greenness of the ECC. The main variables of this study were: type and size of aggregates (local crushed or standard silica sand), type of SCMs (fly ash "FA" or slag), SCM/cement ratio of 1.2 or 2.2, three fatigue stress levels and number of fatigue cycles up to 1 million. The study showed that ECC mixtures produced with crushed sand (with high volume of fly ash and slag) exhibited strain hardening behavior (under static loading) with deformation capacities comparable with those made with silica sand. Class F-fly ash combined with crushed sand was the best choice (compared to class CI fly ash and slag) in order to enhance the ECC ductility with slag–ECC mixtures producing lowest deflection capacity. FA–ECC mixtures with silica sand developed more damage under fatigue loading due to higher deflection evolution than FA–ECC mixtures with crushed sand.

Keywords: engineered cementitious composite; supplimentary cementing materials; polyvinyl alcohol fiber; fatigue flexure; strain hardening; micro-cracking

1. INTRODUCTION

For decades, normal cementitious materials have proven to be a suitable material in infrastructure constructions and have been successfully implemented in numerous projects around the world. However, the deteriorating condition of the infrastructure in North America and elsewhere due to inherently brittle failure resulting from tensile and impact loading in such cementitious materials has motivated authorities and researchers to seek new technology for enhancement of concrete material properties. Therefore, polymeric fibers were employed as a reinforcing material in cementitious materials in order to enhance the resistance and toughness in tension and shear. Four thousand years ago, fiber reinforcement in construction materials started to be used in the Far East regions when horsehair or hays were mixed into the mud for bricks for housing construction. Modern fiber reinforced cementitious materials have been studied since 1960s [1].

In the last decade, a new type of composite called Engineered Cementitious Composite (ECC) has been developed. ECC is a special type of high performance fiber-reinforced cementitious composite featuring high ductility and damage tolerance under mechanical loading, including tensile and shear loadings [2,3,4]. By employing micromechanics-based material optimization, tensile strain capacity in excess of 3% under uniaxial tensile loading can be attained with only 2% polyvinyl alcohol (PVA) fiber content by volume [2,5,6]. Unlike most vinyl polymers, PVA is not prepared by polymerization of the corresponding monomer. The monomer, vinyl alcohol, is unstable with respect to acetaldehyde. PVA instead is prepared by first polymerizing vinyl acetate, and the resulting polyvinylacetate is converted to the PVA.

The characteristic ECC strain hardening after matrix first cracking is accompanied by sequential development of multiple micro-cracking and the tensile strain capacity is 300–500 times greater than that of normal concrete. The formation of multiple micro cracking is necessary to achieve high composite tensile ductility. Even at ultimate load, the crack width remains on the order of 50 to 80 micrometer. This tight crack width is self-controlled and, whether the composite is used in combination with conventional reinforcement or not, it is a material characteristic independent of rebar reinforcement ratio.

In contrast, normal concrete and fiber reinforced concrete (FRC) rely on steel reinforcement for crack width control. By suppressing cracks in the presence of large imposed structural deformations, ECC can offer structural durability improvements in addition to water-tightness and other serviceability enhancements. These properties, together with a relative ease of production including self-consolidation casting [7,8] and shotcreting [9], make ECCs suitable for various civil engineering applications. Currently, self-consolidating ECC is emerging in full scale structural applications [10]. The ingredients and mix proportions of ECC are optimized through micromechanics based material design theory to satisfy strength and energy criteria to attain high composite tensile ductility [2,3,4,11]. The type, size and amount of fiber, matrix ingredients and interface characteristics are tailored for multiple cracking and controlled crack width in ECCs.

Aggregates typically occupy an important volume fraction in conventional concrete, and therefore have important effects on different aspects of material properties. In addition to their role as economic filler, aggregates help to control dimensional stability of cement-based materials, which may be considered to consist of a framework of cement paste with relatively large shrinkage movements restrained by aggregates. However, the presence of coarse aggregates in a paste tends to increase the tortuosity of the fracture path, and lead to a tough matrix which delays crack initiation and prevents steady-state flat-crack propagation in ECC, resulting in loss of tensile ductility. Moreover, the introduction of aggregates with a particle size larger than the average fiber spacing leads to balling and greater interaction of fibers between the large aggregate particles, and the effect becomes more pronounced as the maximum size of aggregate particles increase. Therefore, an increase in aggregate size makes it more difficult to achieve a uniform dispersion of fibers. The greater the size of aggregates, the more clumping and interaction of fibers would occur. Therefore, the size of the aggregates is expected to have a significant influence on the properties of composite. Hence, in spite of positive effects of aggregates on dimensional stability and economy of fiber reinforced cement composites, there are limits on aggregate size beyond which problems with fiber dispersability, fresh mix workability and matrix toughness may start to damage the composite material performance characteristics [12].

Therefore, instead of coarse aggregate, standard ECC incorporates fine aggregate (generally silica sand) with an aggregate to binder ratio (A/B) of 0.36 by mass to maintain adequate stiffness and volume stability [13]. The binder system is defined as the total amount of cementitious material, *i.e.*, cement and SCMs, generally fly ash (FA), in ECC. The silica sand has a maximum grain size of 250 μm and a mean size of 110 μm. Another purpose of using fine silica sand is to obtain the optimum gradation of particles to produce good workability [14].

Due to environmental and economic reasons, there is a growing trend to use industrial wastes or by-products as supplementary materials or admixtures in the production of cementitious composite. Among the various supplementary materials, fly ash (FA) and slag (SL) are the most commonly available SCMs. Because of several potential benefits, fly ash and slag have increasingly found use in high performance concrete in the last few decades [15].

In the past few decades, substitution of SCMs, such as FA and SL, has been of great interest and gradually applied to practical applications of ECC [16,17,18,19,20]. FA is a by-product of the coal power plant, whereas SL is a by-product in the manufacture of pig iron. Both of these waste materials from industrial processes are usually available in large quantities and at a fraction of the price of cement. The absence of coarse aggregate in ECC results in a higher cement content. Partial replacement using FA or SL reduces the environmental burden. Further, it has been found that incorporating high amount of FA, especially class F fly ash, can reduce the matrix toughness and improve the robustness of ECC in terms of tensile ductility. Additionally, un-hydrated FA particles with small particle size and smooth spherical shape serve as filler particles resulting in higher compactness of the fiber/matrix interface transition zone that leads to a higher frictional bonding. This aids in reducing the steady-state crack width beneficial for long-term durability of the structure [18,19,21,22]. The leaking expansion joints are a major source of deterioration of multi-span bridges in Canada. Expansion joints can be replaced by flexible link slabs made with ECC forming a joint-free multi-span bridge and hence, solving the problem of premature deterioration [23,24]. Research studies showed significant enhancement of ductility and crack width control in ECC link slabs confirming that the use of ECC can be effective in extending the service life of bridge deck systems.

Since the increase in aggregate size leads to an increase in the matrix toughness while the use of SCMs leads to reduce the matrix toughness, locally available aggregate could successfully be used in conjunction with high volume SCMs in the production of ECC. Very limited information is currently available in the published literature revealing the influence of aggregate size on the performance (ductility and mechanical behavior) of ECC [12]. Accordingly, one of the current research goals is to design a new class of ECCs with a matrix incorporating locally available aggregates that can show similar mechanical properties compared to standard ECC mixtures containing silica sand.

The main objective of the current research is to explore the development of greener ECCs and analyze the mechanical properties such as flexural strengths and fatigue loading performance of ECCs. Flexure fatigue performance of ECC is very important for its construction applications, for example, link slabs applications in bridge decks.

The present study contributes the existing knowledge of ECC by incorporating locally available crushed sand instead of silica sand and by employing different cement replacement rates up to 70%. It is well known that commercially available silica sand is relatively expensive and difficult to obtain when compared with commonly available sands, such as crushed sand. This paper describes the results of an experimental investigation studying the effects of silica sand (as a control material), local crushed sand and different SCM/C ratio on the flexural properties and fatigue performance as well as crack development characteristics of ECC mixtures. Overall, the recommendations of this paper will be beneficial to engineers, designers and local industries engaged in manufacturing and using cost-effective and greener ECC mixtures for construction applications.

2. EXPERIMENTAL INVESTIGATIONS

Two phases of tests were carried on prismatic prism (beam) samples made of two groups of twelve ECC mixtures designed and selected for this study.

Phase I: This phase consisted of two parts. The first part was static flexural strength test conducted at 28, and 56 days for both ECC groups. The second part was general fatigue flexural test conducted at 28 days for 50,000 cycles at 4 Hz cyclic loading rate and 55% fatigue flexural stress level. At the end of the fatigue flexural tests, static flexural tests were applied on the fatigued ECC specimens to calculate the fatigue residual energy. The best ECC mixtures identified from this phase were included in the second phase (Phase II)—slag mixtures were omitted at the end of this phase because of their inferior multiple cracking behavior.

Phase II: In this phase, special fatigue flexure tests were conducted by using two different approaches. The first approach was applying different fatigue flexural stress levels namely 40%, 55% and 70% of the average maximum stress for 50,000 cycles at 4 Hz cyclic loading rate and the second approach was applying different fatigue cycles namely 200,000, 300,000 and 1,000,000 cycles at 4 Hz cyclic loading rate and 55% fatigue stress level.

2.1. Materials and ECC Mixture Proportions

The material used in the production of standard ECC mixtures were Portland cement (C) Type I general use (GU); two different classes of fly ash (FA): Class-F, and Class-CI with calcium content 5.57%, and 14.30% respectively; Blast furnace slag (SL) with calcium content of 40.80%; silica sand (SS) with a maximum grain size of 0.40 mm; polyvinyl alcohol (PVA) fibers; water; and a polycarboxylic-ether type high-range water-reducing admixture (HRWRA). The chemical composition and physical properties of Portland cement, class-F; class-CI fly ash and slag are presented in Table 1. The commercially available silica sand is relatively expensive and difficult to obtain when compared with commonly available sands [12]. For this purpose, locally available crushed lime stone sand with maximum grain size of 1.18 mm was used in the production of nonstandard ECC mixtures. The grain size distributions of silica sand and crushed sand are given in Table 2. The PVA fibers with a diameter of 39 μm and a length of 8 mm are purposely manufactured with a tensile strength (1620 MPa), elastic modulus (42.8 GPa), and maximum elongation (6.0%) matching those needed for strain-hardening performance. Additionally, the surface of the PVA fibers was coated with a proprietary oiling agent 1.2% by mass to tailor the interfacial properties between fiber and matrix for strain-hardening performance [25].

Table 1. Chemical properties of Portland cement, fly ashes and slag.

Chemical Composition (%)	Cement (C)	Fly Ash (F)	Fly Ash (CI)	Slag (SL)
Calcium Oxide CaO	61.40	5.57	14.30	40.8
Silicon Dioxide SiO_2	19.60	59.5	41.57	35.4
Aluminium Oxide Al_2O_3	4.90	22.2	26.12	13.00
Ferric Oxide Fe_2O_3	3.10	3.90	8.44	0.50
Magnesium Oxide MgO	3.00	–	3.40	8.00
Sulfur Trioxide SO_3	3.60	0.19	1.55	0.10
Alkalis as Na_2O	–	2.75	0.71	0.20
Loss on ignition LOI	2.30	0.21	1.49	–
Sum (SiO_2 + Al_2O_3 + Fe_2O_3)	27.60	85.6	76.13	48.9
Physical properties	–	Fly Ash (F)	Fly Ash (CI)	Slag (SL)
Residue 45 μm (%)	3.00	9.60	17.50	1.00
Density (g/cm^3)	3.15	2.18	2.43	2.11
Blaine fineness (m^2/kg)	410	306	–	430
Air Content (%)	7.79	–	–	–
Initial Setting time (min)	113	–	–	–
Compressive Strength (MPa) 1 day	19.41	–	–	–
Compressive Strength (MPa) 3 day	30.35	–	–	–
Compressive Strength (MPa) 28 day	41.47	–	–	–

Table 2. Sieve analysis of silica sand and crushed sand.

U.S. Sieve #	Opening (mm)	% Retained crushed sand	% Retained silica sand
16	1.18	0.00	–
20	0.841	6.0	–
30	0.60	17.50	–
40	0.42	–	0.00
50	0.30	60.0	2.20
70	0.21	–	14.70
100	0.15	90.25	47.50
140	0.105	–	28.80
200	0.074	98.75	6.40
270	0.053	–	0.40

In order to investigate the influence of aggregate type and size, and different supplementary cementing materials on the mechanical properties of ECC, two main groups of 12 ECC mixtures have been designed and selected. The first group was composed of 6 ECC mixtures which were produced using silica sand. The second group was composed of the same first group ECC mixtures but was produced using local crushed sand. The supplementary cementing materials (SCMs) used in both groups were Class-CI, Class-F fly ash and Slag. These SCMs were used as Portland cement replacement at ratios of 1.2 and 2.2 (55% and 70% by mass), respectively. Mixture proportions and designations for both groups of ECC mixtures are given in Table 3.

Table 3. Engineered cementitious composites (ECC) Mixture proportions.

First Group-Silica Sand									
Mixture ID	Ingredients Per 1 Part of Cement by Mass							FA/C or SL/C	W/B
	Water	Cement	FA	Slag	Sand	PVA kg/m³	HRWRA kg/m³		
CI_1.2_SS	0.60	1	1.2	–	0.80	26	5.4	1.2	0.27
CI_2.2_SS	0.85	1	2.2	–	1.15	26	4.2	2.2	0.27
F_1.2_SS	0.60	1	1.2	–	0.80	26	5.4	1.2	0.27
F_2.2_SS	0.85	1	2.2	–	1.15	26	4.2	2.2	0.27
SL_1.2_SS	0.67	1	–	1.2	0.80	26	4.50	1.2	0.30
SL_2.2_SS	0.96	1	–	2.2	1.15	26	3.80	2.2	0.30
Second Group-Crushed Sand									
Mixture ID	Ingredients Per 1 Part of Cement by Mass							FA/C or SL/C	W/B
	Water	Cement	FA	Slag	Sand	PVA	HRWRA		
CI_1.2_CS	0.60	1	1.2	–	0.80	26	5.4	1.2	0.27
CI_2.2_CS	0.85	1	2.2	–	1.15	26	4.2	2.2	0.27
F_1.2_CS	0.60	1	1.2	–	0.80	26	5.4	1.2	0.27
F_2.2_CS	0.85	1	2.2	–	1.15	26	4.2	2.2	0.27
SL_1.2_CS	0.67	1	–	1.2	0.80	26	4.50	1.2	0.30
SL_2.2_CS	0.96	1	–	2.2	1.15	26	3.80	2.2	0.30

HRWRA: High range water reducing admixture; C: Cement; FA: fly ash; CI: class-CI Fly Ash; F: class-F Fly Ash; SL: Slag; W/B: water to binder ratio (binder = C + SCMs).

The water/binder (W/B) ratio was kept in the range of 0.27 for both class CI and F fly ash and 0.30 for slag. The variable parameters in these mixtures were the aggregate type and size, and SCMs cement replacement rate (SCM/C of 1.2 and 2.2). In both groups, the amount of aggregate and SCMs to binder were held constant.

As shown in Table 3, the ECC mixtures are labeled such that the ingredients are identifiable from their Mix IDs. The first letter in the mixture designations indicates the SCM type (CI, class-CI fly ash; F, class-F fly ash; SL, slag). The numbers after the letter indicate the SCM (CI, F or SL)—cement ratio and last letters represent aggregate type (SS or CS).

2.2. Specimen Preparation

Hobart type mixer with 20-liter capacity was used in preparing all ECC mixtures. All materials of ECC mixtures were weighed separately before starting mixing. All dry materials except PVA fiber such as cement, SCMs (FA or SL), and silica or crushed sands were thrown into the mixer for the dry mix. After 30 s dry mix, 90% of water was added into the mixer for another 2 min. Then the remaining 10% of water and HRWRA were added within the next 1 min. The final step of mixing procedure was to add the PVA fiber into the mixer and the mix was stirred for another 3 min. It should be noted that HRWRA content was not kept constant which means slight adjustment in the amount of HRWRA in each mixture was performed to achieve better fiber dispersion and workability based on past experience.

In the fresh state, workability of ECC mixtures was determined by slump flow test according to ASTM C 1611 [26]. Cube specimens measuring 50 mm × 50 mm × 50 mm were cast to determine compressive strength at the age of 28 days. Although the scope of the paper is limited to the behavior of ECC under flexural loading, the results of slump flow and compressive strength tests are summarized in Table 4. Slump flow values of ECC mixtures ranged between 415 and 540 mm. All ECC mixtures were found flowable and showed no bleeding. The 28 day compressive strength varied from 45 to 69 MPa. ECC mixtures with class CI fly ash exhibited slightly higher compressive strength than Class F while those with Slag showed higher compressive strength than those with fly ash CI or F—the higher compressive strength was attributed to the higher content of CaO. The compressive strength decreases with the increase of fly ash content while decreased with the replacement of silica sand by crushed sand.

355 mm × 50 mm × 76 mm prism (beam) specimens were prepared for the four-point bending test. The total number of all prismatic specimens prepared for this study was 168 beams. All specimens were demolded after 24 h and moisture cured in plastic bags at 95% ± 5% relative humidity (RH), 23 ± 2 °C. The specimens were kept in the curing room until the day of testing.

2.3. Test Procedures

Three prismatic samples of 355 mm × 50 mm × 76 mm were prepared to measure the flexural strength of ECC (Modulus of rupture), for ages 28 and 56 days by performing four-point bending test under displacement control condition at a loading rate of 0.005 mm/s on a closed-loop servo-controlled loading system. The span length of flexural loading was 304.8 mm with a 101.6 mm center span length. During the flexural tests, the load and the mid-span deflection were recorded on a computerized data recording system. The schematic of the test setup with a photo is presented in Figure 1. Flexural strength of the specimens was calculated in accordance with ASTM C 78 [27].

Table 4. Average flexural strength and ultimate deflection results at ages 28 and 56 days.

Mix No.	Mix Designation	Slump Flow (mm)	28-Day Compressive Strength (MPa)	Flexural Properties			
				28 Days		56 Days	
				Strength (MPa)	Deflection (mm)	Strength (MPa)	Deflection (mm)
1	CI_1.2_SS	540	62 (±2)	12.68 (±0.5)	3.17 (±0.2)	13.83 (±0.5)	2.67 (±0.15)
2	CI_2.2_SS	530	53 (±2)	11.98 (±0.5)	3.96 (±0.2)	13.27 (±0.5)	3.56 (±0.15)
3	F_1.2_SS	535	61 (±2)	11.80 (±0.5)	3.91 (±0.2)	12.38 (±0.5)	3.75 (±0.15)
4	F_2.2_SS	530	52 (±2)	10.84 (±0.5)	4.45 (±0.2)	12.25 (±0.5)	3.98 (±0.15)
5	SL_1.2_SS	550	69 (±2)	15.44 (±0.5)	1.79 (±0.2)	15.81 (±0.5)	1.74 (±0.15)
6	SL_2.2_SS	485	67 (±2)	14.08 (±0.5)	2.04 (±0.2)	14.38 (±0.5)	1.86 (±0.15)
7	CI_1.2_CS	520	60 (±2)	12.20 (±0.5)	2.86 (±0.2)	12.92 (±0.5)	2.41 (±0.15)
8	CI_2.2_CS	500	47 (±2)	11.42 (±0.5)	3.14 (±0.2)	12.64 (±0.5)	2.56 (±0.15)
9	F_1.2_CS	490	59 (±2)	11.20 (±0.5)	3.66 (±0.2)	12.12 (±0.5)	3.26 (±0.15)
10	F_2.2_CS	515	45 (±2)	10.48 (±0.5)	4.27 (±0.2)	11.68 (±0.5)	4.02 (±0.15)
11	SL_1.2_CS	420	69 (±2)	14.93 (±0.5)	1.73 (±0.2)	15.24 (±0.5)	1.61 (±0.15)
12	SL_2.2_CS	415	67 (±2)	13.62 (±0.5)	1.80 (±0.2)	13.87 (±0.5)	1.64 (±0.15)

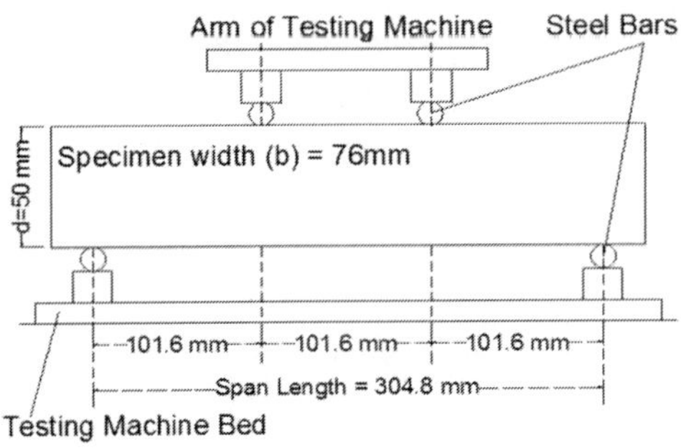

Figure 1. Four point bending test setup.

Two approaches were adopted for investigating the ECC performance under fatigue flexural test: the first was applying different fatigue stress levels and the second was applying different fatigue cyclic loading. Accordingly, four prismatic samples of 355 mm × 50 mm × 76 mm were prepared for the age of 28 days. The four-point bending test was conducted under both static and fatigue loading. Two out of four samples were prepared for static loading tests as control specimens which were carried out under displacement control conditions, while the other two samples were prepared for fatigue loading tests which were performed under load control conditions. Specimens were simply supported on a span of 304.8 mm and subjected to two-point loads at one-third of the span as shown in Figure 1.

In the first approach, static flexural tests were conducted before fatigue flexural tests at the constant rate of 0.005 mm/s. The static flexural strengths were determined by averaging the flexural strength results of ECC control specimens. Based on their static flexural averages, the different maximum fatigue stress levels were determined as 40%, 55% and 70% and the tests were conducted at 50,000 cycles and 4 Hz cyclic loading rate. In the second approach, 55% of fatigue stress level value was fixed and tests were conducted at different fatigue cycles namely 200,000 cycles, 300,000 cycles and 1,000,000 cycles at the same 4 Hz cyclic loading rate as well.

Fatigue flexural tests were performed under load control conditions. The ratio between minimum and maximum flexural stress was set equal to 0.30 for all specimens in order to avoid any impact and slip of specimens during testing. At the first cycle of each specimen, load was gradually applied to the maximum stress level at 0.50 kN/min static loading rate in order to avoid any sudden collapse in the specimen. The cyclic fatigue loading was then applied. The fatigue testing technique mentioned above was adopted in accordance with Suthiwarapirak *et al.* [28]. During the fatigue flexural tests, the mid-span deflection evolutions were recorded on data sheet and at the end of the fatigue flexural tests; static flexural tests were conducted on the fatigued ECC specimens to calculate the fatigue residual values for both strength and mid-span deflection.

3. ANALYSIS OF BEHAVIOR UNDER STATIC FLEXURAL LOADING (PHASE I)

3.1. Flexural Strength

The test results in terms of flexural strength, modulus of rupture (MOR), and ultimate mid-span deflection are given in Table 4 at ages 28 and 56 days, and the average flexural strength-mid span deflection curves of ECC mixtures at age of 28 days are shown in Figure 2. Each result in Table 4 is the average of three measurements.

As seen from Table 4, the average ultimate flexural strengths vary from 10.48 to 15.81 MPa and the mid-span deflection (deflection capacity) of the ECC beams vary from 1.61 to 4.45 mm depending on fly ash class CI, F and slag content.

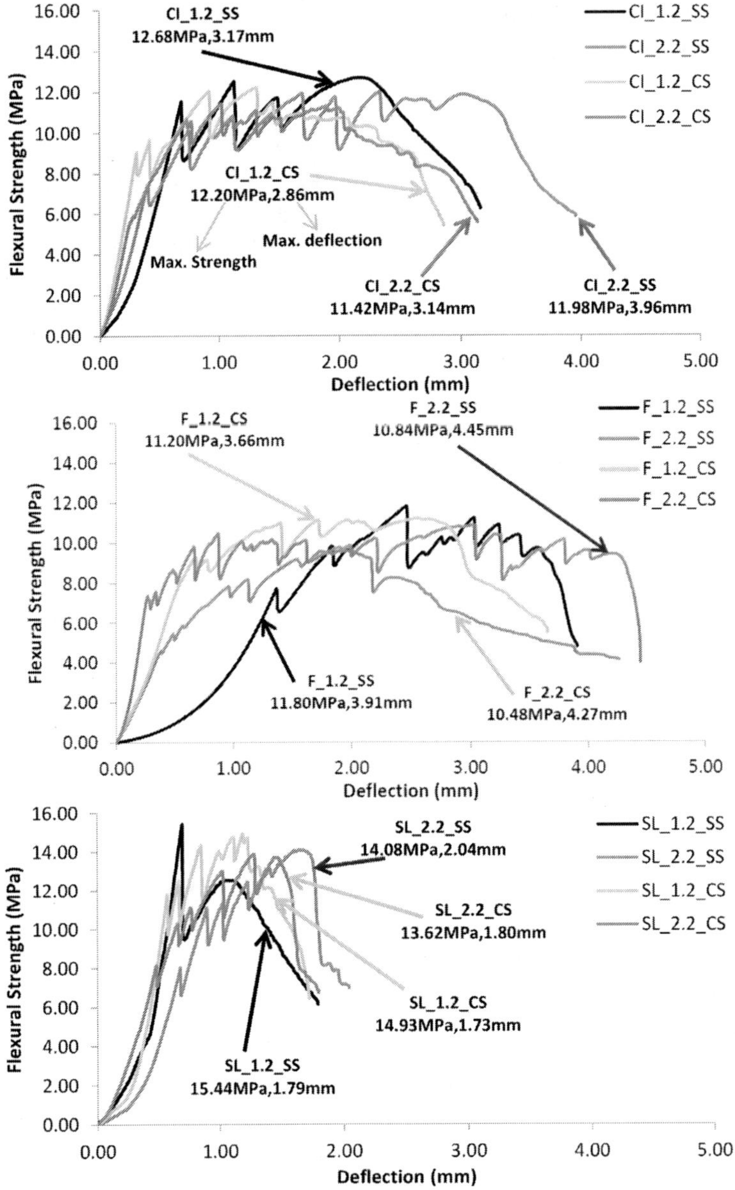

Figure 2. Average flexural strength-mid span deflection curves of ECCs at 28 days.

3.2. ECC's Deflection and Flexural Strength vs. SCMs Cement Replacement Rate

Test results show that the increase of fly ash to cement ratio (FA/C) from 1.2 to 2.2 improved the deflection capacity of ECC mixtures (Table 4 and Figure 2). The improvement in the deflection capacity with the increase in the FA content can be attributed to the fact that the increase in the FA content tends to reduce the PVA fiber/matrix interface chemical bond and matrix toughness, while increasing the interface frictional bond, in favor of attaining high deflection capacity [12,18]. The slag–ECC mixes showed significantly lower deflection capacity when compared to that of the FA–ECC mixes. The reduced deflection capacity can possibly be caused by the higher fracture energy, bond strength and friction between the slag–ECC matrix and the fibers compared with FA–ECC mixes. Previous microstructural studies of slag–ECC mixtures showed that almost all slag–ECC particles participated in hydration process and no un-hydrated particles could be observed after the process of hydration [1]. This means that the fiber surfaces in slag–ECC have more attached matrix material compared to FA–ECC. For a fiber to be pulled out of the slag–ECC matrix, de-bonding at the fiber interface or fracture of the hydrate product is required to overcome interlocking of the hydrate product [1]. It should be noted that the flexural strength values of ECC mixtures incorporating class-CI FA were slightly higher than those of ECC mixtures which were produced with class-F FA while mid-span beam deflection values of the former were lower than the later.

No significant influence of the FA or slag replacement rate from 1.2 to 2.2 (55% to 70%) on the flexural strength values especially at the ages of 28 days or later in the tested range. However, even at about 70% replacement of cement by FA or slag, the flexural strength of ECC at 28 days was significantly higher than that of conventional concrete and fiber reinforced concrete.

3.3. ECC's Deflection and Flexural Strength vs. Aggregate Size

Adverse effect of increased size of aggregates on deflection capacity performances of ECC mixtures was observed. The increase in aggregate size up to 1.18 mm slightly reduces the total mid-span beam deflection of ECC mixtures as shown in Table 4. This negative effect of increasing aggregate size may be attributed to the adverse effect on the uniform dispersion of fibers. The balling of fibers encouraged by coarser sands at constant sand content prevents sufficient coating of fibers by the matrix, and thus reduces the fiber-to-matrix bonding, which is an important factor influencing ductility [12]. Moreover, for ECC with the larger aggregate size, a higher degree of aggregate interlock is expected, resulting in higher matrix toughness and work-of-fracture during crack propagation. According to the micromechanical model of steady state cracking, which is essential to achieving strain hardening behavior, high matrix fracture toughness reduces the margin to develop multiple cracking [13]. However, aggregates within the size range studied, as long as they do not interfere with the uniform dispersion of fibers, do not negatively influence the ductility of ECC [12].

As in the case of cement replacement rate case, the aggregate particle size had no or only a minor effect on the flexural strength test results. Simultaneously, flexural strength and mid-span deflection obtained with crushed sand is within the permissible limit of silica sand and might be similar or even better than the mechanical properties of ECC mixtures made with silica sand [12].

It could be concluded that the most important feature of ECC, high deflection capacity with multiple cracking behaviors, was protected and is not sacrificed by replacing cement with a maximum of 70% FA or by replacing silica sand with local crushed sand.

4. ANALYSIS OF GENERAL FATIGUE FLEXURE PERFORMANCE (PHASE I)

The first part of the Phase I aimed at investigating the flexural strength characteristics of the two ECC groups by applying static flexural testing at 28 and 56 days. The second part involved fatigue flexural tests which performed under load control conditions at 28 days where specimens were subjected to 4 Hz sinusoidal cyclic loading rate for 50,000 cycles at the fixed fatigue stress level of 55% of the averaged flexural strength of ECC control specimens. At the end of the fatigue flexural tests, static flexural tests were applied on the fatigued ECC specimens to calculate the fatigue residual energy.

Table 5 shows the test results (for 12 ECC mixtures) of both static flexural testing without fatigue flexural loading and after fatigue flexural loading tests. By using these results, the percentage of fatigue residual energy of the fatigued ECC specimens for both residual stress and deflection could be calculated.

Table 5. Fatigue flexural test results (phase I).

Mix ID.	Mix Designation	SCM/C (%)	Static Flexural Testing		Static Flexural after Fatigue Testing		Residual Energy (%)	
			Stress (MPa)	Deflection (mm)	Stress (MPa)	Deflection (mm)	Residual Stress (%)	Residual Deflection (%)
1	CI_1.2_SS	55%	12.68	3.17	11.77	3.92	92.82	123.66
3	F_1.2_SS		11.80	3.91	11.16	5.29	94.58	135.29
7	CI_1.2_CS		12.20	2.86	10.72	3.32	87.87	116.08
9	F_1.2_CS		11.20	3.66	10.75	4.81	95.98	131.42
2	CI_2.2_SS	70%	11.98	3.96	7.91	2.41	66.03	60.86
4	F_2.2_SS		10.84	4.45	7.04	3.09	64.94	69.44
8	CI_2.2_CS		11.42	3.14	8.13	2.58	71.19	82.13
10	F_2.2_CS		10.48	4.27	11.88	5.55	113.39	129.98
5	SL_1.2_SS	55%	15.44	1.79	8.91	0.94	57.73	52.57
11	SL_1.2_CS		14.93	1.73	11.12	1.17	74.45	67.80
6	SL_2.2_SS	70%	14.08	2.04	6.33	0.82	44.98	40.10
12	SL_2.2_CS		13.62	1.80	9.35	0.96	67.44	53.11

As seen from Table 5, FA–ECC mixtures with 55% cement replacement (FA/C = 1.2) showed superior performance relative to higher volume fly ash mixtures with 70% cement replacement (FA/C = 2.2) with respect to both fatigue flexural strength and deflection. The fatigue flexural strength remains equal or slightly lower than static flexural strength.

It was reported that conventional plain concrete with class-C fly ash mixture having 15% cement replacement showed superior performance relative to high-volume fly ash mixture with 50% cement replacement with respect to compressive strength and flexural fatigue strength [29]. Moreover, Tse *et al.* [30] indicated that concrete with equivalent or higher compressive and fatigue strength could be obtained with cement replacement of 25% by weight of low-calcium fly ash (class-F) or 50% by weight of high-calcium fly ash (class-C). The reason of ordinary performance of FA–ECC mixtures with 70% cement replacement relative to 55% cement replacement in both fatigue flexural strength and deflection can be attributed to the fixation of fatigue stress level at 55%. In addition, Suthiwarapirak *et al.* [28] indicated that the evolution of mid-span deflection depends on the fatigue stress level. Namely, the mid-span deflection increased to more than twice as much under high fatigue stress levels S = 0.8–0.9 compared to that under low stress levels S = 0.5–0.6. However, as shown in the Table 5, fly ash concrete mixtures with cement replacement 55% and 70% showed almost the same results when flexural fatigue strength was expressed as percentage of the flexural static strength.

Fatigue test results showed that F_2.2_CS ECC mixture with crushed sand had a remarkable performance with respect to both fatigue flexural strength and deflection capacity. The performance of F_2.2_CS ECC mixture was almost equal to that of all other FA–ECC mixtures with 55% cement replacement in term residual deflection. The higher residual strength and deflection capacity F_2.2_CS ECC mixture compared to other ECC mixtures (as illustrated in Table 5) may be attributed to high water demand of class F fly ash (causing higher degree of hydration) leading to denser matrix and increased friction among crushed sand particles and PVA fivers.

Slag–ECC mixtures with 55% and 70% cement replacement exhibited inferior performance with respect to both flexural fatigue strength and deflection. It was found out that the compressive fatigue limit ratios at two million cycles in water were 41% for the concrete with blast furnace slag, 44% for the silica fume concrete, and 31% for the plain cement concrete at 50% probability of survival fatigue life [31]. Higher chemical bond strength between slag particles and fibers compared with FA–ECC results in significant lower deflection capacity and inferior performance in residual fatigue flexural strength and deflection capacity of slag–ECC mixtures. Although slag–ECC mixtures exhibit lower deformation capacity compared with FA–ECC, their flexural deflection capacity is around or more than 2 mm at 28 days of age. The 2 mm deflection capacity remains almost 150 times higher than that in normal concrete and conventional fiber reinforced concrete.

The 70% cement replacement group rather than 55% was chosen for investigation in Phase II due to number of reasons. In general, fly ash has been used to replace cement in ECC mixture because of the absence of coarse aggregate in ECC results in higher cement content. Partial replacement using FA reduces the environmental burden. Further, it has been found that the addition of FA at high volumes improves the fresh and durability properties, reduces the drying shrinkage and matrix toughness, and improves the robustness of ECC in terms of tensile ductility [18,19]. Additionally, un-hydrated FA particles with a small particle size (<45 μm) and smooth spherical shape serve as filler particles resulting in higher compactness of the fiber/matrix interface transition zone that leads to a higher frictional bond, which aids in reducing the steady-state crack width beneficial for the long-term durability of the structure [18,19].

Accordingly, fly ash-to-cement ratio (FA/C) of 2.2 was chosen in phase II to satisfy the above requirements. The ECC mixtures selected to be involved in the second phase were CI_2.2_SS, F_2.2_SS, CI_2.2_CS and F_2.2_CS.

5. ANALYSIS OF SPECIAL FATIGUE FLEXURE PERFORMANCE (PHASE II)

In this phase, fatigue flexure tests were conducted on 28 day specimens in two different approaches; the first approach applied different fatigue flexural stress levels namely 40%, 55% and 70% of the average maximum static stress for 50,000 cycles at 4 Hz cyclic loading rate and the second approach applied different fatigue cycles namely 200,000 cycles, 300,000 cycles and 1,000,000 cycles at 4 Hz cyclic loading rate and 55% fatigue stress level as well.

5.1. First Approach—Fatigue Stress Levels

5.1.1. Mid-Span Deflection Evolution

The evolution of the average mid-span deflection as the number of cycles increased to 50,000 cycles for ECC specimens with different fatigue stress levels of 40%, 55% and 70% are plotted inFigure 3. Only a typical result at each fatigue stress level is shown in the Figure.

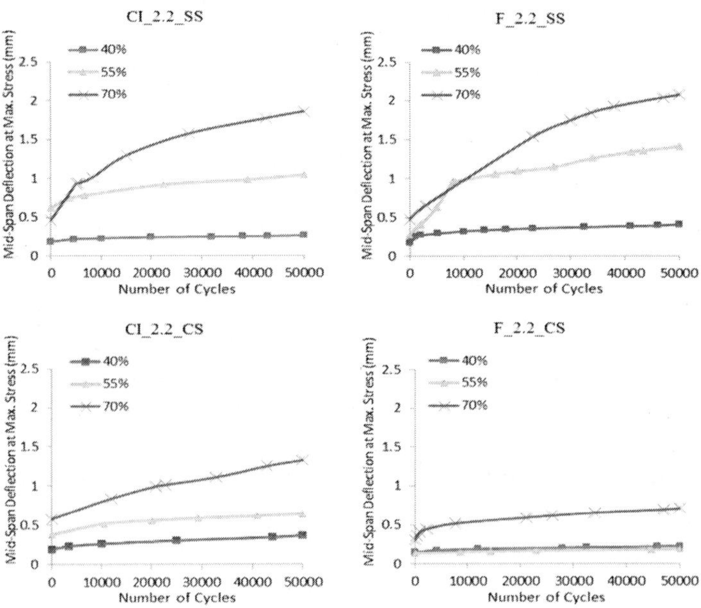

Figure 3. Evolution of mid-span deflection at different fatigue stress level (Phase I).

The results showed that FA–ECC specimens with silica sand (CI_2.2_SS and F_2.2_SS) under fatigue loading developed much more damage than FA–ECC specimens with crushed sand (CI_2.2_CS and F_2.2_CS). The range of displacement evolution of ECC specimens between the maximum stress and minimum stress levels was very large except for F_2.2_CS specimen. Under high stress level 70%, the displacement at the maximum stress evolved up to about 2.1 mm mid-span deflection in silica sand ECC mixtures while the CI_2.2_CS specimen evolved up to 1.35 mm compared to 0.7 mm for the F_2.2_CS specimen. The evolution of the mid-span deflection was found to depend on the fatigue stress level. The mid-span deflection increased to around or more than twice at high fatigue stress level 70% compared to that under low fatigue stress level 55%. It should be noted that FA–ECC mixtures with silica sand exhibited slightly higher deformation capacity than FA–ECC mixtures with crushed sand under static flexural loading while exhibited much higher deformation capacity under fatigue loading. It is also noticed that the CI_2.2_CS mixture exhibited higher deformation capacity than the F_2.2_CS mixture under fatigue loading. This is contrary to the fact that the former had lower deformation capacity than the latter under static flexural loading.

The fatigue flexural tests in this approach were applied for 50,000 cycles at 4 Hz cyclic loading rate. 4 Hz means 4 cycles per second; therefore, the time required to finish 50,000 cycles is 12,500 s or 208 min. According to this, the mid-span evolution speed rate could be calculated at different fatigue stress levels, 40%, 55% and 70% of maximum static stress as shown in Table 6. For example, as shown in Figure 3, at fatigue stress level 40% for CI_2.2_SS mixture, the mid-span evolution started at 180 μm and achieved at 260 μm when the 50,000 cycles were done. The difference between these two values is 80 μm; divided this value by 208 min resulted in 0.39 μm/min as shown in Table 6.

Table 6. Speed rate of mid-span deflection evolution at different fatigue stress levels.

Mix ID.	Mix Designation	Fatigue Stress Levels: Difference between First and Last Values of Mid-Span Beam Deflection (μm)			Mid-Span Evolution Speed Rate (μm/208 min)		
		40%	55%	70%	40%	55%	70%
1	CI_2.2_SS	260 − 180 = 80	1040 − 630 = 410	1860 − 465 = 1395	80/208 = 0.39	2.00	6.70
2	F_2.2_SS	410 − 170 = 240	1410 − 280 = 1130	2080 − 480 = 1600	1.15	5.43	7.70
3	CI_2.2_CS	373 − 193 = 180	645 − 381 = 264	1325 − 580 = 745	0.87	1.27	3.58
4	F_2.2_CS	227 − 150 = 77	192 − 143 = 47	711 − 306 = 405	0.37	0.24	1.95

Figure 4 shows the speed rate of mid-span evolution for all ECC mixtures. As seen from the Figure and as was expected, FA–ECC mixtures with silica sand have the highest speed rate of the evolution of mid-span deflection than FA–ECC mixtures with crushed sand while F_2.2_CS has the lowest speed rate in the two groups. This means that silica sand mixtures are developing much more damage under fatigue loading. The results of speed rate mid-span evolution of all ECC mixtures are consistent with the results of mid-span deflection evolution (Figure 3).

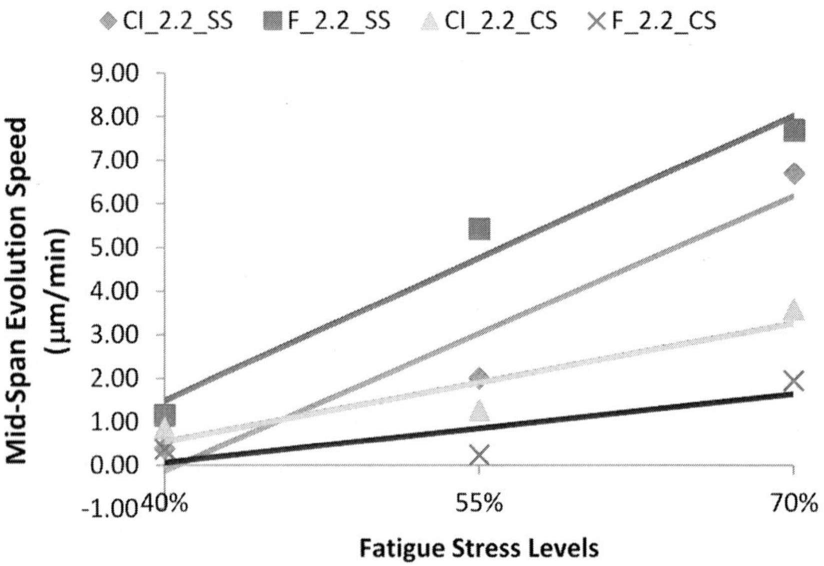

Figure 4. Speed rate of mid-span evolution at different fatigue stress levels.

5.1.2. Number and Width of Cracks

Figure 5 and Figure 6 illustrate the crack numbers and the crack widths of ECC mixtures respectively. It should be noted that the results in this section have been collected when the static flexural loading was applied after the fatigue flexural loading. The width of the crack was measured on the tension surface of the specimens by using a crack-scope.

It was found that more cracks formed at higher fatigue stress levels and unexpectedly few cracks formed at lower fatigue stress levels (Figure 5). At 40% stress level, a small number of cracks formed, meaning that ECCs tend to behave in a similar fashion to single cracking conventional normal concrete. The reason behind this is not clear. Qian and Li [32] suggested that when the fatigue load level decreases, it is more difficult for ECC beams to reach saturated multiple cracking since the corresponding tensile stress at the bottom of the beam may be very close to the cracking strength. Based on general trend (Figure 5), the number of cracks should increase with the increase of fatigue stress level (generally associated more damage).

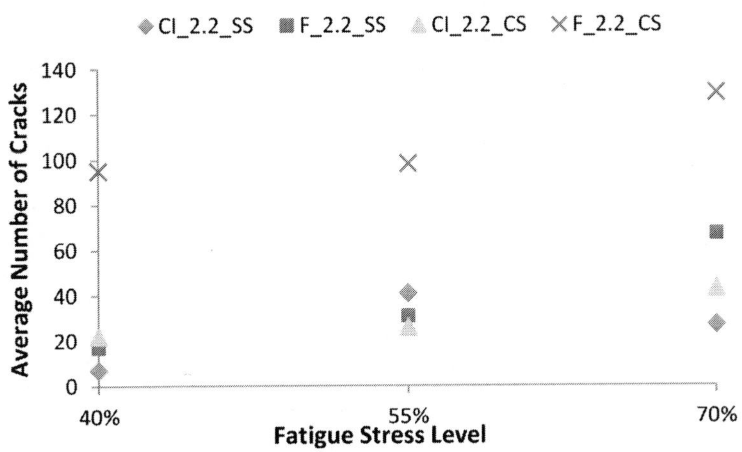

Figure 5. Average numbers of cracks at each fatigue stress level.

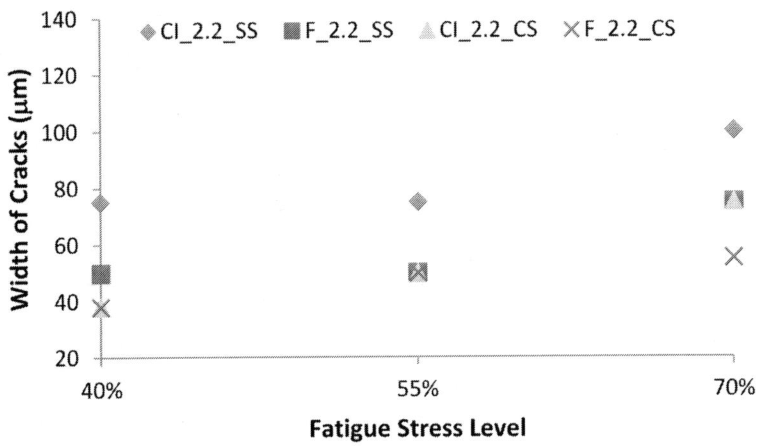

Figure 6. Crack Widths at each fatigue stress level.

It was noticed that the F_2.2_CS mixture had higher number of cracks and smaller crack widths than other ECC mixtures as shown in Figure 5 and Figure 6, respectively. However, it has to be mentioned again, at least at the range of low number of cycles studied in the present approach, that the large number of cracks with small width did not increase the deformation capacity in the case of fatigue as in F_2.2_CS mixture (Figure 3). As was expected, the relation between crack widths and fatigue stress level was directly proportional as shown in Figure 6.

In general, it can be concluded that the final number of multiple cracks of ECC mixtures is depending on a maximum fatigue stress level and also that the multiple-cracking behavior terminates earlier at a lower stress level.

It is essential to confirm whether or not multiple cracks can be formed in ECC mixtures even under fatigue loading, since structural applications can't be realized when the multiple cracking ability of ECC is negated during fatigue loading. Especially when improved durability is expected with ECC, the formation of fine multiple cracks is a necessary condition to minimize the migration of aggressive substances and could be the key to assess the feasibility of the structural applications under fatigue loading [33].

The results of the present study for fatigue testing confirmed that ECC mixtures showed multiple cracking under fatigue flexural loading as well as static flexural loading. Figure 7confirms this fact. As shown in the photograph, the colored cracks resulted under fatigue flexural loading while the new, un colored, cracks resulted under static flexural loading which applied after fatigue loading to get the residual fatigue energy in the tested ECC beam. To highlight even more on the performance of F_2.2_CS and F_2.2_SS ECC mixtures according to multiple-cracking behavior, the results of both mid-span deflection evolution as in Figure 3 and crack measurement data as in both Figure 5 and Figure 6 could be used. It can be concluded that the F_2.2_SS mixture exhibited more damage under fatigue loading than the F_2.2_CS ECC mixture as the evolutions of mid-span deflection was higher, and the total number of cracks of the SS mixture was lower than the CS mixture at the same fatigue stress level. This was caused by the fact that ECC exhibited multiple cracks and that the number of cracks initiated depended on the fatigue stress level. The number of cracks was higher when the specimens underwent high fatigue stress levels.

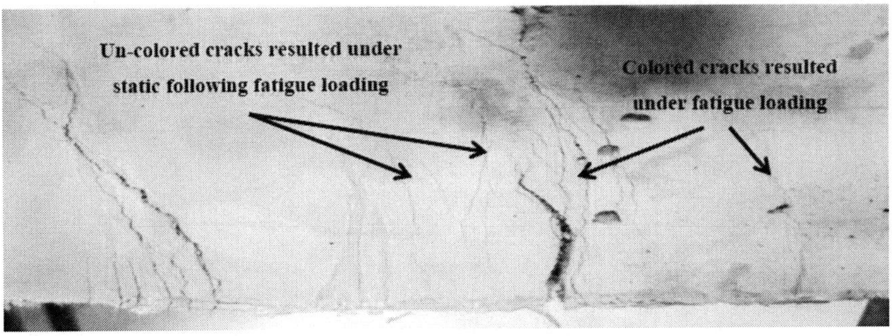

Figure 7. Multiple cracking under flexural fatigue loading.

5.1.3. Static Tests Following Fatigue Loading

At the end of the fatigue flexural tests as performed in the second part of the first phase, static flexural tests were applied on the exhausted and fatigued ECC specimens to calculate the fatigue residual energy for both strength and mid-span deflection. The fatigue flexural tests were expressed as percentages of static flexural tests and the residual strength and deflection are presented in Figure 8 and Figure 9, respectively.

Figure 8 and Figure 9 show that at higher fatigue stress level (70%), all ECC mix designs exhibited higher strength and deformation capacity than lower fatigue stress levels (40% and 55%). It was indicated that when a high fatigue stress level was applied, ECCs exhibited a significantly prolonged fatigue life in comparison with other FRC due to their multiple crack characteristics. Their fatigue life tended to be equivalent to or become shorter than that of FRC at lower fatigue stress levels [28].

As mentioned in the first phase, at 55% of fatigue stress level, FA–ECC mixtures with 70% cement replacement exhibited ordinary performance relative to 55% cement replacement for both residual fatigue strength and deflection. It is found that when comparing ECC mixtures with 70% cement replacement under 70% of fatigue loading stress level to those 55% cement replacement under 55% of fatigue loading stress level, the former showed equal or slightly higher superior performance of the latter with respect to both residual fatigue strength and deflection. For convenience, the first phase results of both fatigue flexural strength and mid-span deflection are presented in Table 7 compared to present phase results. These results confirm that the selection for FA–ECC mixtures with 70% cement replacement to be included in the second phase was a good choice.

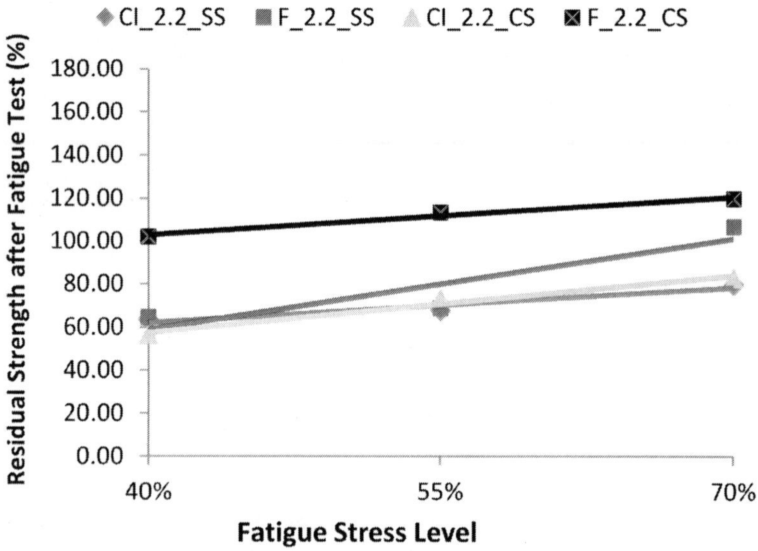

Figure 8. Percentages of residual strength after fatigue test.

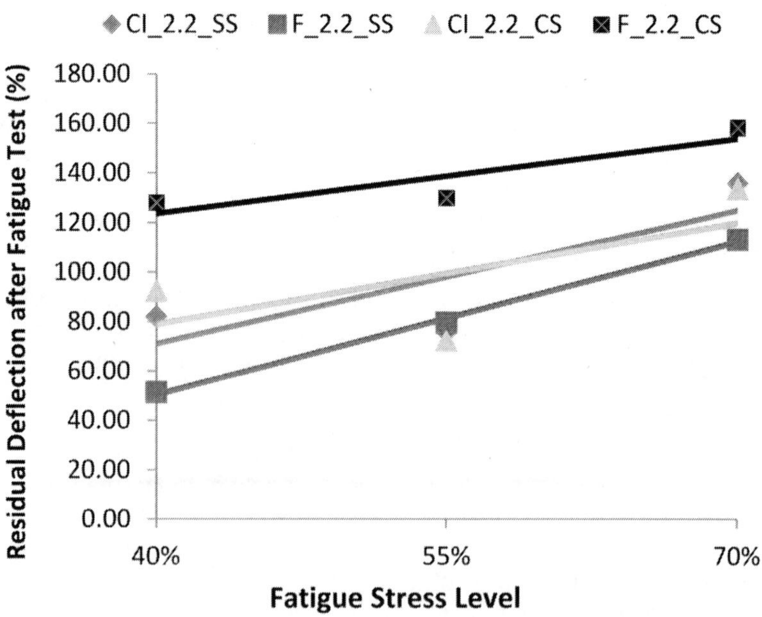

Figure 9. Percentages of residual mid-span deflection after fatigue test.

Table 7. Comparison between second and present Fatigue strength and deflection results.

Mix ID.	Mix Designation	Residual Fatigue Energy at 55% of Stress Level (First Phase)		Residual Fatigue Energy at 70% of Stress Level (Present Phase)	
		Residual Stress (%)	Residual Deflection (%)	Residual Stress (%)	Residual Deflection (%)
1	CI_1.2_SS	92.82	123.66	-	-
3	F_1.2_SS	94.58	135.29	-	-
7	CI_1.2_CS	87.87	116.08	-	-
9	F_1.2_CS	95.98	131.42	-	-
2	CI_2.2_SS	66.03	60.86	79.76	136.11
4	F_2.2_SS	64.94	69.44	106.71	113.26
8	CI_2.2_CS	71.19	82.13	82.69	133.69
10	F_2.2_CS	113.39	129.98	119.76	158.29

It is noticed that at 70% fatigue stress level, the F_2.2_SS mixture still shows high flexural strength capacity (106.71%) even with larger deflections during fatigue loading found in the results of the mid-span deflection evolution (2.082 mm) (Figure 3). Again, it should be noted that the F_2.2_CS mixture shows the highest superior performance consistently at all fatigue stress levels with respect to residual fatigue flexural strength and mid-span deflection as shown in Figure 8 andFigure 9, respectively. In the case of F_2.2_CS, it is noticed that the mid-span deflection evolution is inversely proportional to the residual fatigue flexural strength and mid-span deflection capacity. The lower the mid-span deflection evolution as in Figure 3 the higher the residual fatigue flexural strength and deflection capacity as in Figure 8 and Figure 9, respectively. F_2.2_CS behaved as fiber reinforced concrete (FRC) under fatigue loading. Suthiwarapirak *et al.* [28] mentioned that the evolution of mid-span deflection for FRC was very small. But unlike FRC behavior when static loading was applied after fatigue loading at high fatigue stress levels, the residual energy for both stress and deflection was much larger than both FRC and F_2.2_SS in this study. To highlight the practical application of this finding; Li *et al.* [34] confirmed that the most important properties required for link slab applications are tensile strain capacity (ductility) and crack width control for durability purposes. The minimum ductility required to withstand temperature and drying shrinkage stress, as well as live loads, was computed to be 1.4% using a factor of safety of two. Furthermore, it was confirmed that crack widths should be below 100 μm to minimize water/chloride penetration. These requirements are difficult, if not impossible, to attain for normal concrete, but are easily achievable with fly ash ECC mixtures with silica sand [34]. The current study revealed that the performance of fly ash ECC mixtures with crushed sand was comparable with that of silica sand mixtures under static loading while they exhibited much better performance under fatigue loading in terms of both residual fatigue flexural strength and deflection capacity.

5.2. Second Approach—Fatigue Number of Cycles

The range of fatigue cycles studied in the first approach was somewhat low, only 50,000 cycles. Therefore, further fatigue experiments were needed to be applied with higher range of fatigue cycles to confirm the case of FA–ECC mixtures. In the second approach, the fatigue tests were applied at different fatigue cycles, namely 200,000 cycles, 300,000 cycles and 1,000,000 cycles. The specimens in the second approach were subjected to 4 Hz sinusoidal cyclic loading rate of 55% fatigue stress level.

5.2.1. Mid-span Deflection Evolution

The evolutions of mid-span deflection as function of the number of cycles are plotted in Figure 10 for each ECC specimens with 55% fixed fatigue stress level.

At 1,000,000 cycles and as in fatigue stress level results (first approach), FA–ECC specimens with silica sand (CI_2.2_SS and F_2.2_SS) under fatigue loading developed much more damage than FA–ECC specimens with crushed sand (CI_2.2_CS and F_2.2_CS). The range of displacement evolution of FA–ECC specimens with silica sand was very large compared to FA–ECC specimens with crushed sand. Under stress level of 55%, the displacement evolved up to about 1.10 mm mid-span deflection in silica sand ECC specimens compared to 0.53 mm crushed sand ECC specimens. The evolution of mid-span deflection was found to depend on the number of cycles in FA–ECC specimens with silica sand. The mid-span deflection increased to more than twice at high number of cycles (1,000,000 cycles) compared to low number of cycles (200,000 and 300,000 cycles) as shown in Figure 10.

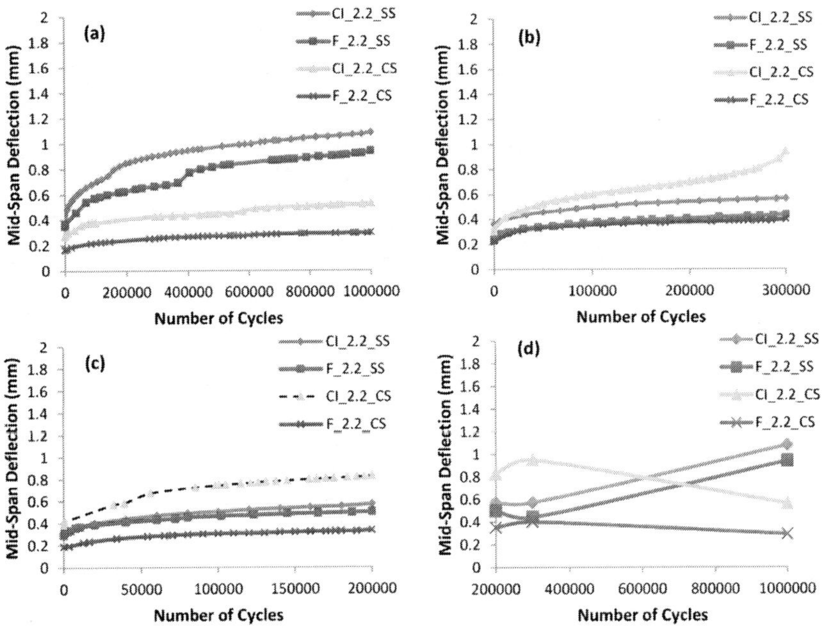

Figure 10. Evolution of mid-span deflection at 55% fatigue stress level and different number of cycles: (**a**) 1,000,000 cycles; (**b**) 300,000 cycles; (**c**) 200,000 cycles; (**d**) Trends showing deflection evolutions.

Un-expectedly, the relationship between the number of cycles and the evolution of mid-span deflection in FA–ECC specimens with crushed sand is inversely proportional. Their behavior for mid-span deflection evolution under high fatigue number of cycles was totally opposite to those with silica sand specimens. The more the number of cycles in FA–ECC with crushed sand the lower evolution of mid-span deflection, at least in the range of number of cycles studied in present study (Figure 10d). The reason behind this is not completely clear, but is likely associated with many un-hydrated spherical FA particles in the interfacial zone of FA–ECC mixtures which will make FA particles difficult to connect with other cement product crystals [1]. This is consistent with the fact that around 25%–30% in the paste with 50% replacement of fly ash did not hydrate and the unreacted fly ash may act like a micro-aggregate [35]. According to this, 70% of cement replacement was used in the present study which means around 30%–40% of un-hydrated fly ash particles did not participate in the hydration process, but could be seen as mere filler material, or aggregate. Therefore, the use of crushed sand in FA–ECC mixtures facilitated to keep the un-hydrated fly ash particles as filler materials along with the PVA fibers due to its high porosity. FA–ECC mixtures with silica sand definitely reduced the chances for fly ash particles to act as a filler material because silica sand has extremely fineness than fly ash particles. Accordingly, when the fatigue tests applied on the crushed sand samples, the friction between cement crystals and the un-hydrated fly ash particles along with the confined PVA fibers in the zones of un-hydrated fly ash particles would be increased due to the fatigue vibration and bouncing. Thus, enhanced friction may increase the chance to reactivate and accelerate the chemical reaction between the three mentioned above parties. Further experimental works are needed to confirm the case of FA–ECC with crushed sand.

Furthermore, Song [1] indicated that the major phases of fly ash used in his study were glass phases containing mainly a SiO_2 with 60% content. The FA glassy phase and the spherical particle shape may slow down the hydration process. This could indicate that FA–ECC mixture may have many un-hydrated particles featured with smooth surfaces, resulting in significant interface between the matrix and the FA particles. In the present study, SiO_2 proportion was around 60% and 40% of fly ash class F and CI respectively. This shows that class F fly ash mixture has more un-hydrated fly ash particles than class CI mixture which may enhance the composite performance under fatigue loading as mentioned above.

According to the fact that the un-hydrated fly ash particles do not participate in the hydration process and SiO_2 content, F_2.2_CS mixture exhibited better performance than CI_2.2_CS mixture. This is signified in Figure 10a–c, respectively. Figure 10 shows that the mid-span deflection evolution of F_2.2_CS was consistently less than CI_2.2_CS mixture at fatigue cycles of 200,000 cycles, 300,000 cycles and 1,000,000 cycles, respectively. Moreover, Figure 10d shows that the trend curve of F_2.2_CS mixture was more flat than CI_2.2_CS. This means that class-F fly ash mixture has a material characteristic independent of the number of cycles than all ECC mixtures mentioned herein at least within the range of fatigue number of cycles studied.

5.2.2. Number and Width of Cracks

To investigate the effect of fatigue loading on FA–ECC specimens, different fatigue number of cycles was applied on them namely 200,000 cycles, 300,000 cycles and 1,000,000 cycles. It should be noticed that the fatigue stress level was maintained at 55% for all of them. Again, it has to be mentioned that the results in this section had been collected when the static loading was applied following the fatigue loading. Crack numbers and widths of FA–ECC specimens are presented inFigure 11 and Figure 12, respectively.

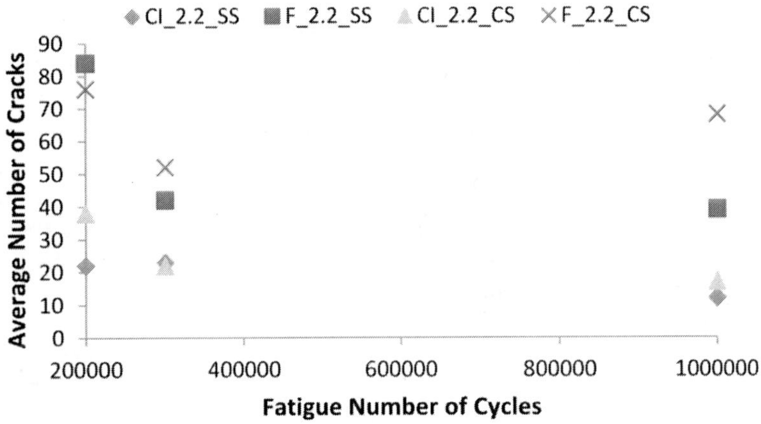

Figure 11. Average numbers of cracks at each fatigue number of cycles.

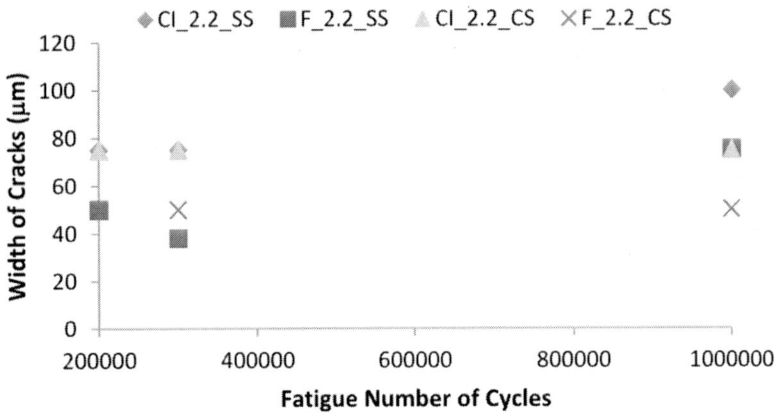

Figure 12. Crack widths at each fatigue number of cycles.

It was found that more cracks formed at lower fatigue number of cycles in FA–ECC mixtures and unexpectedly, few cracks were formed at higher fatigue number of cycles. This might be attributed to the fact that the fiber/matrix interfacial bond stress degradation results in fiber pullout and fiber fatigue results in fiber rupture [36]. Furthermore, Suthiwarapirak *et al.* [28] indicated that when comparing the fatigue specimens and static specimens, it was found that PVA fibers were severely ruptured under fatigue loading.

As in fatigue stress level results, first approach, F_2.2_CS mixture had higher number of cracks and smaller value of crack widths than other ECC mixtures as shown in Figure 11 and Figure 12, respectively. It should be noted that F_2.2_CS and CI_2.2_CS mixtures exhibited the same value of crack width at all fatigue number of cycle tests, 50 μm and 75 μm, respectively.

Although the final number of multiple cracks of FA–ECC mixtures was inversely proportional to the number of cycles, FA–ECC mixtures still showed multiple cracking under fatigue flexural loading as well as static flexural loading as shown in Figure 13. The photograph shows that the colored cracks resulted under fatigue flexural loading while the new, un-colored, cracks resulted under static flexural loading which applied after fatigue loading to get the residual energy in the tested ECC beam.

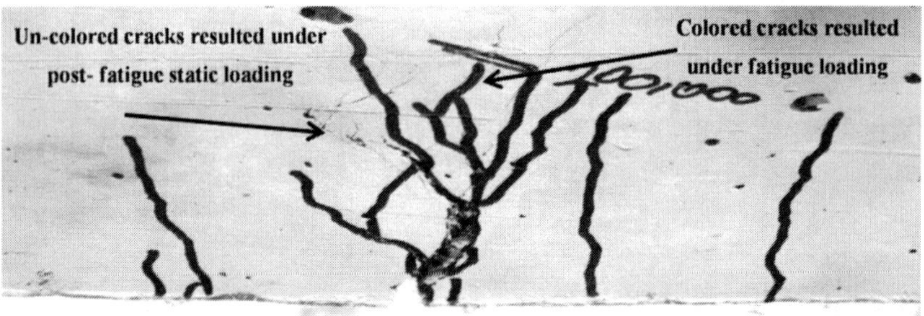

Figure 13. Multiple cracking under flexural fatigue loading.

5.2.3. Static Loading Following Fatigue Loading

Static flexural tests were applied on the fatigued ECC specimens to calculate the fatigue residual energy in terms of strength and mid-span deflection at the end of the fatigue flexural tests. The fatigue flexural tests were expressed as percentages of static flexural tests (in terms of residual strength and deflection) and are presented in Figure 14 and Figure 15, respectively.

Figure 14 and Figure 15 show that at higher number of cycles, FA–ECC mixtures with crushed sand exhibited higher residual energy in terms of flexural strength capacity than ECC mixtures with silica sand, while all ECC mixtures exhibited lower residual energy in terms of deflection capacity. These results are consistent with the mid-span deflection evolution results (Figure 10). It is found that the relation between the residual strength and the mid-span deflection evolution is inversely proportional. The lower the mid-span deflection evolution, the higher the residual fatigue flexural strength and vice versa. As in fatigue stress level results, F_2.2_CS mixture shows the highest performance consistently for all number of cycles with respect to fatigue stress and mid-span deflection as shown in Figure 14 and Figure 15, respectively.

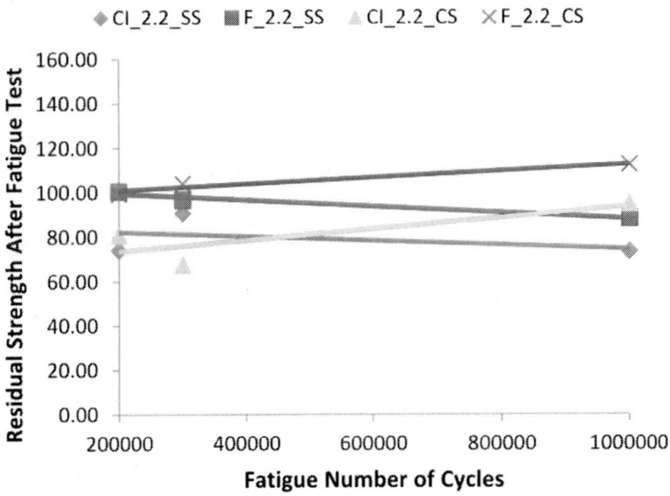

Figure 14. Percentages of residual strength after fatigue test.

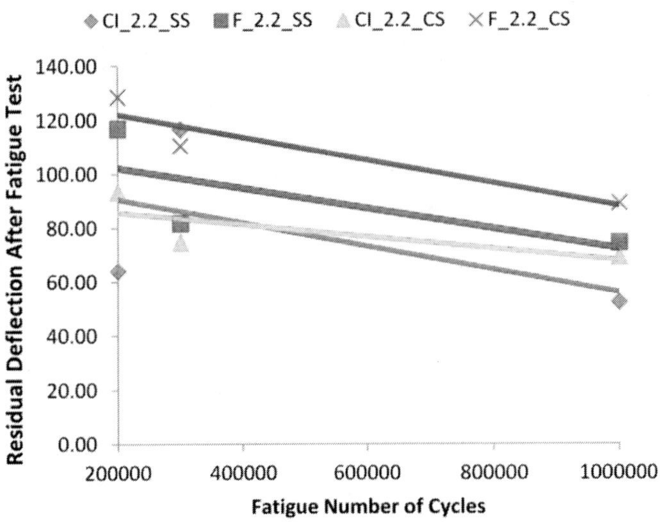

Figure 15. Percentages of residual mid-span deflection after fatigue test.

The good performance in residual energy of crushed sand ECC mixtures even after applied 1,000,000 cycles could be attributed to the decrease in the maximum fatigue stress level. The stress level used herein was somewhat low, 55%. According to this, Awad [37] reported that when normal concrete is subjected to high repeated stresses, a decrease of maximum stress level results in an increase of the number of cycles to failure. In addition to the decrease in fatigue stress level, high volume fly ash concrete subjected to high fatigue number of cycles at lower fatigue stress level as well would improve the performance of concrete. It was indicated that concrete with equivalent or higher compressive and fatigue strength of cylindrical concrete specimens incorporating large quantities of fly ash could be obtained with cement replacement of 25% by weight of low-calcium fly ash (class F) or 50% by weight of high-calcium fly ash (class C) [30]. Furthermore, Ramakrishnan *et al.* [38] revealed that the high-volume fly ash concrete has slightly higher (7%) endurance limit when expressed as a ratio (ratio of flexural fatigue strength to static flexural strength) compared to plain Portland cement concrete. The results further indicated that there was an increase (15% to 30%) in static flexural strength for high-volume fly ash concrete which was previously subjected to four million cycles of fatigue stresses at their respective lower fatigue limit load (10%). In the present study, both the decrease in fatigue stress level, which was 55%, and the production of fly ash ECC mixtures with 70% cement replacement, could be the reasons for the superior performance in residual fatigue energy of crushed sand ECC mixtures. In addition to these two reasons, most researchers supported that during cyclic loading, fatigue of concrete occurs because of the propagation of the micro cracks and macro cracks present in the material, especially in the interface region as well as in the matrix. As an attempt to delay fatigue failure, Naik *et al.* [29] reported the addition of fiber to concrete restrict crack formation and delays crack growth. Therefore, unstable cracks produced during loading are transformed into a slow and controlled growth. The overall tensile rupture strain of concrete is increased due to the introduction of fiber which will lead to the great improvement in the fatigue life of concrete.

5.2.4. Fatigue Stress Life Diagram, *S–N* Curve

The fatigue stress (*S*) to fatigue life *versus* the number of cycles (*N*) relationships for FA–ECC mixtures are plotted on a semi-logarithmic scale as shown in Figure 16. All 28 days specimens reached one million cycles without failure.

Figure 16 shows all average static and fatigue test results of all ECC mixtures. The fatigue number of cycle's axis was drawn by the logarithmic scale to the base 10 while stress axis was drawn in normal scale. It should be noted that further fatigue experiments were done for all ECC mixtures types at 50,000 and 100,000 fatigue cycles in addition to 200,000, 300,000 and 1,000,000 fatigue cycles to construct sufficient *S–N* curve for all fly ash ECC mixtures. In this series of tests, two specimens at each static and fatigue loading were conducted. As shown in Figure 16, it is found that F_2.2_CS and F_2.2_SS mixtures exhibited prolonged fatigue life compared with other CI-ECC mixtures at higher fatigue number of cycles.

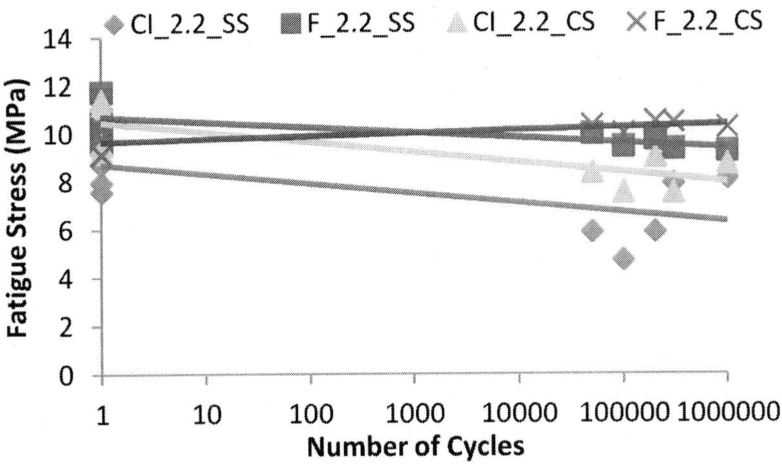

Figure 16. Fatigue stress–life relationships.

The flexural fatigue stress–life relationships of the four ECC mixtures can be expressed as a function of the fatigue stress, *S*, and logarithmic values of the number of cycles, *N*. The developed equations are:

CI_2.2_SS:

$$S = -0.172 \ln (N) + 8.6844, R^2 = 0.38 \quad (1)$$

F_2.2_SS:

$$S = -0.093 \ln (N) + 10.674, R^2 = 0.42 \quad (2)$$

CI_2.2_CS:

$$S = -0.181 \ln (N) + 10.456, R^2 = 0.65 \quad (3)$$

F_2.2_CS:

$$S = 0.054 \ln (N) + 9.6397, R^2 = 0.48 \quad (4)$$

Unfortunately, on the *S–N* curve for ECC mixtures, the author could not determine whether ECC mixtures exhibit a bilinear relation on a semi-logarithmic scale or not when compared to *S–N* curve presented done by Suthiwarapirak *et al.* [28], which is similar to the *S–N* relation of a metallic material. This is due to comparatively small number of fatigue cycles applied; one million cycles, compared to two million cycles by Suthiwarapirak *et al.* [28]. The authors could not have done more than one million cycles due to research time constraints.

In general, all ECC specimens showed superior fatigue strength compared to fatigue tests done by Zhang and Stang [39]. Table 8 shows fatigue strength results as a percentage of ultimate static flexural strength of the present study along with Zhang and Stang's results [39]. Fatigue flexural strength of plain concrete is 60%, and that of fiber reinforced concretes (FRCs) varies from 65% for hooked end (2% fiber content) to 90% for smooth steel (1%). Fatigue strength of polypropylene ECC (PE–ECC) is weaker than smooth steel (1%), but stronger than hooked end (2%). The fatigue strength of PE–ECC reaches as high as 78% of the ultimate static strength and it fails within the observed range of fatigue strength of FRCs [39]. FA–ECC mixtures with silica and crushed sand showed better fatigue strength performance at 1,000,000 cycle than all plain, FRCs and PE–ECC mixture except CI_2.2_SS mixture, which showed lower fatigue strength than FRCs but still a way better than plain concrete performance.

5.2.5. The Case of F_2.2_CS ECC Mixture

As a conclusion to the high performance of ECC mixture incorporating class-F fly ash with crushed sand, the mixture, F_2.2_CS, exhibited a unique performance which was equal to the superior performance of FA–ECC mixtures with 55% cement replacement in fatigue deflection and even better than them in fatigue strength. For 50,000 cycles and 70% higher fatigue stress level, the displacement evolution of F_2.2_CS mixture evolved up to 0.70 mm which was the lowest range of all other ECC mixtures. Accordingly, the speed rate of the mixture was 1.95 μm/min which is again the lowest fatigue speed rate. Higher number of cracks and lower crack widths of F_2.2_CS mixture at higher fatigue stress level, 70%, was remarkable. Test results showed that F_2.2_CS mixture exhibited 120 numbers of crack and 50 μm crack widths. The maximum numbers of crack and crack widths of all other ECC mixtures were 70 numbers of crack and 75 μm crack width, respectively.

Table 8. Comparison between fatigue strength of plain concrete, fiber reinforced concretes (FRCs) [39], polypropylene engineered cementitious composites (PE-ECC) [33] and present study fly ash (FA)-ECC specimens.

Concrete Type	Ultimate Static Strength (MPa)	Fatigue Strength (MPa)	Fatigue Strength (%) of Ultimate Static Strength at 1 Million Cycles
Plain Concrete and FRCs [39], and PE-ECC [33]			
plain concrete	6.94	–	60%
smooth steel 1%	10.15	–	90%
hooked steel 1%	9.88	–	87%
hooked steel 2%	12.82	–	65%
hooked steel 1% + polypropylene 1%	9.46	–	87%
PE-ECC 1.5%	8.12	–	78% failed at 0.88 million
Present study fatigue strength			
CI_2.2_SS	11.04	8.095	73.32%
F_2.2_SS	10.50	9.26	88.19%
CI_2.2_CS	9.20	8.712	94.70%
F_2.2_CS	9.11	10.24	112.40%

At high fatigue number of cycles (1,000,000 cycles), F_2.2_CS mixture exhibited 0.30 mm displacement evolution range which was again the lowest value of all ECC mixtures. The average number of cracks of F_2.2_CS mixture at 1,000,000 cycles and at 55% stress level was around 70 whereas the crack width was 50 μm which was consistently remained fixed for all fatigue number of cycles (started at 200,000 through 1,000,000 cycles). It can be said that the F_2.2_CS mixture has material characteristics independent of the fatigue number of cycles of all FA–ECC mixtures, which means that the more this material is subjected to fatigue loading the more it gets stronger and more flexible at the same time, at least within the range of number of cycles studied.

Suthiwarapirak *et al.* [28] and Matsumoto *et al.* [40] reported that the age of specimens at fatigue flexural testing should be at least two months to alleviate the effect of initial hydration development. According to this, and due to the consistent superior performance in the case of F_2.2_CS mixture, the author preferred to conduct another flexural fatigue test for 1,000,000 cycles to the mentioned mixture at the age of 56 days to confirm this superiority. The mixture's specimen was subjected to similar fatigue loading conditions employed in the present study. It was subjected to 4 Hz sinusoidal cyclic loading rate at 55% fatigue stress level of the maximum static test result.

For comparison purposes, the evolution of mid-span deflection of this mixture was compared with all ECC mixtures mid-span deflection evolution. It can be found that the displacement evolved up to 0.6 mm mid-span deflection in F_2.2_CS_56 days compared to 0.3 mm in F_2.2_CS_28 days. The mid-span deflection increased to twice the value of 28 days mixture at high number of cycles (1,000,000 cycles) as shown in Figure 17. The Figure shows that mid-span deflection evolution of F_2.2_CS_56 days still exhibits less evolution than 28 days fly ash ECC mixtures with silica sand, even at the age of 56 days.

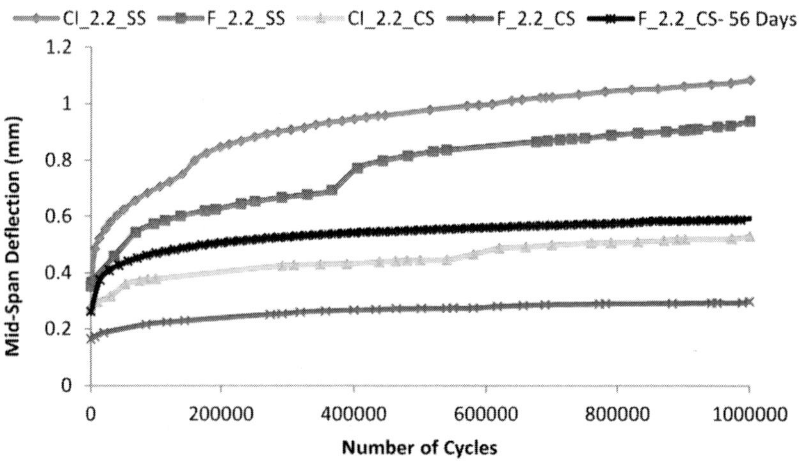

Figure 17. Evolution of mid-span deflection at 55% fatigue stress level and 1,000,000 cycles.

Table 9 shows the fatigue loading characteristics of fly ash ECC mixtures studied in the second approach of fatigue performance including the characteristics of F_2.2_CS at the age of 56 days. The presented fatigue characteristics summarized in Table 9 are for fatigue flexural loading test results at one million cycles only. It should be mentioned again that the results in this section had been collected when static loading was applied after fatigue loading. Crack numbers, crack widths and residual fatigue energy expressed as percentages of static flexural tests (in terms of residual strength and deflection) of all FA–ECC mixtures mentioned above are presented in Table 9.

Table 9. Fatigue loading characteristics of fly ash ECC mixtures at one million cycles.

Mix Designation	Age of Specimen	Static after Fatigue Loading		Residual Fatigue Energy at 55% Fatigue Stress Level	
		Crack Numbers	Crack Width (μm)	Residual Stress (%)	Residual Deflection (%)
CI_2.2_SS	28	12	100	73.32%	52.22%
F_2.2_SS	28	39	75	88.19%	74.46%
CI_2.2_CS	28	17	75	94.70%	69.09%
F_2.2_CS	28	68	50	112.40%	89.09%
F_2.2_CS	56	45	50	99.22%	81.20%

As shown in Table 9, the F_2.2_CS mixture at 56 days still has higher number of cracks and smaller value of crack widths than fly ash ECC mixtures with silica sand and CI_2.2_CS mixture. It should be noted that both F_2.2_CS at 28 and 56 days exhibited the same value of crack width at 1,000,000 fatigue cycles test which was around 50 μm. Furthermore, Table 9 shows that F_2.2_CS mixture at 56 days exhibited better performance at 1,000,000 fatigue number of cycles with respect to residual fatigue flexural strength and mid-span deflection of all fly ash ECC mixtures except for the F_2.2_CS mixture at 28 days.

6. CONCLUSIONS

This paper describes the performance engineered cementitious composites (ECCs) under static and fatigue loading conditions. The variable parameters are: type and size of aggregates, type of supplementary cementing materials (SCMs), SCM/cement ratio of 1.2 or 2.2 (55% or 70% by mass), fatigue stress levels (40%, 55% and 70%), number of fatigue cycles (50,000 to 1,000,000) and age (up to 56 days). Twelve ECC mixtures produced from different combinations of mineral admixtures (both class CI and F fly ash as well as slag) and sands (silica sand and local crushed with maximum grain sizes of 0.42 mm and 1.18 mm, respectively) are used. The aggregate/binder ratio is fixed at 0.36 and water/binder ratio is kept in the range of 0.27 for both class CI and F fly ash (FA) and 0.30 for slag ECC mixtures. The non-fatigue, fatigue and post-fatigue performances of ECC mixtures are determined based on flexural strength/deflection capacity, evolution of deflection and crack formation with crack width characteristics. The following conclusions are drawn from the study.

- The use of coarser local crushed sand in place of finer silica sand does not affect the deflection capacity and flexural strength of ECC mixtures for a particular SCM as long as polyvinyl alcohol (PVA) fibers are distributed evenly during the mixing process. The combination of sands (either silica or crushed) with different SCMs can affect both flexural strength and deflection capacity. The use of high volumes (up to 70%) of SCMs produced ECC mixtures with enhanced bending ductility. Class F-fly ash was the best choice when combined with crushed sand (compared to class CI-fly ash and slag) in order to enhance the ECC ductility. Test results revealed that slag–ECC mixtures had lower deflection capacity under static loading. However, the deflection capacity under static loading of slag–ECCs remained 150 times greater than normal and conventional fiber reinforced concrete although they showed slightly higher flexural strength than FA–ECC mixtures.

- Under low fatigue cycles (50,000) at constant stress level of 55%, low volume (55%) FA (both CI and F) ECC mixtures with both silica and crushed sands produced higher residual fatigue strength and deflection capacity compared to those made with high volume FA (70%). Only exception was high volume class F FA–ECC mixture with crushed sand which showed similar behavior compared to those made of lower volume FA with both silica and crushed sands. However, slag–ECC mixtures showed inferior performance in terms of lower residual flexural strength and deflection capacity under fatigue loading compared to those with class F and CI fly ash.
- In both cases of high fatigue stress level (70%) and high fatigue cycles (1,000,000 cycles), high volume FA–ECC mixtures with silica sand developed more damage in deformation capacity, bigger crack width and lower number of cracks than high volume FA–ECC mixtures with crushed sand. Silica sand ECCs developed crack width in less than or equal to 100 μm whereas in general crack width was found to be reduced with the crushed sand (<50 μm). It is interesting to note that the use of crushed sand seemed to have increased the resistance of ECC mixtures against fatigue damage (especially at higher number of cycles) due to capacity enhancement in terms of generating more internal energy. At high fatigue cycles up to 100,000, crushed sand-ECC mixtures showed the best performance (even at high stress level of 70% and at higher age of 56 days) exhibiting higher residual strength and deflection capacity compared to other mixtures with all combinations of both types of fly ash and sands.
- Fatigue tests at higher stress level (70%) produced more multiple cracks and bigger crack width in FA–ECC mixtures than those conducted at low stress level (40%). In addition, more cracks were formed during lower fatigue number of cycles (200,000) and during

higher fatigue number of cycles (1,000,000 cycles), few cracks were developed.

- It is concluded that the most important feature of ECC, high deflection capacity with multiple cracking behaviors, was protected and is not sacrificed by replacing cement with a maximum of 70% FA or by replacing silica sand with local crushed sand.
- The results confirmed that the use of crushed sand not only provide cost effective ECC mixtures but also better mechanical properties in terms of enhanced ductility, fatigue resistance and post-fatigue mechanical characteristics. Such ECC mixtures have great potential to be used in the construction building and bridge structures. However, more investigations are needed to explore the effect of crushed sand on durability performance of ECC mixtures. Researches are in progress on the short and long term properties as well as structural performance of joint-free bridge decks with ECC link slabs.

ACKNOWLEDGMENTS

The authors gratefully acknowledge the financial assistances of Ministry of Transportation of Ontario (MTO) Highway Infrastructure Innovations Funding Program and National Science and Engineering Research Council (NSERC), Canada. Supports from Lafarge Canada and St. Marys Cement in terms of materials are highly appreciated. Special thanks to Hanna Schell, David Rhead and Clifford Lam of MTO as well as Abdurrahman Lotfy (Lafarge, Canada) and Wafiq Taha (St. Marys Cement, Canada)—their supports are gratefully acknowledged.

AUTHOR CONTRIBUTIONS

This work was carried out in collaboration among all three authors. Mohamed A.A. Sherir carried out experiments, data analyses and prepared the draft of the manuscript. Khandaker M.A. Hossain designed the study, supervised the research, involved in experimental data analyses and prepared the manuscript for submission. Mohamed Lachemi was involved in supervision of the research and preparation of the manuscript. All three authors proof-read and approved the paper.

REFERENCES

1. Song, G. Matrix Manipulation to Study ECC Behaviour. Master's Thesis, University of Stellenbosch, Matieland, South Africa, 2005.
2. Li, V.C.; Kanda, T. Engineered cementitious composites for structural applications. *ASCE J. Mater. Civ. Eng.* **1998**, *10*, 66–69.
3. Li, V.C.; Wang, S.; Wu, C. Tensile strain-hardening behavior of polyvinyl alcohol engineered cementitious composite (PVA-ECC). *ACI Mater. J.* **2001**, *98*, 483–492.
4. Li, V.C. On engineered cementitious composites (ECC): A review of the material and its applications. *J. Adv. Concr. Technol.* **2003**, *1*, 215–230.
5. Lin, Z.; Li, V.C. Crack bridging in fiber reinforced cementitious composites with slip-hardening interfaces. *J. Mech. Phys. Solids* **1997**, *45*, 763–787.
6. Lin, Z.; Kanda, T.; Li, V.C. On interface property characterization and performance of fiber reinforced cementitious composites. *Concr. Sci. Eng.* **1999**, *1*, 173–184.
7. Kong, H.J.; Bike, S.G.; Li, V.C. Development of a self-consolidating engineered cementitious composite employing electrosteric dispersion/stabilization. *Cem. Concr. Compos.* **2003**, *25*, 301–309.
8. Kong, H.J.; Bike, S.G.; Li, V.C. Constitutive rheological control to develop a self-consolidating engineered cementitious composite reinforced with hydrophilic poly(vinyl alcohol) fibers. *Cem. Concr. Compos.* **2003**, *25*, 333–341.
9. Kim, Y.Y.; Kong, H.; Li, V.C. Design of engineered cementitious composite suitable for wet-mixture shotcreting. *ACI Mater. J.* **2003**, *100*, 511–518.
10. Lepech, M.D.; Li, V.C. Large scale processing of engineered cementitious composites. *ACI Mater. J.* **2008**, *105*, 358–366.
11. Yang, E.; Li, V.C. A Micromechanical Model for Fiber Cement Optimization and Component Tailoring. In Proceedings of the 10th International Inorganic-Bonded Fiber Composites Conference, San Paulo, Brazil, 15–18 November 2006; pp. 1–13.
12. Sahmaran, M.; Lachemi, M.; Hossain, K.M.A.; Ranade, R.; Li, V.C. Influence of aggregate type and size on ductility and mechanical

properties of engineered cementitious composites.*ACI Mater. J.* **2009**, *106*, 308–316.

13. Li, V.C.; Mishra, D.K.; Wu, H.C. Matrix design for pseudo-strain-hardening fibre reinforced cementitious composites. *Mater. Struct.* **1995**, *28*, 586–595.

14. Fischer, G.; Wang, S.; Li, V.C. Design of engineered cementitious composites (ECC) for processing and workability requirements. In *Brittle Matrix Composites 7*; Brandt, A.M., Li, V.C., Marshall, M.H., Eds.; Woodhead Publishing: Cambridge, UK, 2003; pp. 29–36.

15. Mehta, P.K. Influence of fly ash characteristics on the strength of portland-fly ash mixtures.*Cem. Concr. Res.* **1985**, *15*, 669–674.

16. Kim, Y.Y.; Kim, J.S.; Kim, H.S.; Kim, J.K.; Ha, G.J. Uniaxial Tensile Behavior of High Ductile Fiber Reinforced Mortar Designed with Ground Granulated Blast Furnace Slag. In Proceedings of the KSCE Annual Conference, Pyung-Chang, Korea, 20–22 October 2004; pp. 546–551.

17. Kim, J.K.; Kim, J.S.; Ha, G.J.; Kim, Y.Y. Tensile and fiber dispersion performance of ECC (engineered cementitious composites) produced with ground granulated blast furnace slag.*Cem. Concr. Res.* **2007**, *37*, 1096–1105.

18. Wang, S.; Li, V.C. Engineered cementitious composites with high-volume fly ash. *ACI Mater. J.* **2007**, *104*, 233–241.

19. Yang, E.-H.; Yang, Y.; Li, V.C. Use of high volumes of fly ash to improve ECC mechanical properties and material greenness. *ACI Mater. J.* **2007**, *104*, 303–311.

20. Zhou, J.; Qian, S.; Sierra Beltran, M.G.; Ye, G.; Breugel, K.; Li, V.C. Development of engineered cementitious composites with limestone powder and blast furnace slag. *Mater. Struct.* **2010**, *43*, 803–814.

21. Lepech, M.; Li, V.C. Water Permeability of Cracked Cementitious Composites. In Proceedings of the 11th International Conference on Fracture (ICF11), Turin, Italy, 20–25 March 2005; p. 4539.

22. Lepech, M.; Li, V.C. Durability and long term performance of engineered cementitious composites (ECC). In *International RILEM Workshop on High Performance Fiber Reinforced Cementitious Composites in Structural*; Fischer, G., Li, V.C., Eds.; RILEM Publications SARL: Bagneux, France, 2006; pp. 165–174.

23. Caner, A.; Zia, P. Behavior and design of link slabs for jointless bridge decks. *PCI J.* **1998**, *43*, 68–80.

24. Hossain, K.M.A.; Lachemi, M.; Kayes, M.A.; Mavani, M.B. Influence of ECC Mixture on the Structural Performance of Link Slabs in Bridge Decks. In Proceedings of the 3rd Specialty Conference on Material Engineering & Applied Mechanics, Montréal, Québec, Canada, 29 May–1 June 2013.

25. Li, V.C.; Wu, C.; Wang, S.; Ogawa, A.; Saito, T. Interface tailoring for strain-hardening polyvinyl alcohol-engineered cementitious composite (PVA-ECC). *ACI Mater. J.* **2002**, *99*, 463–472.

26. American Society for Testing and Materials. *Standard Test Method for Slump Flow of Self-Consolidating Concrete*; ASTM C1611 2014; American Society for Testing and Materials: West Conshohocken, PA, USA, 2014.

27. American Society for Testing and Materials. *Standard Test Method for Flexural Strength of Concrete (Using Simple Beam with Third-Point Loading)*; ASTM C78 2002; American Society for Testing and Materials: West Conshohocken, PA, USA, 2002.

28. Suthiwarapirak, P.; Matsumoto, T.; Kanda, T. Multiple cracking and fiber bridging characteristics of engineered cementitious composites under fatigue flexure. *J. Mater. Civ. Eng. ASCE* **2004**, *16*, 433–443.

29. Naik, T.; Singh, S.; Ye, C. *Fatigue Behavior of Plain Concrete Made with or without Fly Fsh*; The Department of Civil Engineering and Mechanics, University of Wisconsin-Milwaukee for Electric Power Research Institute (EPRI): Palo Alto, CA, USA, 1993.

30. Tse, E.W.; Lee, D.Y.; Klaiber, F.W. Fatigue Behavior of Concrete Containing Fly Ash. In Proceedings of the Second International Conference on the Use of Fly Ash, Silica Fume, Slag and Natural Pozzolans in Concrete, Madrid, Spain; Malhotra, V.M., Ed.; American Concrete Institute Special Publication: Madrid, Spain, 1986; pp. 273–290.

31. Ozaki, S.; Sugata, N. *Fatigue of Concrete Composed of Blast Furnace Slag or Silica Fume under Submerged Conditions*; Malhotra, V.M., Ed.; American Concrete Institute (ACI): Detroit, MI, USA, 1992; pp. 1509–1524.

32. Qian, S.; Li, V.C. Durable pavement with ECC. In *International Conference on Microstructure Related Durability of Cementitious*

Composites; Sun, W., van Breugel, K., Miao, C., Ye, G., Chen, H., Eds.; RILEM Publications: Bagneux, France, 2008; pp. 535–543.

33. Matsumoto, T. Fracture Mechanics Approach to Fatigue Life of Discontinuous Fiber Reinforced Composites. Ph.D. Thesis, Department of Civil and Environmental Engineering, University of Michigan, Ann Arbor, MI, USA, 1998.

34. Li, V.C.; Lepech, M.; Wang, S.; Weimann, M.; Keoleian, G. Development of green ECC for sustainable infrastructure systems. In Proceedings of the International Workshop on Sustainable Development and Concrete Technology, Beijing, China, 20–21 May 2004; pp. 181–192.

35. Termkhajornkit, P.; Nawa, T.; Nakai, M.; Saito, T. Effect of fly ash on autogenous shrinkage.*Cem. Concr. Res.* **2005**, *35*, 473–482.

36. Matsumoto, T.; Suthiwarapirak, P.; Asamoto, S. Model development of ECC fatigue analysis.*Proc. Jpn. Concr. Inst.* **2002**, *24*, 237–242.

37. Awad, M.E. Strength and Deformation Characteristics of Plain Concrete Subjected to High Repeated and Sustained Loads. Ph.D. Thesis, University of Illinois at Urbana, Champaign, IL, USA, 1971.

38. Ramakrishnan, V.; Malhotra, V.M.; Longley, W.S. Comparative evaluation of flexural fatigue behavior of high volume fly ash and plain concrete. *ACI Spec. Publ.* **2005**, *229*, 351–368.

39. Zhang, J.; Stang, H. Fatigue performance in flexure of fiber reinforced concrete. *ACI Mater. J.***1998**, *95*, 58–67.

40. Matsumoto, T.; Suthiwarapirak, P.; Kanda, T. Mechanisms of multiple cracking and fracture of DFRCC under fatigue flexure.*J. Adv. Concr. Technol.* **2003**, *1*, 299–306.

5
CHAPTER

Chemical and Physical Properties of Polyimides: Biomedical and Engineering Applications

Anton Georgiev[1], Dean Dimov[2], Erinche Spassova[2], Jacob Assa[2], Gencho Danev[2] and Peter Dineff[3]

[1] University of Chemical Technology and Metallurgy, Department of Organic Chemistry, Sofia, Bulgaria

[2] Institute of Optical Materials and Technologies "Acad. Jordan Malinovski",Department of "Nanostructured Materials and Technology", Sofia, Bulgaria

[3] Technical University,Sofia, Bulgaria

1. INTRODUCTION

The development of chemistry and solid matter physics lead to improved technologies in producing polymers and nanosized films such as vacuum deposition and solid state reactions. Over the past 20 years because of intensive research in chemistry and physics of organic materials, place the development of the new area in materials chemistry - organic electronics and photonics (polytronics). Polymer materials are becoming more widely used to improve the quality of human life: from social to high-tech - household appliances, textiles, insulation materials in industry and construction, medical implants, materials for optoelectronics, the formation of nanosized films and more over. Important classes of polymeric materials responded to certain technical requirements are polyimides (PI). They contain in molecules functional group $CO-NR_2$ called imide. The presence of n-π conjugation between non-pair electron of nitrogen atom and π electrons of the carbonyl group makes them resistant to chemical agents and moisture. Mainly the type of hydrocarbon residues (arenes, aliphatic) and the presence of other functional groups (Cl, F, NO2, OCH3, etc) determine their physical properties and their application in practice.In this chapter will be discussed preparation of PI, the mechanisms of reactions, possible competitive reactions, physical properties, their application as biomedical materials (implants, functionalized nucleic acids) and their application as optical materials for the producing of nanocomposite layers matrix including metallic or dielectric clusters as "guest". One of the achievements in obtaining PI layer is the application of microwave (MW) irradiation combined with low temperature treatment compared to the application of pure thermal imidization. This approach is particularly effective imidization for inclusion of PI layers in multilayer lithography or optical systems in which the formed nanocomposite films are not stable in thermal imidization condition.

1.1. Classification of Polyimides

Depending on the polymer chain, the type of hydrocarbon residues and the presence of other functional groups, polyimides can be classified as follows:

1.1.1. Linear

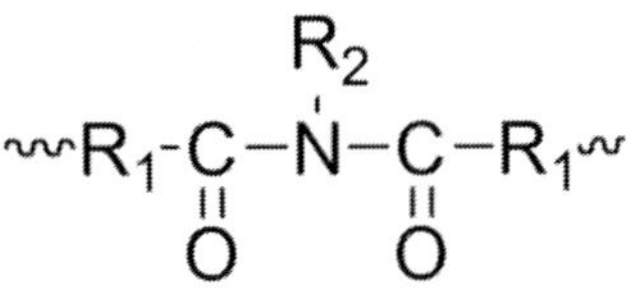

1.1.2. Cyclic

a. Aromatic (main-chain)

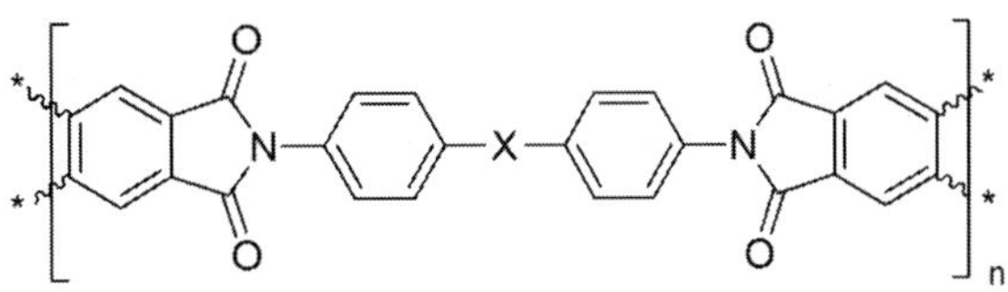

X = -O-, >C=O, >CH_2, -N=N-, -SO_2- и др.

b. Aliphatic – aromatic (main-chain)

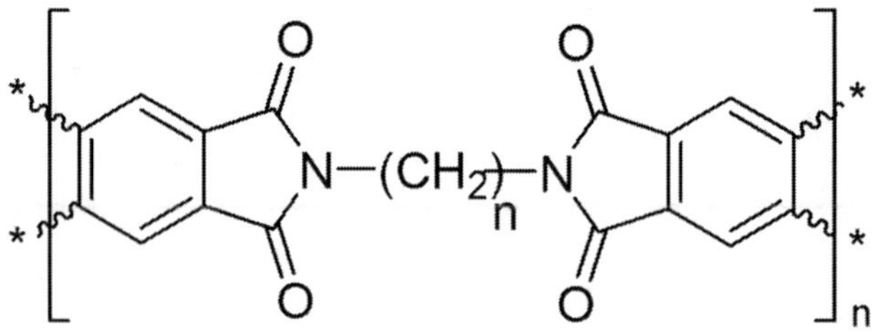

1.1.3. Side-chain

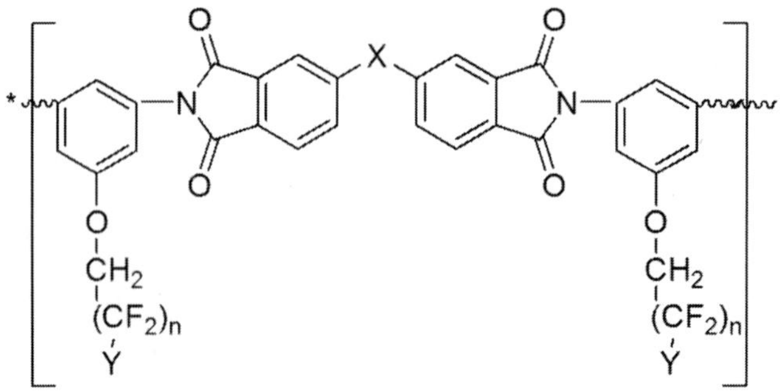

n = 1,2,3,4,5; Y= F, H

1.1.4. Functionalizing

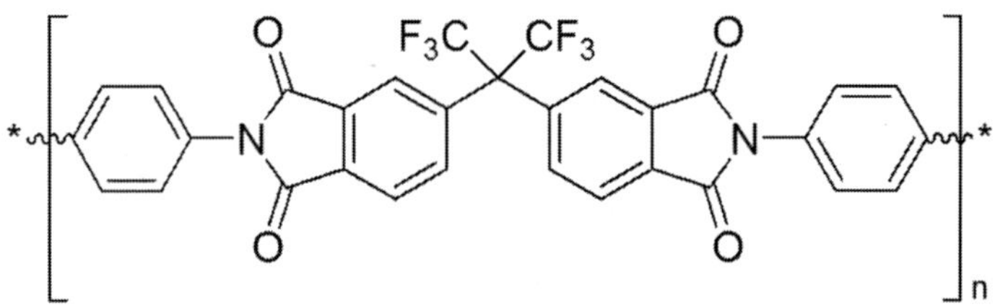

Many polyimide investigations have mainly been concentrated on aromatic polyimides, and little information is available about aliphatic polyimides that are also potential candidates for engineering and biomedical applications.

2. CHEMICAL PROPERTIES OF POLYMIDES

2.1. Preparation and Structure of Polyimides

Aromatic polyimides generally prepared by a two-step procedure from aromatic diamines and aromatic tetracarboxylic dianhydrides. The chemistry of polyimides is a specific area with large variety of monomers available and several methodologies for synthesis [1,2,3]. The subtle variations in the structure dianhydride and diamine components have tremendous effect on the properties of the final polyimide. The most widely practiced procedure in polyimide synthesis is two-step process via poly(amic acid). The reaction between dianhydride (*pyromellitic dianhydride* PMDA) and diamine (*4,4'-oxydianiline* ODA) is required ambident conditions in dipolar aprotic solvents, such as N-methylpyrrolidone (NMP) or N,N-dimethylacetamide (DMAc). The next polycyclodehydration reaction of poly(amic acid) depending on ratio of precursors lead to final polyimide with different molar mass (Scheme 1) [4,5].

Reactions between cyclic anhydrides and primary diamines run as $S_{N2}Ac$ mechanism (bimolecular nucleophillic acyl substitution). The reaction running in two steps, the first is attaching of nucleophillic reagent to electrophillic carbonyl C-atom. The intermediate poly(amic acid) is formed by the nucleophilick attack of the amino group on the carbonyl carbon of the anhydride group. Thus reaction is irreversible, because the amino group is strong nucleophillic agent, consequence is not good leaving group than the hydroxyl group from the carboxylic acid [1,2,5,6]. In additions, anhydride cycle have not good resonance stability and charge delocalization, because oxygen atoms have equal electronegativity and electron structure. The second step is nucleophillic ring closure due to dehydration and imide ring formation. One of the disadvantages of this method is unavoidable presence of solvents and need for their removal [7]. During the polycyclodehidration side reaction is run that defy stereoregular control [1]. In Figure 1 shows the structural formulas of the side products of the reaction for obtaining PI.

These side compounds caused defects in the synthesized PI film. They are unwanted impurity in production of nanostructured films with chromophores as a "guest" in the matrix, since they lead to low quality of dielectric and optical properties of layers. At temperatures above 200 °C by intramolecular rearrangement isoimide convert to imide (Scheme 2) [9,10].

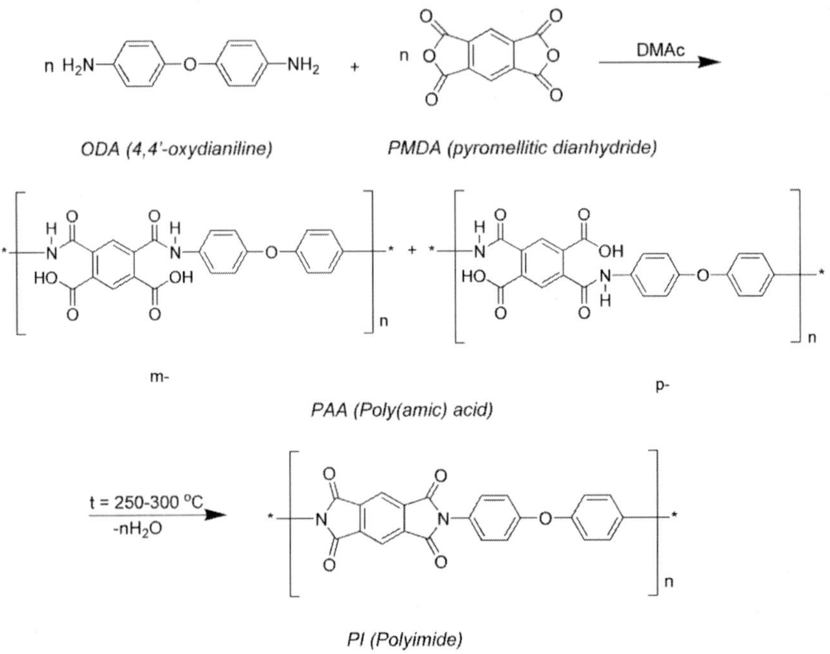

Scheme 1. Reaction between ODA and PMDA to PAA with following cyclodehidration to PI.

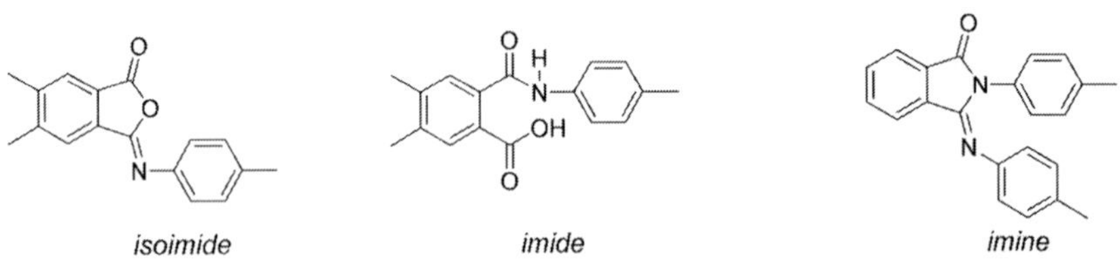

Figure 1. Structural formulas of side compounds of the reaction for preparation of PI.

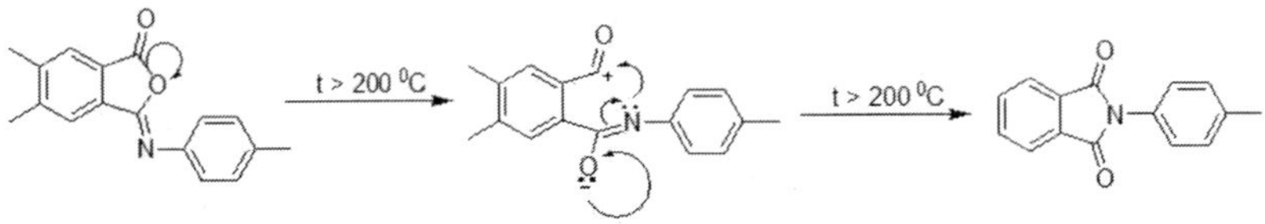

Scheme 2. Intramolecular rearrangement of isoimide to polyimide.

Another way for synthesis of polyimides from Nylon-Salt-Type Monomers has been reported from Imai [4]. The method based on the melt polycondensation of diamine and pyromellitic acid half diester (pyromellitic acid diethyl ester) (Scheme 3).

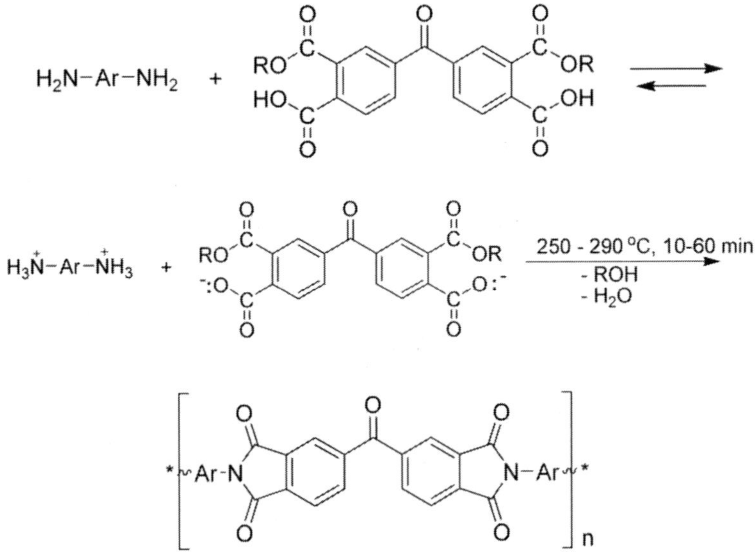

Scheme 3. Synthesis of PI from pyromellitic acid half diester and diamine.

These salt monomers have been prepared as white crystalline solids by dissolving an equimolar amount of each individual diamine and tetracarboxylic acid half diester in hot ethanol (or methanol), and subsequently cooling the resultant solution. The author has been found that imidization (polycondensation) of salt run to polyimide for 10 min at

250 °C. The high-pressure polycondensation of the salt monomer has been applied. The pressure affect on the temperature and reaction time that directly afforded high molecular-weight polyimide. This method is useful for the synthesis of the polyimides having well defined structures, compared with the other synthetic methods [4].

2.2. Vapour Deposition and Solid State Reaction

The development of the chemistry of polymers and their application in nanotechnology many researchers have been seen alternative forms for obtaining nanosized films. Our studies have focused primarily on vapour deposition of precursors and solid state imidization reaction [7,10,11,12]. Vapour deposition processes of organic layers play an important role in polytronics. They allow the construction of systems without solvents, based on principle of bottom-up and have significant role in the formation of intermediate and protective layers. Important advantages of vapour processes are follows: (i) vacuum deposition is basically a cleaning process from impurities and resulting deposited layers have a much higher purity. This reduces problems with local anisotropy properties, polymorphism, etc.; (ii) vapour process allows the run of additional activated or modified processes as resulting deposited films are changed and improved properties: higher physical density, polymerization, high chemical purity and others. This is related take plasma processes, electronic flow, microwave irradiation or photon interaction. All these additional processes generate enough number of active particles as free radicals and ions in the gas phase. The disadvantage of these so-called vapour assistance deposition processes is the need for very precise control [7,13].

In the vacuum deposition of the precursors PMDA and ODA at temperature of 120-145 °C reaction of polycondensation to PAA with opening of the anhydride ring of PMDA takes place (Ac-S_N2 -reaction). These processes are to great extent accelerated and controlled in the thermal treatment of the condensed solid phase which represents PAA, with regard to their transformation to PI by means of reaction of polycyclodehydration in solid state to linear PI [7,10,11]. The FTIR spectra of individual films of PMDA, ODA and PAA are shown in Figure 2.

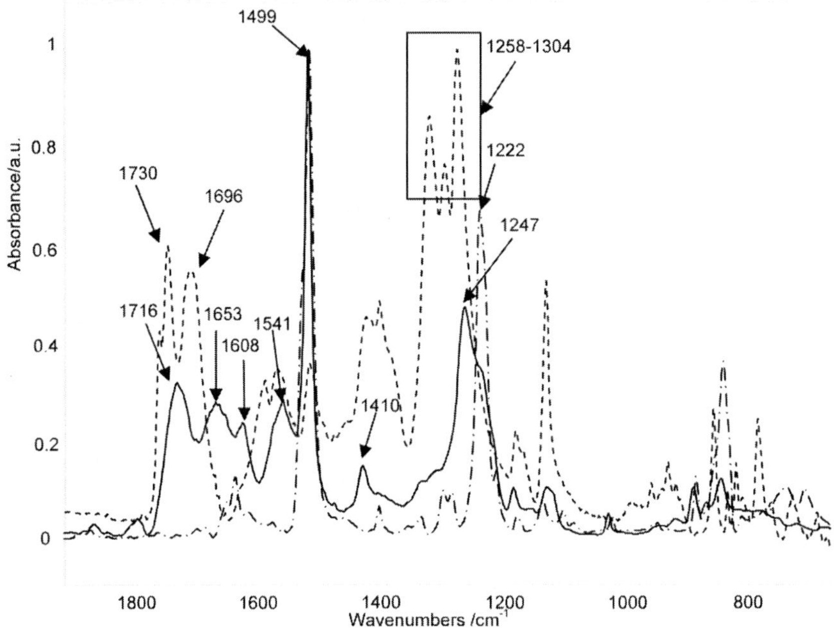

Figure 2. FTIR spectra of the vacuum deposited films of: - - - - - - - PMDA; — · — · — · ODA; --------simultaneously deposited both precursors in a mole ratio of PMDA : ODA= 1:1(PAA).

The band ν 1716 cm^{-1} for the >C=O group in the PAA spectrum related to acid. The presence of hydroxyl group of the acid (C-OH) is corresponded at δ 1247 cm^{-1}. The amide bond is identified by the bands at ν 1653 cm^{-1} (>C=O amide I band) and δ 1608 cm^{-1} (N-H amide II band).

Compared to the classical methods for producing films from PAA, in which the acid is preliminarily obtained in a solution and after that deposited as a thin film usually by spin coating, in the vacuum deposition method this process is performed in only one step. More and more researcher accept that the method of vacuum deposition provides for a greater degree of purity in the thin film production, opportunity for controlling and computerizing the processes of heating, imidazion and layer formation designed for obtaining standard PI or nanocomposite products of reproducible composition, thickness, structure and properties [12,14]. In Figure 3 are compared the spectra of PAA and solid imidization film of PAA to PI obtained after thermal treatment.

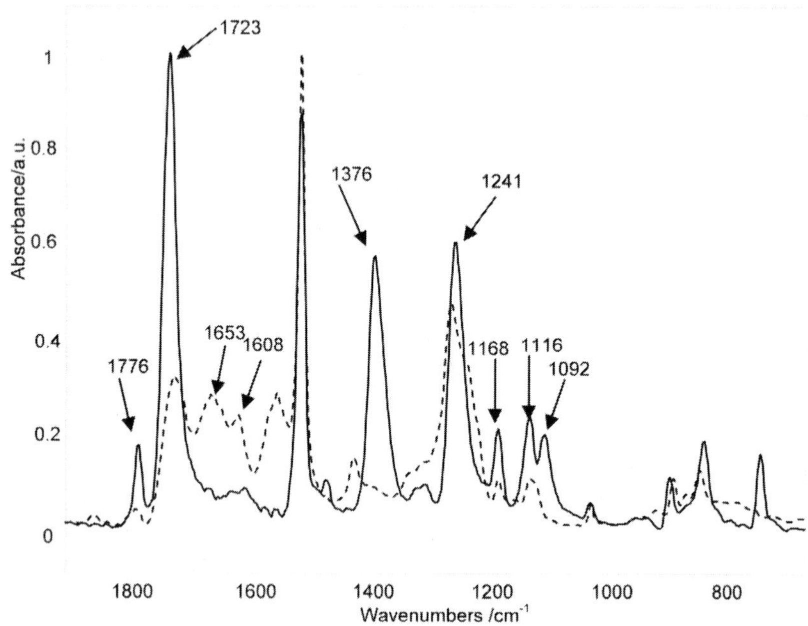

Figure 3. FTIR spectra of vacuum deposited films with thickness 200 nm and PMDA:ODA(PAA)=1:1 - - - - - - - - thermally untreated; ------------- thermally treated 1 h at 300 ºC.

The bands at ν_s 1776 and ν_{as} 1723 cm^{-1} characterize the >C=O groups of the imide ring. The imidization confirmed by the $\nu_{C\text{-}N}$ 1376 cm^{-1} imide III band where a minimum is observed in the spectrum of the untreated layer in this frequency. The exits of new bands in the area deformation vibrations for C-O and C-N bonds respectively at 1241, 1168, 1116 and 1092 cm^{-1} is observed [7,9,10].

Similarly, the vacuum deposition gives the opportunity to syntheses of azo-polyimide, type "main-chain". In this case, as starting precursors were used PMDA and DAAB (*4,4 '-diaminoazobenzene*), which are then vacuum deposition in a thin layer subjected to thermal and microwave treatment for imidization reaction to polyimide (Scheme 4) [7,15]. As is seen the azo-chromophore is covalently bonded in the polymer matrix. So a preparation of colored polymer is realized without to incorporate in the matrix an additional chromophore.

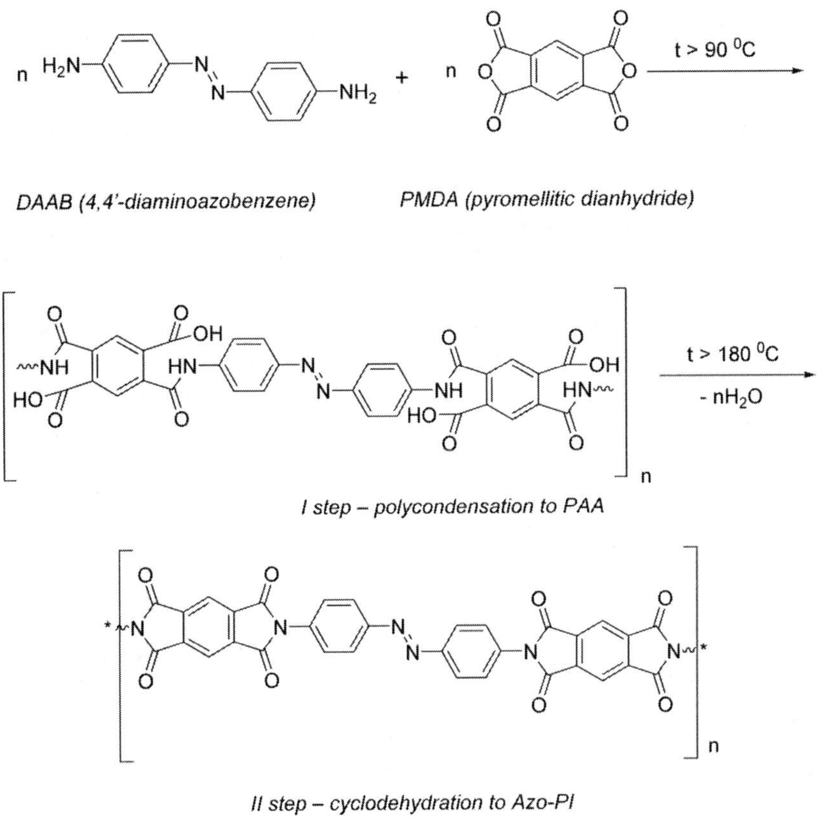

Scheme 4. Reaction between PMDA and DAAB to Azo-PI

The resulting films were characterized by FTIR and UV-VIS spectroscopies before and after imidization. It was studied kinetics imidization at different temperatures. The results show that optimal conditions for imidization are temperature 300 °C for 1 h. It is known that azobenzene derivatives possess optical properties associated with optical anisotropy due to photoisomerisation photoorientation of azochromophore, perpendicular to the direction of the polarized beam. These properties of azobenzene derivatives are important for their application in nonlinear optics and nanotechnology, optical modulators, optical recording media and other devices [16,17,18,19].

2.3. Microwave Synthesis

One of the achievements of polymer chemistry is development of microwave synthesis. It is new and promising methods in organic synthesis, that time of the reaction manifold shorten. The literature describes that reaction occurs within hours or days, by microwave synthesis reaction time is greatly reduced - from 5 to 30 min [19,20,21]. Microwaves are distributed evenly in the reaction mixture, which makes the temperature field uniform (homogeneous) and radio frequency radiation provides the energy required for the reaction, and molecules get more energy (Ea) and the reaction speed increases manifold [20,21,22,23].

Our investigation showed that after combined treatment - MW and thermal, the imidization reaction take place for 5 to 15 min that is confirmed by FTIR spectroscopy. The quality of the film is identical with the one of PI obtained only after a thermal treatment for 1h at 300 ^{0} C. The method carried out allowed for PI production to be optimized by the involving of MW treatment of vacuum deposited films [10, 24].

It was found that the prerequisite for obtaining films of PI without mechanical defects is the establishment impact energy. It was investigated and established optimum conditions of MW and thermal treatment. Our suggestion that the type of energetic treatment power and time can be reduced to the substitution of different parameters: time, power and intensity of impact is not confirmed, i.e. it was not prepared the PI films with the same imidization degree in terms variation of treatment. Experimentally proved that 5 hours additional heat treatment at 170 ^{o}C is not offset one hour treatment at 250 °C. By applying the DTA (Differential Thermal Analysis) and FTIR spectroscopy was found to be necessary to achieve clearly defined temperatures for better imidization to PI. These results are also valid in MW treatment. Only the combination of both MW and heat, leading up to the high imidization. It was found that higher imidization degree achieved for 15 min. after application of MW and thermal treatment at 250 °C compared with only thermal treatment [24,25,10]. Therefore, the mechanism of the reaction according Scheme 1 and 6 requirements not only affect energy and same temperature achievement in the layer, but also requires strict established parameters: the type of treatment, power and time of action to achieve a high imidization degree.

2.4. Kinetic of the Imidization

The kinetics of imidization is measured by the degree of transformation of PAA into PI by given output parameters - temperature and time of reaction.

Appropriate method for determination of imidization degree is FTIR spectroscopy. The band at 1376 cm^{-1} is called imide III (C-N-C vibrations) band and using for qualitative and quantitative estimations of imidization degree in the present of internal standard at 1500 cm^{-1} ($C\text{-}C_{Ar}$ vibrations) [26,27]. The imidization degree of PAA to PI was determined using eq 1.

$$Degree\ of\ imidization = \frac{(peak\ area\ at\ 1376cm^{-1})time/(peak\ area\ at\ 1500cm^{-1})time}{(peak\ area\ at\ 1379cm^{-1})300\,^{\circ}C \big/ (peak\ area\ at\ 1500cm^{-1})300\,^{\circ}C} \quad (1)$$

Yang et al. was used Seo model to correlate of imidization with time [28]. In Seo's approach the rate constant was proposed as *k(t) = b x sech(-at).* Inserting into the first-order rate equation, the relationship between the degree of imidization and curing time is obtained as:

$$-\ln(1-x) = -\frac{2b}{a}\tan^{-1}e^{-at} + const. \quad (2)$$

The constant in eq. (2) was obtained by fitting with the experimental data to the original Seo model. However the constant in eq. (2) was found to be 0.785 by fitting the experimental data at the initial condition ($t = 0, x = 0$). Hence, the kinetic equation was expressed as eq (3)

$$-\ln(1-x) = -\frac{2b}{a}\tan^{-1}e^{-at} - 0.785. \quad (3)$$

The parameters *a* and *b* were expressed in the form of Arrhenius expressions in this study, which are shown as eqs (4) and (5)

$$a = A_a \exp(-E_a/RT) \quad (4)$$

$$b = A_b \exp(-E_b/RT) \quad (5)$$

The constants *Aa, Ab* and *Ea* were calculated from Arrhenius plots of the parameters *a* and *b*. The assumption that the parameter *b* in the original Seo model is independent of temperature conflict obviously with the fact that the rate constant of first-order reaction *k(t)* is a function of temperature since *k(t)* approaches *b* at the initial time, $t \rightarrow 0$. The corresponding activation energy obtained from parameter *a* should be the energy barrier for the transition state and parameter *b* represents the rate constant for the imidization reaction.

3. PHYSICAL PROPERTIES OF POLYIMIDES

3.1. Intramolecular Interaction

The formation of charge transfer complex (CTC) has been studied from many researches [29,30]. They reported that CTC formated between dianhydride and diamine groups in polyimides being an important reason for high glass transition temperature (Tg) of polyimides [29,30]. The increased interchain attractive forces due to such interactions were proposed as effectively increasing the chain rigidity and hence the Tg. It was also proposed that the presence of any bridging group in the dianhydride had a strong influence on the glass transition as it changed its electron affinity and hence promoted the possibility of CTC formation. Figure 4 shows the idealized form of such an interaction between the dianhydride and diamine groups.

In terms of electronic structure of PI is a prerequisite for strong intermolecular interactions and charge transfer. The nitrogen atom is electrondonor to carbonyl group which is electronacceptor, this leads to electron move and charge interactions. On the other hand they are interactions between aromatic rings with their π electronic sextet, which lead to parallel and planar orientation of individual chains to each other. Many of the properties of the PI's dependent on intramolecular interactions due to the use of vapour processes and solid state reactions to obtain essential to create packing of the polymer chain and effective CTC.

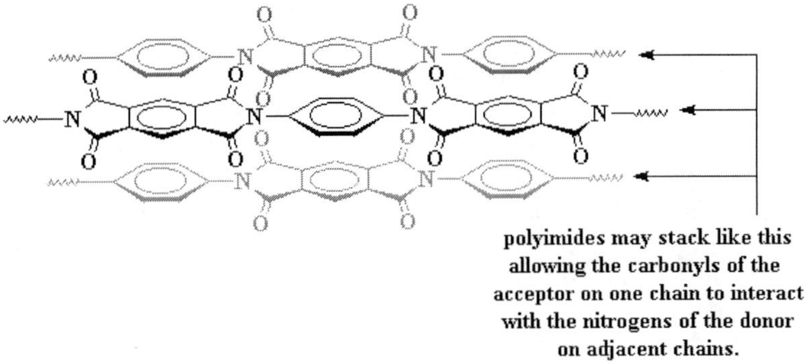

Figure 4. Idealized charge transfer complex formation between dianhydrides and diamines

3.2. Thermal Stability

Thermo-Gravimetric Analysis (TGA) is an appropriate method for estimation of thermal stability of polymers. In Figure 5 TGA curves presented our experiments after thermal treatment of ODA and PMDA mixture [11].

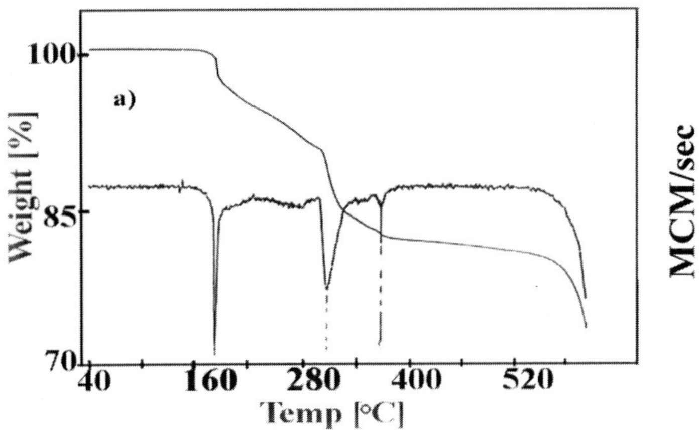

Figure 5. TGA curves of ODA and PMDA powder mixture (1:1 ratio)

Two drastically changes of the TGA curves are observed at about 180⁰C and about 280⁰C, which indicate relatively weight loss of 3% and 9%. These values correspond to one or three mols of water. It is well known that the polycondenzation reaction of ODA and PMDA (Sheme 1), takes place by separate of water. Therefore is quite logical to accept that the weight loss is a consequence of the process starting at lower (180⁰C) temperature till the moment the PAA is obtained. With increasing the temperature the interaction process continues and at higher (280⁰C) temperature the dehydratation leads to imidization and PI formation. It is also clearly seen (Fig. 5), that the rest of the sample remains stable up to 520⁰C. As the rest most probably is consisting of PI, it can be underline that the PI considering the destruction or weight loss is stable at temperatures higher than 500⁰C.

3.3. Optical Transmittance

High light transmittance for a wide spectral range in combination with dielectric and chemical resistance properties is an attractive feature for potential applications as encapsulating layers, protective films and intermediate for micro- and optoelectronics. In Figure 6 presents a series of spectra of PI obtained under different conditions. It is seen that the conditions for the PI films do not significantly affect the values of transmittance in the range 80 to 92 % in a relatively wide spectral range λ 420-900 nm.

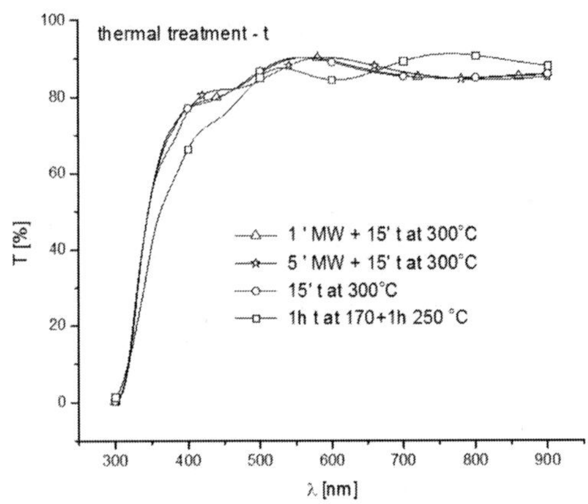

Figure 6. Spectral dependence of transmittance (T) for PI (PMDA-ODA) films 0,22 μm prepared under different conditions.

3.4. Electrical Properties

The possibility of treatment condensation of the precursors in Ar atmosphere during vapour deposition can be interpreted as incorporation of pores, i.e. change the density or porosity of the condensed films of ODA and PMDA respectively, and the influence of pores on the permittivity PI layers. There was also depending on the reduction of dielectric constant with increasing imidization degree which is higher for higher temperatures of treatment in view of running imidization reaction [31,32,33]. Tables 1to 3 are summarized data for conductivity, dielectric constant and capacity of the PI films with different composition, thickness, condition of the substrate or energy treatment and influence the conditions for obtainig PI or composite layers with embedded nanosized carbon particles.

Impressive ability to drastically change the conductivity by incorporation of carbon particles. The measured differences in conductivity depending on the concentration of "guests" - carbon 0,55% and 1,4% vol, respectively filling factor can be interpreted as a condition of embedded clusters. Therefore, there is the possibility of one side by vacuum deposition of the precursors and conditions of thermal treatment to influence the parameters examined. On the other, through the incorporation one type particles created opportunities to change individual parameters in a wide range in identical composition, for example different conduction with or without embedded carbon [30,31].

Table 1. Thermal dependence of polyimide permittivity.

Thickness d [μm]	Thermal treatment [°C]	Permittivity [ε]
0.1	1 h 170° C + 1 h 250° C	3.0 - 3.2
0.1	1 h 170° C + 1 h 350° C	2.8
0.22	1 h 170° C + 1 h 250° C	3.1 – 3.3
0.22	1 h 170° C + 1 h 350° C	2.9

Table 2. Polyimide capacity (C) end permittivity (ε) depending on the dynamic state of the substrates during the vapour depositions of precursors.

Vacuum deposited layer 500 nm	Capacity [pF]	Permittivity [ε]
Planetary rotating substrates	9	3.2
Linearly moving substrates	11	3.4

Table 3. Polyimide conductivity (σ) depending on the type of layer formation

Type of layer formation	Conductivity (σ), [$ohm^{-1}m^{-1}$]
Without Ar (residual pressure 10^{-4}Pa)	$1.9x10^{-8}$
With Ar (residual pressure 10^{-2}Pa)	$0.7x10^{-14}$
With 0.55 vol.% Carbon	$2.0x10^{-5}$
With 1.4 vol.% Carbon	$4.0x10^{-1}$

4. BIOMEDICAL AND ENGINEERING APPLICATION

4.1. Biomedical

Polymer materials have been established as excellent materials in chemistry, automotives and electronics to interconnect different components, to electrically insulate conductors and to survive harsh corrosive environments. Polyimides are the most common material class for substrate and insulation materials in combination with metals for interconnection wires and electrode sites. Therefore, it is quite natural that the medical device industry has focused its attention to polyimides for medical devices in general and especially for encapsulation and insulation of active implants. Neural implants are technical systems that are mainly used to stimulate parts and structures of the nervous system with the aid of implanted electrical circuitry or record the electrical activity of nerve cells. Their application in clinical practice has given rise to the fields known as "neuromodulation" and "neuroprosthetics" or neural prostheses.

Neuromodulation, namely the stimulation of central nervous system structures to modulate nerve excitability and the release of neurotransmitters, alleviates the effects of many neurological diseases. Deep brain stimulation helps patients suffering from Parkinson's disease to suppress tremor and movement disorders. It is also a treatment option for severe psychiatric diseases like depression and obsessive-compulsive disorder. Neural prostheses aim to restore lost functions of the body, either sensory, motor or vegetative. All neural implants have to fulfill general requirements to become approved as a medical device - they must not harm the body and should stay stable and functional over a certain life-time which is in most cases in the range of decades. Generally polymides are used as an insulation or passivation layer, polyimides provide protection for underlying circuitry and metals from effects such as moisture absorption, corrosion, ion transport, and physical damage. Furthermore, it acts as an effective absorbcr for alpha particles that can be emitted by ceramics, and as a mechanical stress buffer. Key properties are thermoxidative stability, high mechanical strength, high modulus, excellent insulating properties, and superior chemical resistance. Devices made of polyimide have elicited only mild foreign body reactions in several applications in the peripheral and central nervous system showing good surface and structural biocompatibility. They have proven to be biostable and functional for months in chronic *in vitro* and *in vivo* studies. Most often used polyimide is BPDA-PPD as biomaterial and commercially available under the trademark of DuPont's PI2611 or UBE's U-Varnish-S (Figure 7) [34]

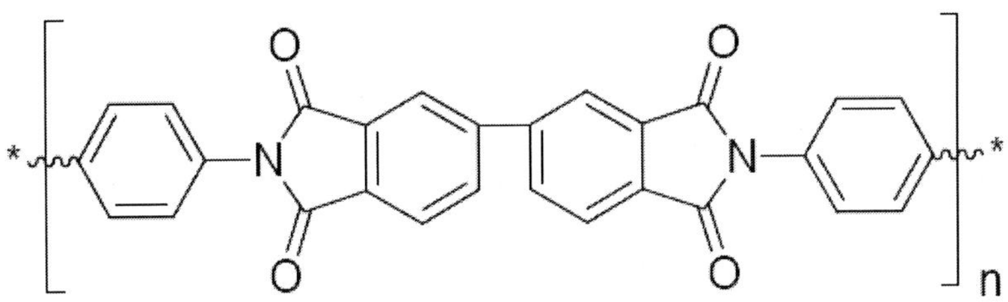

Figure 7. Structure of BPDA-PPD polyimide.

Advanced microtechnologies offer new opportunities for the development of these active implants. Biocompatible materials such as titanium and polyimide are potential candidates in encapsulating implant devices. A. Main et al. have been tested laser joined titanium-polyimide samples for bond strength, andwas observed that the laser bonding parameters such as laser power and feed speed affect the bond strength of all material systems

[35]. The applicability of laser joining for a specific material combination and the selection of the appropriate approach depends on the optical properties of the materials. The need to join dissimilar materials occurs as the encapsulation includes functional elements such as electrodes used for neural stimulation or modules with defined micropores for the effusion of drugs. Currently, the implants such as pacemakers and cochlea implants are being used to treat cardiac and hearing disorders, respectively. Examples of other implants include neural and muscular stimulators for the treatment of seizures caused by epilepsy, and implantable drug delivery systems to locally effuse chemicals and drugs such as pain medication, hormones and other pharmaceutical compounds. Research is underway to develop subretinal implant devices that will replace damaged photoreceptor cells. These damaged cells fail to send electrical signals to the rest of the eye and eventual the brain, that results in blindness [36]. The study of Humayun et al. has showed successful results to restore vision by permanently implanting a retinal prosthesis in the blind eye. These active implants utilize microelectronic devices that are developed on biologically toxic materials such as silicon. Jong-Mo Seo et al. have been development biocompatibilities of polyimide microelectrode array for retinal stimulation [37]. They development a platinum electrode- embedded polyimide film and tested in feline eye. The authors used flexible polyimide selected as the substrate material of gold microelectrode array to minimize the damage during ophthalmic surgery and to get better contact to retina. To evaluate the feasibility of polyimide microelectrode array as a retinal prosthesis, in vitro and in vivo biocompatibility have been shown that retinal pigment epithelial on gold electrode-embedded polyimide film in a monolayer after 10 days of culture, and showed good affinity to it. Electroretinography revealed no difference between the transplanted eye and the healthy eye. Gold electrode-embedded polyimide film showed good biocompatibility in vitro and in vivo test and was suitable as a candidate biomaterial for the fabrication of retinal stimulator in visual prosthesis system.

Development of nanotechnology and chemistry of polyimide have been applied in biology and medicine as nano-biosensor device, biochips and thin film DNA immobilize. Biochips, particularly those based on DNA, are powerful devices that integrate the specificity and selectivity of biological molecules with electronic control and parallel processing of information. Examples of current applications of DNA chips include genomic analysis to screen and identify single nucleotide polymorphisms or to sequence gene fragments, pathogen identification, and gene expression profiling. Other possible features of a DNA chip surface include the presence of microelectrodes based on PI matrix thin films, that can be used as sensing devises and to generate electric fields that promote the migration of oligonucleotides, hybridization and covalent binding [39]. Forster et.al.

have been purposed the Nanochip™ with a layered polyimide structure that supports the controlled electrophoretic transport of oligonucleotides [40]. They shown show that the biased electrodes preserve the integrity of the DNA by performing an electronic reverse-dot blot hybridization assay after electrophoretic transport of the target oligonucleotides [40].This nanodevice in future can serve as the basis for a low cost specific biomolecular detection tool in clinical diagnostics.

4.2. Engineering

The attractiveness and applicability of the PI's are determined not only from single their characteristics like high chemical and thermal stability, high optical transmission, high electrical resistance, but most important is their saving at combination of the cited properties [41,42]. The combination of the high optical transparency by simultaneously high chemical and thermal stability make the PI very attractive material for applying as capsulating or protective layers, films for laser microstructuring in micro- or optoelectronic, etc. [11,42]. Additionally, it could be noted the influence of the fast unlimited choice of the "guest" to enlarge the number of composites for many different purposes – from conductive or insulating transparency or colored films until insulating substrates, optical, chemical and thermal stable coatings [15,43]. Recently, much research interest focuses on polyimide films containing azobenzene fragments. These films are interest because of their potential application as photoactive materials for optical recording, biosensors, reversible optical switch, liquid crystal, photosensitive elements, optical information storage, holographic effect etc. An intensive research effort is being undertaken to use holographic techniques for optical information storage and optical information processing. Holographic storage is technologically very promising because information storage capacity, that can be reached with this technique, is much higher compared with other techniques (the storage capacity of CD is 0.7 GB, of DVD is from 4.7 to 17 GB, of PAP DVD is about 40 GB, of HOLO CD could be 1 TB) [19,44]. Development of holographic technology depends on the properties of recording materials. For such practical applications materials besides of specific requirements viz the principal the presence of photochromic moieties, the thermal stability of orientation alignment, the high optical non-linearity, high damage threshold, chemical resistance, mechanical endurance, they should exhibit the high optical quality ability and feasibility of device fabrication which are determined in wide range by their solubility. Significant efforts have been made to improve solubility of polyimides by designing their structure [45].

5. CONCLUSIONS

In this chapter, we have discussed methods for obtaining of polyimides, chemical properties and physical parameters that are related with obtaining nanosized films by vapour deposition. It was discussed possibilities for the solid state synthesis of polyimides in thin film and applied microwave synthesis. The studies show that can be obtained homogeneous films without defects on the surface and volume of layers with controlled density, thickness and dielectric properties. The developments of polymer chemistry produce polyimide films with covalent bonded chromophore to the chain. One of these achievements is our development a method for solid state synthesis of azo-polyimide.

Polyimide layers are suitable matrix for incorporation of metal, salts, chromophores as nanoscale particles to obtain of nanocomposite materials. It was discussed the possibility of use polyimides in materials chemistry and nanomaterials, one of these applications is the use for making biomedical implants for neurology, ophthalmology, biosensor device and chips which are a powerful tool in clinical diagnostics. Another important trend is use in electronics and optoelectronics such as dielectric substrates and intermediate barrier layers, creating nanocomposite films with various nanosized particles such as dyes, metal, dielectric and other clusters.

ACKNOWLEDGEMENT

The financial support of the National Fund of Ministry of Education and Science, Bulgaria – contract № DO-02/254 – 18.12.2008 is gratefully acknowledged.

REFERENCES

1. Y Strunskus, M Grunze, 1994Polyimides-fundamentals and applications. In: Crosh M, Mittal K, editors. New York: Marcel Dekker. 187205
2. J. M Marc, Abadie Alexander L. Rusanov (2007Practical Guide to Polyimides. Smithers Rapra Technology Limited. 4577
3. M. I Bessonov, M. M Koton, V. V Kudryavtsev, and L. A Laius, 1987Polyimides: Thermally Stable Polymers. Plenum, New York, 2-nd edition. 56187

4. Yoshio Imai (1999Rapid Synthesis of Polyimides from Nylon-Salt-Type MonomersIn: Progress in Polyimide Chemistry I. H.R. Kricheldorf editor, Springer. 320

5. H Rohit, P Vora, Santhana Gopala Krishnan, S. Veeramani and Suat Hong Goh (2003Poly(amic acid)s and their ionic salt solutions: Synthesis, characterization and stability study. In: Polyimides and Other High Temperature Polymers, 2K.L. Mittal editor. VSP. 1435

6. F. W Harris, 1990Polyimides. Wilson D., Stenzenberger, H.D., Hergenrother, P.M., Chapman and Hall editors. New York. 2396

7. Anton GeorgievErinche Spassova, Jacob Assa and Gencho Danev (2010Preparation of Polyimide Thin Films by Vapour Deposition and Solid State Reactions. Polymer Thin Films. Abbass A Hashim editor. InTech. 7192

8. D Sek, A Wanic, 2000High-temperature polycondensation of six membered dianhydrides with o-substituted aromatic diamines 1. Model compounds investigations. Polymer. 4123672378

9. D Yu, S. N Likhachev, Chavlin, Yu. A. Zubov, R.N. Nurmukhamedov and I. Ye. Kardash (1991Effect of chemical structural defects on morphology of polyimide films. Polymer Science.33918851894

10. Anton GeorgievIlyana Karamancheva, Dejan Dimov, Ivailo Zhivkov, Erinche Spassova (2008FTIR study of the structures of vapor deposited PMDA-ODA film in presence of copper phthalocyanine. Journal of Molecular Structure. 888214233

11. E Spassova, 2003Vacuum deposited polyimide thin films. Vacuum. 70551561

12. KiRyong Ha and John LWest (2002Studies on the photodegradation of polarized UV-exposed PMDA-ODA polyimide films. Journal of Applied Polymer Science. 8630723077

13. K.S. Sree Harsha (2006) Principles of Physical Vapor Deposition of Thin Films Elsevier, San Jose State University, CA, USA, first edition. pp. 11, 367, 961.

14. Mitchell AnthamattenStephan A. Letts, Katherine Day, Robert C. Cook, Anthony P. Gies, Tracy P. Hamilton and William K. Nonidez (2004Solid-state amidization and imidization reactions in vapor-deposited poly(amic acid). Journal of Polymer Science: Part A: Polymer Chemistry. 4259996010

15. A Georgiev, I Karamancheva, D Dimov, E Spassova, J Assa, G Danev, 2008Polyimide coatings containing azo-chromophores as

structural units. Journal of Physics: Conference Science. 113: 012032.

16. Masashi TakahashiTakashi Okuhara, Tomohiro Yokohari, Koichi Kobayashi (2006Effect of packing on orientation and cis-trans isomerization of azobenzene chromophore in Langmuir-Blodgett film. Journal of Colloid and Interface Science. 296212219

17. Cristina Cojocariu and Paul Rochon (2004Light-induced motions in azobenzene-containing polymersPure Applied Chemistry. 7614791497

18. Kiyoaki UsamiKenji Sakamoto, Norio Tamura, Akihiko Sugimura (2009Improvement in photo-alignment efficiency of azobenzene-containing polyimide films. Thin Solid Films. 518729734

19. Ewa Schab-BalcerzakLukasz Grobelny, Anna Sobolewska and Andrzej Miniewicz (2006Cycloaliphatic-aromatic polyimides based on diamines with azobenzene unit. European Polymer Journal. 4228592871

20. Brittany L Hayes (2002Microwave Synthesis: Chemistry at the Speed of LightCEM Publishing. 11

21. D Michael, P Mingos, 2005Theoretical aspects of microwave dielectric heating. Microwave Assisted Organic Synthesis. Tierney J., P., & Lidstroem P. editors. Blackwell Publishing Ltd. 63179

22. C. Oliver Kappe, Doris Dallinger and S. Shaun Murphree (2009) Practical Microwave Synthesis for Organic Chemists. WILEY-VCH Verlag GmbH & Co. KGaA, Weinheim. pp. 11, 87, 161, 203.

23. Yumin LiuY. Xiao, X. Sun, D. A. Scola (1999Microwave irradiation of nadic-end-capped polyimide resin (RP-46) and glass-graphite-RP-46 composites: cure and process studies. Journal of Applied Polymer Science. 731223912411

24. D Dimov, A Georgiev, E Spassova, I Karamancheva, Y Shopov, and G Danev, 2007Microwave assisted processes for producing thin layer materials in the field of nanotechnologies. Journal of Optoelectronics and Advanced Materials. 92494499

25. Richard Hoogenboom and Ulrich SSchubert (2007Microwave-Assisted Polymer Synthesis: Recent Developments in a Rapidly Expanding Field of Research. Macromolecular Rapid Communication. 28368386

26. M. B Saeed, Mao-Sheng Zhan (2006Effects of monomer structure and imidization degree on mechanical properties and viscoelastic

behavior of thermoplastic polyimide films. European Polymer Journal. 4218441854

27. M. B Saeed, Mao-Sheng Zhan (2007Adhesive strength of partially imidized thermoplastic polyimide films in bonded joints. International Journal of Adhesion & Adhesives. 27919

28. Chang-Chung YangKuo Huang Hsieh and Wen-Chang Chen (2003A new interpretation of the kinetic model for the imidization reaction of PMDA-ODA and BPDA-PDA poly(amic acid)s". In: Polyimides and Other High Temperature Polymers, 2K.L. Mittal editor. VSP. 3745

29. M Fryd, 1984Structure-Tg relationships in Polyimides: Synthesis, Characterization and Properties, 1Mittal K.L. editor. Plenum New York. 377384

30. StClair, T.L. (1990In Polyimides. Wilson D., Stenzenberger, H.D., Hergenrother, P.M., Chapman and Hall editors. New York. 187208

31. D Dimov, E Spassova, J Assa, and G Danev, 2009Ion beam assisted physical deposition of polyimide. Journal of Optoelectronics and Advanced Materials.111014361439

32. V Strijkova, D Dimov, A Paskalevaa, I Zhivkov, E Spassova, J Assa, G Danev, 2005Electrical properties of a thin layer polyimide matrix Journal of Optoelectronics and Advanced Materials. 7313191322

33. F. -Y Tsai, Y. -H Kuo, and D. R Harding, 2006Properties and structure of vapor-deposited polyimide upon electron-beam irradiation. Journal of Applied Physics 99: 064910.

34. Christina HasslerTim Boretius, Thomas Stieglitz (2011Polymers for Neural Implants. Journal of Polymer Science: Part B: Polymer Physics. 491833

35. A Mian, G Newaz, l Vendra, N Rahman, D. G Georgiev, G Auner, R Witte, H Herfurth, 2005Laser bonded microjoints between titanium and polyimide for applications in medical implants. Journal of Materials Science: Materials in Medicine. 16229237

36. Keekeun LeeAmarjit Singh, Jiping Heb, Stephen Massia, Bruce Kima, Gregory Rauppc (2004Polyimide based neural implants with stiffness improvement. Sensors and Actuators B. 1026772

37. M. S Humayun, de Juan Jr, E.; Weiland, J. D.; Dagnelie, G.; Katona, S.; Greenberg, R.; Suzuki, S (1999Pattern electrical stimulation of the human retina. Vision Research. 3925692576

38. Jong-Mo SeoaSung June Kimb, Hum Chunga,b, Eui Tae Kimb, Hyeong Gon Yua, Young Suk Yua (2004Biocompatibility of

polyimide microelectrode array for retinal stimulation. Materials Science and Engineering C. 24185189

39. F Fixe, A Faber, D Gongalves, D. M. F Prazeresl, R Cabea, V Chu, G Ferreira, and J. P Conde, 2002Thin film micro arrays with immobilized DNA for hybridization analysis. Materials Research Society. 723125130

40. H Anita, Forster, Michael Krihak, Paul D. Swanson, Trevor C. Young and Donald E. Ackley (2001A laminated, flex structure for electronic transport and hybridization of DNA. Biosensors & Bioelectronics 16187194

41. G Danev, E Spassova, J Assa, J Ihlemann, D Schumacher, 2000Excimer laser structuring of bulk polyimide material. Applied Surface Science. 168162165

42. G Danev, E Spassova, J Assa, 2005Vacuum deposited Polyimide- A Perfect Matrix for Nanocomposite Materials. Journal of Optoelectronics and Advanced Maters. 7313791390

43. M Hasegawa, and K Horie, 2001Photophysics, Photochemistry, and optical properties of polyimides. Progress in Polymer Science. 26295335

44. E Grabiec, E Schab-balcerzak, D Sek, A Sobolewska, A Miniewicz, 2004New polyamides with azo-chromophore groups. Thin Solid Films. 453-454: 367-371.

45. M Moniruzzaman, P Zioupos, G. F Fernando, 2006Investigation of reversible photo-mechanical properties of azobenzene-based polymer films by nanoindentation. Scripta Materialia. 54257261

6

CHAPTER

Preparation of Nanofibers with Renewable Polymers and Their Application in Wound Dressing

Ying Zhao, Yihui Qiu, Huanhuan Wang, Yu Chen, Shaohua Jin, and Shuseng Chen

School of Material Science and Engineering, Beijing Institute of Technology, Beijing 100081, China

ABSTRACT

Renewable polymers have attracted considerable attentions in the last two decades, predominantly due to their environmentally friendly properties, renewability, good biocompatibility, biodegradability, bioactivity, and modifiability. The nanofibers prepared from the renewable polymers can combine the excellent properties of the renewable polymer and nanofiber, such as high specific surface area, high porosity, excellent performances in cell adhesion, migration, proliferation, differentiation, and the analogous physical properties of extracellular matrix. They have been widely used in the fields of wound dressing to promote the wound healing, hemostasis, skin regeneration, and treatment of diabetic ulcers. In the present review, the different methods to prepare the nanofibers from the renewable polymers were introduced. Then the recent progress on preparation and properties of the nanofibers from different renewable polymers or their composites were reviewed; the application of them in the fields of wound dressing was emphasized.

1. INTRODUCTION

Renewable polymers, including the polymers directly extracted from biomass and proteins, for example, starch, chitosan, cellulose, alginate, gelatin, and collagen [1], have attracted considerable attentions in the last two decades, predominantly due to their environmentally friendly properties, renewability, good biocompatibility, biodegradability, bioactivity, and modifiability [2]. The nanofibers prepared from the renewable polymers can combine the excellent properties of the renewable polymer and nanofiber [3], such as high specific surface area, high porosity, excellent performances in cell adhesion, migration, proliferation, differentiation, and the analogous physical properties of extracellular matrix (ECM). They have been widely used in the fields of wound healing, hemostasis, skin regeneration, and treatment of diabetic ulcers. In the present review, we will focus on the preparation, composition, and properties of nanofibers prepared with renewable polymers and their applications in the wound dressing. The research progress and development directions were also discussed.

2. METHODS FOR THE PREPARATION OF NANOFIBERS WITH RENEWABLE POLYMERS

As nanofibers are becoming increasingly important, the high throughput preparation of high quality nanofibers has drawn remarkable attentions. Currently, electrospinning is the most widely used method for the preparation of nanofibers. Other methods including Bubbfil spinning, centrifugal spinning, and freeze-drying have also been reported for their potential applications in the biological field.

2.1. Electrospinning

Electrospinning is the most commonly used method for the fabrication of nanofibers due to its simplicity, adaptability, cost-efficiency, and versatility [4].

In a typical electrospinning process, a high voltage is applied to a liquid polymer, leading to the ejection of a continuous jet strand from the spinneret towards a grounded collector. The surface tension of the polymer droplet is overcome by the applied electric field. The droplet then elongates to form a cone known as "Taylor cone" and then is extruded from the cone to form a fiber jet. The solvent in the jets is evaporated as the fiber jets travel through the atmosphere and solid polymer fibers are deposited on the metal collector as nonwoven web [4].

Based on the preparation process of the polymer, the electrospinning process can be classified into two groups including solution electrospinning and melt electrospinning. Solution electrospinning is limited mainly by its low productivity, extra solvent extraction process, and requirement of toxic solvent. Yet it is much more thoroughly reviewed than melt electrospinning by researchers due to more concerns on melt electrospinning, such as difficulties inherent in the finer fiber formation, higher viscosity of polymer melts, and electrical discharge issues associated with the application of high voltage [5]. Thus, several setups including multijets from single needle, multijets from multineedle, and needleless systems have been proposed for solution electrospinning to increase its productivity.

2.1.1. Multijets from Single Needle Electrospinning

For single needle electrospinning (SNE) process, a single jet is initiated from the Taylor cone formed by the application of electric field. Li et al. prepared poly(lactic acid) (PLA)/chitosan core-shell structure nanofibers from a homogeneous solution using SNE and found that this method was simpler and more effective in producing core-shell structures from homogeneous solutions than other methods [6]. Ayaz et al. successfully prepared composite nanofibers with SNE and coaxial (double needle) electrospinning and compared their performances. The polymer matrix and filler were mixed before being infused into the single needle for SNE. Therefore, the filler was uniformly distributed in the whole nanofiber. In contrast, polymer matrix and filler were infused separately into two nozzles axially centered to each other for coaxial electrospinning, which restricted the filler in the center of nanofiber. In addition, the average diameter of the nanofibers prepared by SNE is smaller than that of the nanofibers prepared by coaxial electrospinning. However, nanofibers prepared through SNE showed lower breaking elongation than those prepared by the coaxial method [7].

2.1.2. Multijets from Multineedle Electrospinning

As discussed above, SNE provides very low throughput productivity [8]. Multineedle electrospinning (MNE) system is a simple way to improve the productivity and a simple technique to fabricate composite fibers from the polymers that cannot be easily dissolved in common solvents. However, the needle configuration, number of needles, and needle gauge need to be optimized for the MNE design. For example, needles can be arranged in one- or two-dimensional configuration. However, charged jets could interfere with each other and restrain jets injected from the spinneret, leading to low production efficiency [9, 10]. The interference between the charged jets can be resolved by several methods.

Nuryantini et al. prepared PVA/chitosan nanofiber membranes using an MNE system with a drum collector. Compared with conventional electrospinning setups, MNE setup can significantly improve the throughput while maintaining the uniformity of the size and thickness of the stacked nanofibers [11]. Yu et al. fabricated nanofibers using a novel MNE setup with assistant sheath gas as shown in Figure 1 [9]. The assistant sheath gas introduced to the spinneret improved the stretching process of the charged jet and realized multijet injection. During the electrospinning, assistant gas was pumped into the chamber through gas input channels, which were uniformly distributed and thus contributed to each spinneret evenly. Sheath gas at a pressure of 50 kPa decreased the average diameter of the nanofibers from 600 nm to 472.8 nm and decreased the diameter distribution region from 249 nm to 188 nm. The assistant sheath gas also promoted the stability of the electrospinning and the uniformity of the nanofibers due to its ability to reduce the required voltage for jet injection and jet surface charge density [9].

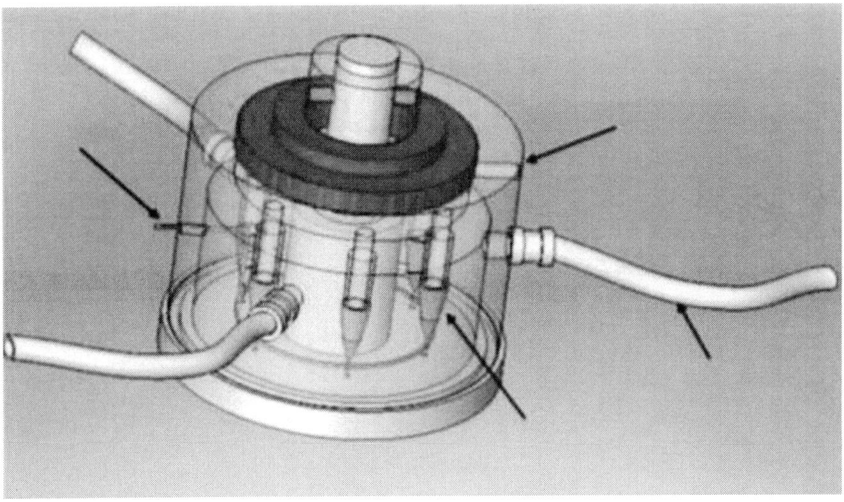

Figure 1. Setup of MNE spinneret assisted with sheath gas [9].

2.1.3. Multijets from Needleless Electrospinning

MNE system usually needs a big space and optimization of the relative distance between needles to avert strong charge repulsions between the jets. Therefore, needleless electrospinning has been developed and widely used due to its notable ability to increase the production of nanofibers. For the needleless electrospinning, an electric field is applied on a large area, producing numerous polymeric jets from free liquid surface [5, 8].

Sasithorn and Martinová prepared silk nanofibers by both needle electrospinning and needleless electrospinning and found that the mass production rate of needleless electrospinning was much higher than that of needle electrospinning. In addition, the production rate of needleless electrospinning could be increased significantly with the increase of solution concentration and applied voltage. However, nanofibers produced by needle electrospinning were much smaller and had a narrow diameter distribution [12].

Wang et al. prepared uniform nanofibers using a needleless electrospinning system with a rotating spiral wire coil as the spinneret (Figure 2). The polymer solution was charged with a high voltage and loaded onto the coil surface by slowly rotating the coil at 40 rpm. They found the production of nanofiber was improved by increasing the length () and diameter () of the coil or decreasing the spiral distance () and the diameter () of the wire. However, these factors showed no significant effect on the average diameter of the nanofibers. Compared with the needle electrospinning, the needless electrospinning with coil spinneret showed better performance, such as producing thinner nanofibers and higher productivity [13].

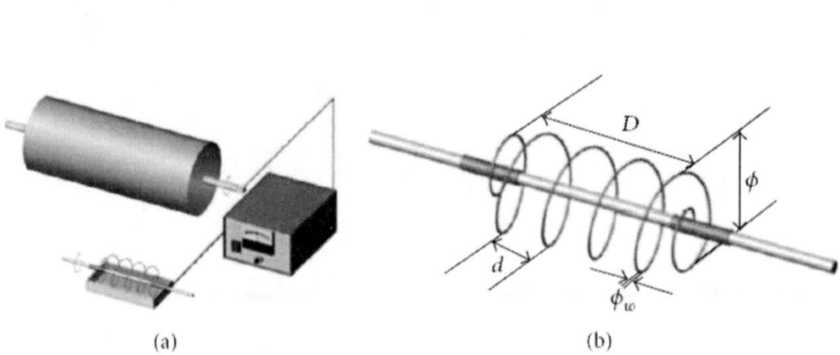

Figure 2. (a) Schematics of spiral coil electrospinning setup; (b) magnified view of the coil [13].

2.2. Bubbfil Spinning

It is known that the increases in the molecular weight and viscosity of polymers lead to higher surface tension of their Taylor cone, which cannot be used to fabricate nanofibers by the conventional electrospinning methods. Therefore, higher electrostatic force is required for the electrospinning to fabricate desired nanofibers [14]. Other drawbacks of electrospinning including low productivity and uncontrollability of the spinning process call for the development of other technologies [15].

Bubbfil spinning including bubble electrospinning and blown bubble spinning is one of main newly developed methods for the preparation of nanofibers.

2.2.1. Bubble Electrospinning

Bubble electrospinning has attracted considerable attention due to its high throughput. Unlike conventional electrospinning, bubble electrospinning uses electrostatic force rather than Taylor cone to overcome the surface tension of a polymer bubble [16, 17]. The electrospinnability of bubble electrospinning depends geometrically on the sizes of the obtained bubbles, which avoid using high voltages and thus make it the best candidate in the green production. Figure 3 shows a typical bubble electrospinning setup.

In a typical bubble electrospinning process, bubbles are formed on the surface of a polymer solution by a controllable syringe pump as the gas pump slowly turned on. The number of the bubbles is reduced with the decrease of the tube diameter. Multiple jets are then ejected from the bubbles to the grounded collector when the applied voltage surpasses a critical value. The fluid jets are usually solidified into fibers in less than 0.001 seconds [5, 14].

Jiang et al. mass-produced nanofibers with an improved microbubble electrospinning system where a pyramid-shaped copper spinneret was used as the generator. They found that the increase of bubble size from 29.6 ± 8.2 μm to 152.7 ± 22.7 μm led to an increase in the average diameter of nanofibers from 139 ± 39 to 165 ± 26 nm. The microbubble electrospinning produced nanofibers with average diameter of 147 ± 28 nm with the bubble sizes of 59.9 ± 10.8 μm, which are similar to those produced by the conventional electrospinning with the same processing parameters. The throughput of microbubble electrospinning reached 8.9 g/h, which was much higher than that of the conventional electrospinning (5.2 g/h) [18].

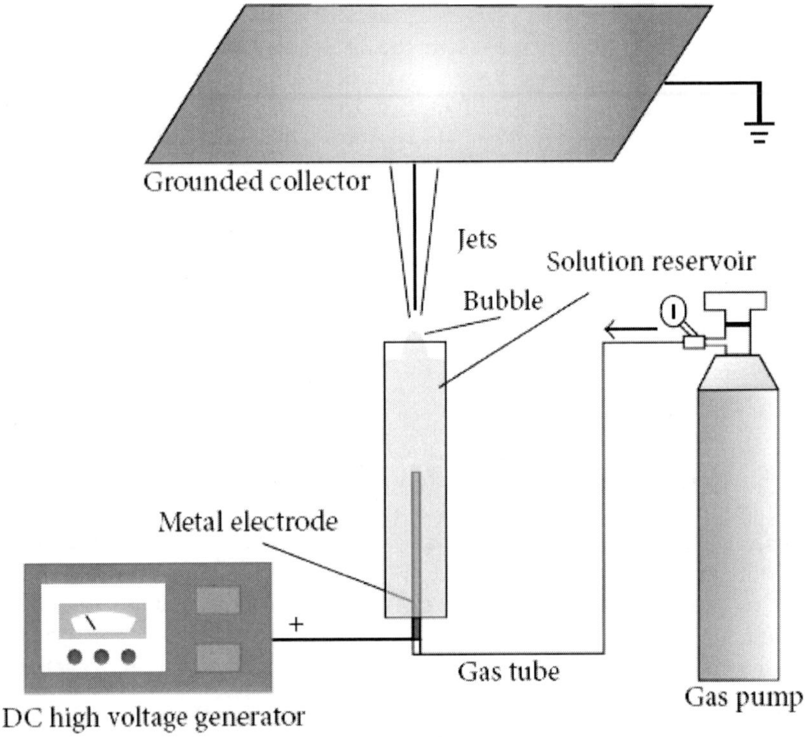

Figure 3. Bubble electrospinning setup for tunable bubble size [14].

2.2.2. Blown Bubble Spinning

Blown bubble spinning produces nanofibers using blowing air and has been used commercially as a one-step process to convert polymer resin directly into a nonwoven mat of fibers. Figure 4 shows an experimental setup of blown bubble spinning.

Blown bubble spinning process is very much similar to bubble electrospinning except that it uses blowing air instead of electronic force to produce fine fibers [19]. However, a weak electronic field can significantly improve the performance of the blown bubble spinning, which actually turns into an electrostatic-field-assisted blown bubble spinning (Figure 5). It can also be called as air-jet assisted bubble electrospinning, depending on which one is the major functional force [14].

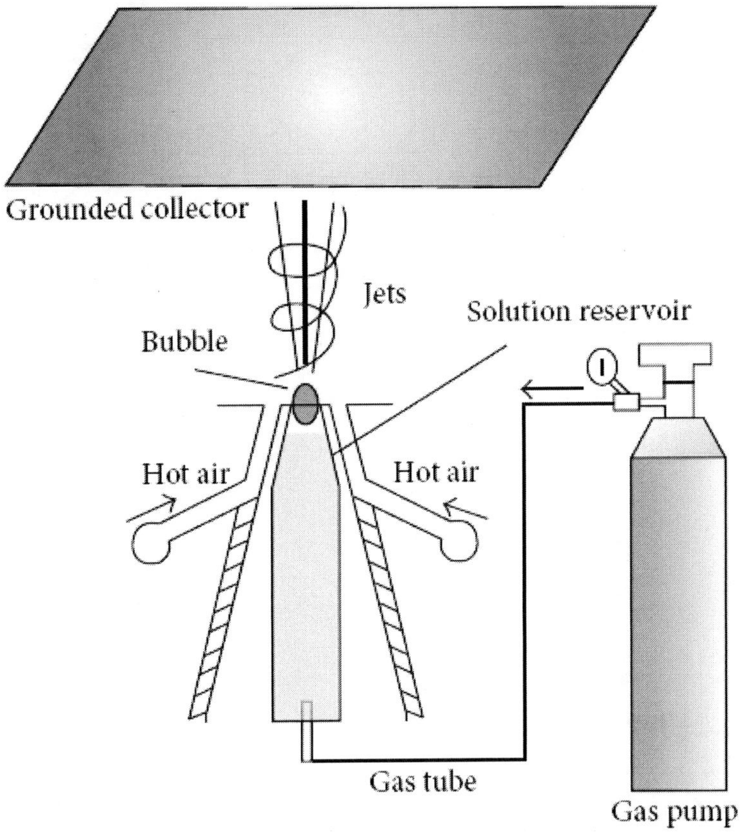

Figure 4. Blown bubble spinning (the only acting force is blowing air) [14].

Li et al. prepared silk fibroin through blown bubble spinning and found that the direction of blowing air, both perpendicular (setup I in Figure 6(a)) and parallel (setup II in Figure 6(b)) to the ejected direction of broken polymer bubble, could affect the morphology, structure, and arrangement of the produced fibers. For example, random oriented nanofibers with an average diameter of 524.25 ± 142.41 nm were formed through setup I and well organized fiber bundles were obtained with setup II [16].

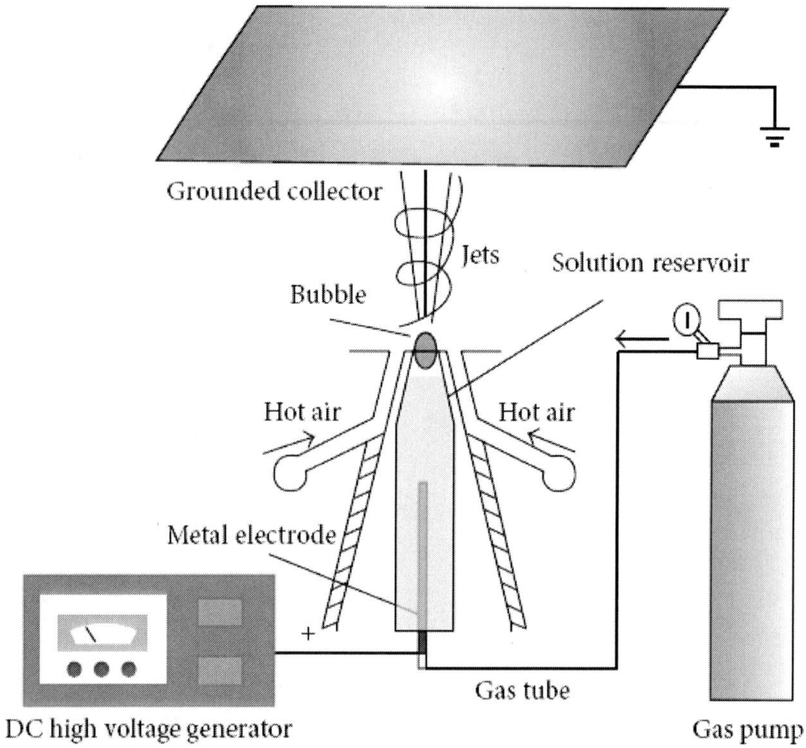

Figure 5. Air-jet assisted bubble electrospinning or electrostatic-field-assisted blown bubble spinning [14].

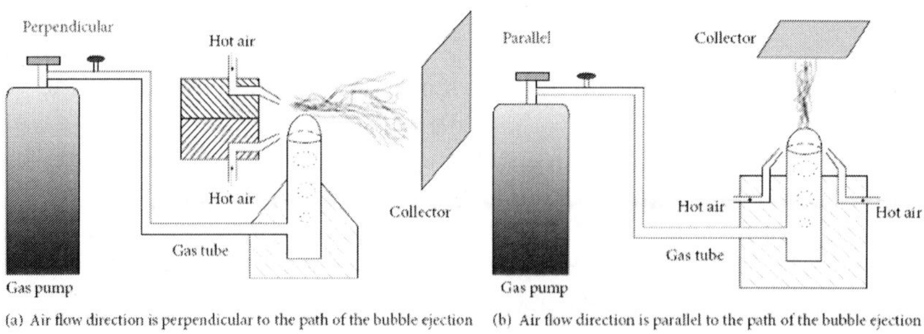

Figure 6. The schematic of the experimental setup of two airflow directions: (a) perpendicular flow and (b) parallel flow [16].

2.3. Centrifugal Spinning

Centrifugal spinning, also called rotary spinning or rotational jet spinning, has been well developed. Since no high voltage is required by the centrifugal spinning, the system can alleviate safety-related concerns. In addition, centrifugal spinning can remarkably improve the production efficiency by increasing the rotational speed, allowing the fast and large-scale fabrication of nanofibers. Compared with the electrospinning process, it also allows nanofiber fabrication from the polymers at much higher concentrations, which decreases the use of solvent and lowers the production cost [20].

During a typical spinning process, a polymer solution is introduced into a highly rotating spinneret and then ejected as jets at a high rotating speed which is used to overcome the surface tension. The jets are then deposited on the collectors to form dried nanofibers as the solvent evaporated during the stretching process. Figure 7 shows a typical setup for centrifugal spinning [20].

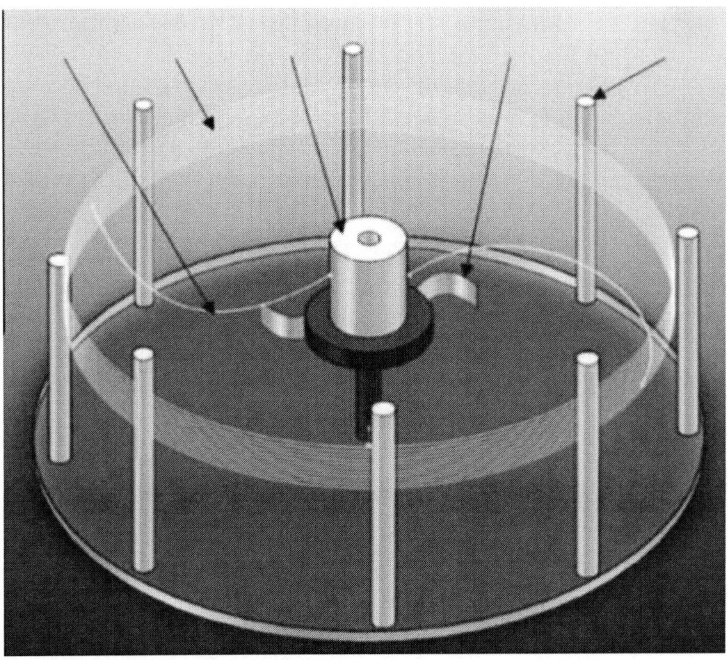

Figure 7. Setup of centrifugal spinning [20].

Hammami et al. prepared nanoscale fibers using centrifugal force spinning and found that centrifugal force spinning also required certain chain overlap or entanglement. However, needles with smaller inner diameters were used for the centrifugal force spinning, which significantly improved the spinnability of the nanofiber [21].

Erickson et al. fabricated highly aligned chitosan/polycaprolactone (PCL) nanofibers via centrifugal electrospinning system. Compared with those produced by the conventional electrospinning, the nanofibers produced by centrifugal electrospinning showed more uniform diameters and were aligned vertically. The diameter distribution of the nanofibers produced by centrifugal electrospinning was from 100 nm to 275 nm, which was much narrower than that of the nanofibers produced by the conventional electrospinning (25–450 nm) [22].

2.4. Freeze-Drying

Freeze-drying (FD) can fabricate porous structures with controllable sizes directly from polymers, such as chitin, without structure-directing additives or pretreatments needed, which cannot be achieved by other techniques, such as self-assembly and electrospinning [23]. In addition, no high temperature and further leaching step are required for the freeze-drying process [24]. Therefore, it has drawn increasing attention in the fabrication of nanofibers. Starting with solution, emulsion, or dispersion, freezing leads the solute or solids to be excluded by ice front in the interstitial spaces between ice crystals. Then porous structures are created by the following sublimation [23].

Ma et al. prepared chitosan (CS)/sodium hyaluronate (SH) polyelectrolyte complex (PEC) fibers by FD method. The SEM images indicate that the equimolar ratio of CS to SH benefits the formation most. The in vitro tests indicate that the freeze-dried fibers have no cytotoxicity and good compatibility, which are essential for its application in aqueous systems, especially for biological and environmental applications [25].

Wu and Meredith prepared nondirectional and aperiodic chitin nanofibers (CNFs) by FD method. The prepared structures successfully mimicked the size and interconnectivity of white beetle structure with improved porosity. In all, the fine and porous structure can be realized by more facile water-based FD method [23].

3. RENEWABLE POLYMERS FOR THE PREPARATION OF NANOFIBERS

3.1. Chitosan

Chitosan (CS) is a natural cationic polyelectrolyte copolymer that can be obtained by the deacetylation of its parent polymer chitin, a naturally occurring source derived from the exoskeleton of insects, crustaceans, and certain fungi. The abundance as a renewable resource and excellent properties, such as hemostasis, biodegradability, biocompatibility, nontoxicity, and chelation with metals, render chitosan suitable for biomedical applications. Moreover, chitosan has also shown good performances in inhibiting the growth of a wide variety of yeasts, fungi, and bacteria as well as gas and aroma barrier properties in dry condition. All these positive features of chitosan have provided ample opportunities for further development in biomedical and other industrial fields [26–28].

It has been reported that pure chitosan fibers can be produced in nonaqueous solvents. For example, electrospinning of pure chitosan was successfully achieved in trifluoroacetic acid (TFA) by Ohkawa et al. [29]. Sencadas et al. improved the homogeneity of the nanofibers by adding dichloromethane to the chitosan/TFA mixtures and systematically studied the processing parameters to set the ground for a reproducible way to obtain chitosan nanofibers for specific applications [30].

However, electrospinning individual chitosan from its aqueous solution remains a challenge. First, the rigid D-glucosamine repeat unit, high crystallinity, and ability to form strong hydrogen bond of chitosan make it difficult to be dissolved in most solvents. In addition, its polycationic property also leads to high viscosity, which hinders the formation of sufficient chain entanglements and results in the formation of nanobeads rather than nanofibers [26, 31]. However, the solvents for the spinning of the pure chitosan are environmentally harmful [32]. The electrospinning for chitosan can only be achieved at concentrations in the range between 2% and 8% (wt/v). This urges us to develop new versatile methods for the electrospinning of chitosan at higher concentrations to ensure the mechanical stability and retain the bioactivity of its nanofiber products [33].

Therefore, special attention needs to be paid to improving the spinnability of chitosan. Attempts have been made by blending chitosan with synthesized polymers, such as polyethylene oxide (PEO), polyvinyl alcohol (PVA), and PLA, and renewable polymers such as gelatin, alginate, and silk fibroin and nanoparticles to improve the mechanical strength, antibacterial activity, and antiadhesive properties towards bacteria of its nanofibers [31, 34].

PVA is a typical nontoxic and water-soluble synthesized [illegible] biocompatibility and hydrophi[illegible] mechanical properties [35]. Liu et al. explored the preparation [illegible] PVA/chitosan hydrogel nanofibers by solution blowing method with ethylene glycol diglycidyl ether (EGDE) as cross-linker at various concentrations (as shown in Figure 8) and found that CS/PVA hydrogel nanofiber mats possessed the properties of both hydrogel and nanofiber mats. In addition to the similar excellent exudate absorption property to that of hydrogel films, hydrogel nanofiber mats allowed gaseous exchange, which decreased with the increase of its cross-linking degree (DCL). Moreover, CS/PVA hydrogel nanofiber mats demonstrated good antibacterial rates of over 81% against E. coli, which was not significantly affected by DCL. The findings suggest that the CS/PVA nanofiber hydrogel is a promising material for wound dressing [36].

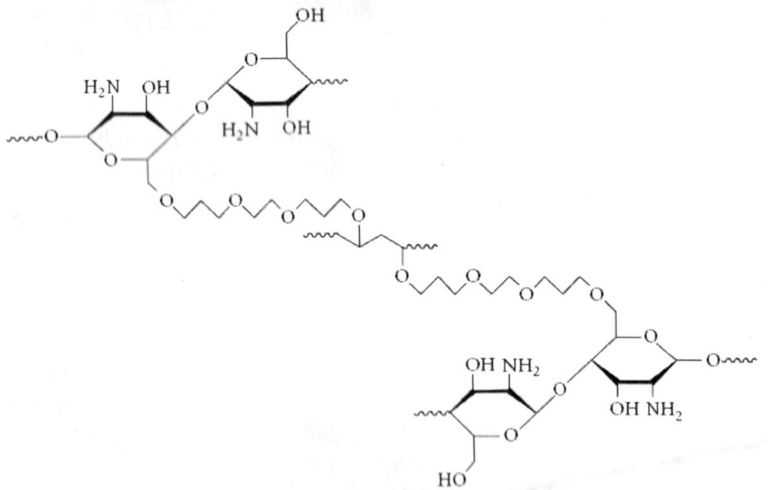

Figure 8. The structure of CS/PVA hydrogel nanofibers cross-linked by EGDE.

Compared with synthetic polymers, natural polymers have better biocompatibility and lower immunogenicity. Sericin is one of the biodegradable and biocompatible natural polymers with good antioxidant, moisture absorption, and antibacterial and UV resistance properties and has been widely used in cosmetics and fabrics. Zhao et al. prepared chitosan/sericin composite nanofibers by blending binatural polymers and electrospinning to enhance their biological performance. The nanofibers had diameters ranging from 240 nm to 380 nm and continuous and uniform diameter distribution. Bioassays indicated that these novel composite nanofibers showed noncytotoxicity and remarkably enhanced cell proliferation and antibacterial activities against both Gram-negative bacteria Escherichia coli and Gram-positive bacteria Bacillus subtilis [34].

Ag nanoparticles (Ag-NPs) have shown superior antibacterial properties. Ag-NPs can attach to cells to interrupt their permeability and intermembrane exchange. In addition, Ag-NPs have no adverse effects on viable cells and cause no significant microbial resistance. Therefore, Ag-NPs have been widely used as additives to various materials and in wound dressings against microbial contaminations [35, 37]. Abdelgawad et al. successfully prepared chitosan/PVA/Ag-NPs nanofiber mats with chitosan as capping polymer at high concentrations and glucose as reducing agent for antimicrobial applications. Ag-NPs were able to improve the electrospinnability of the blend, reduce the diameter of fibers, and more importantly enhance the antibacterial performances of the electrospun fibers, which made them viable to wound dressing [37].

3.2. Cellulose

Cellulose is one of the most abundant biodegradable materials and can be extracted from different natural sources, such as wood and cotton pulps, via chemical and/or physical methods [38].

Bacterial cellulose (BC) is natural cellulose synthesized by acetobacter xylinum [38] and has high biocompatibility and transparency, nontoxicity, and excellent mechanical strength. Moreover, BC fibers have a very high surface area per unit mass, which, combining with its high hydrophilicity, leads to a very high liquid loading capacity. Thus, BC has been applied in various fields [39]. Costa et al. carried out electrospinning of acetylated BC nanofiber mats to produce artificial symmetric nanoporous structures. The SEM images revealed that nanopores were more uniformly distributed throughout the electrospun nanofiber mat than those casted with BC mats. In addition, the electrospinning of modified BC nanofibers was easier than that of unmodified BC nanofibers, which could be explained by the dissolution mechanism of cellulose in DMA/LiCl solvent system [40].

However, BC shows poor antibacterial, antioxidant, and conducting and magnetic properties, which limits its application in biomedical and electronic fields [41]. One of the solutions is to blend BC with other substances to promote its properties. Carbon nanotubes (CNTs) have been of great interest in various fields since their discovery. They are promising fillers for polymer matrices to prepare CNTs-reinforced composite materials because of their extraordinarily high aspect ratio, elastic modulus, and high axial strength. Chen et al. prepared multiwalled CNTs-embedded BC nanofibers (MWCNTs/BC nanofibers) by electrospinning. The TGA analysis showed that the initial temperature of degradation followed the increasing order of BC < electrospun BC nanofibers < electrospun MWCNTs/BC nanofibers, which was attributed to the crystalline polymorph transformation and orientation as well as the embedded MWCNTs. In addition, the mechanical properties including tensile strength and Young's modulus and electrical conductivity of the products were significantly enhanced due to the well-dispersed and aligned MWCNTs. The finding indicates that MWCNTs/BC nanofibers may be potentially used in medical, mechanical, and electrical fields [42].

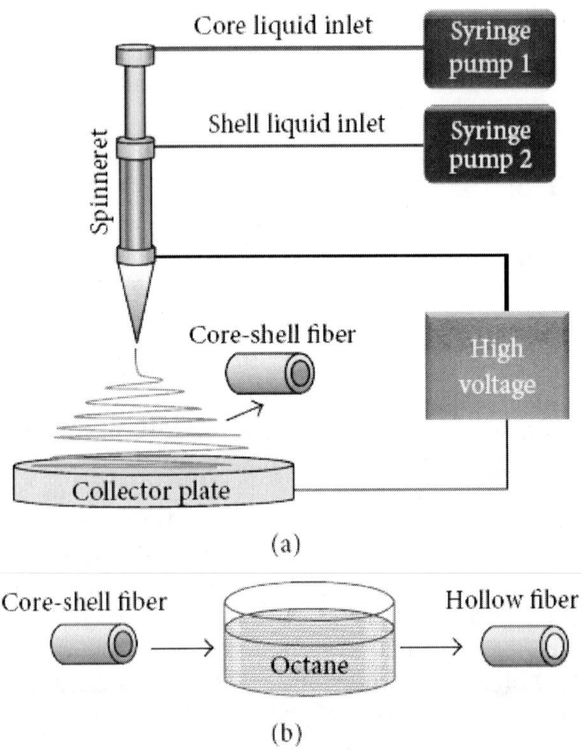

Figure 9. The preparation process of CA/PCL core-sheath nanostructures [44].

Cellulose acetate (CA) is a derivative of cellulose and has good stability and solubility in organic solvent. It can also selectively absorb low-level organic compounds and toxins. However, CA has poor mechanical properties [43]. Khalf et al. fabricated a CA/polycaprolactone (PCL) core-sheath nanostructure by coaxial electrospinning to promote its mechanical and biological properties. The preparation process is shown in Figure 9. Their results indicate that the loading of PCL was able to improve the elastic elongation of CA solid and hollow fibers. Cell attachment test suggested that the nontoxicity and higher cell attachment performance of CA nanofibers blended with PCL were attributed to its higher surface area per volume ratio [44].

3.3. Alginate

Alginate is a natural linear polysaccharide copolymer and can be obtained from brown sea weed. Sodium alginate (SA) has been well studied on its biomedical applications due to its superior properties including biocompatibility, biodegradability, nontoxicity, hydrophilicity, and low cost. Unlike chitosan, alginate is highly soluble in water [45].

To improve the structural integrity of the prepared nanofibers, alginate-based materials are usually cross-linked with divalent ions, such as calcium and barium. In addition, double cross-linking with glutaraldehyde, hexamethylene diisocyanate (HMDI), epichlorohydrin, and adipic acid (ADA) hydrazide has also been proposed (as shown in Figure 10). Unlike the non-cross-linked electrospun nanofibers, the cross-linked electrospun nanofibers are highly stable and relatively insoluble in both water and simulated body fluid [46–48].

However, pure SA nanofibers have been rarely fabricated via electrospinning. Nie et al. [49] prepared smooth and uniform alginate nanofibers by introducing glycerol as a cosolvent and simply adjusting the volume ratio between glycerol and water. Fang et al. [50] reported that pure alginate had been successfully electrospun by introducing Ca^{2+} cations to SA aqueous solutions. They demonstrated that high intermolecular interactions and low surface tension were the critical factors to improve the electrospinnability of SA solutions.

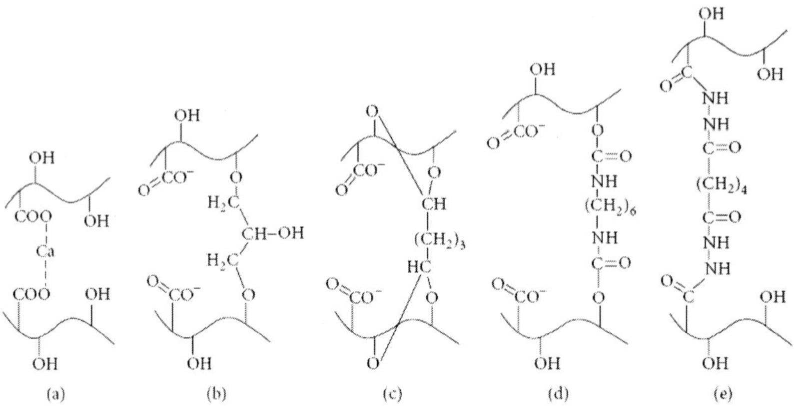

Figure 10. Molecular network structures formed between alginate chains in nanofibers after cross-linking with (a) $CaCl_2$, (b) epichlorohydrin, (c) glutaraldehyde, (d) HMDI, and (e) ADA [46].

Alginate has poor properties including low mechanical strength, high degradation rates, and processing difficulties, which are usually overcome by compositing alginate with other polymers such as PEO [51] and PVA [45]. It can be attributed to the intermolecular hydrogen bonds formed between the polymers and alginate. Surfactants, such as lecithin [52], and cosolvents have also been proposed to facilitate its processing for the preparation of nanofibers [53].

3.4. Gelatin

Gelatin (GT) is a renewable polymer obtained by the partial hydrolysis of collagen, a most abundant natural substance in connective tissues. GT has been widely used as a bioengineering material due to its high biocompatibility, biodegradability, and low cost. GT can also prevent fluid loss, which plays an important role in wound healing [54–56]. However, the electrospun gelatin nanofibers are soluble in water and have poor mechanical strength [56, 57]. Therefore, cross-linking treatments have been adopted to increase their water resistance [13] and mechanical properties [58, 59]. The cross-linking can be realized by physical processes, such as drying, heating, γ-ray, electron beam, and UV light exposure and chemical treatments by using cross-linker, such as formaldehyde, glutaraldehyde (GA), epoxy compounds, tannic acid, carbodiimide, acyl azide, and transglutaminase [60], through dissolving gelatin in 2,2,2-trifluoroethanol [61], formic acid dope solution [62], and water-based cosolvent composed of ethyl acetate and acetic acid [63]. Among all chemicals, GA is by far the most widely used due to its high efficiency, easy accessibility, and low cost. However, glutaraldehyde at high concentrations is cytotoxic and can disrupt the electrospun fiber morphology [59, 64]. Ko et al. [60] demonstrated that the gelatin nanofibers cross-linked with 0.5% (w/v) genipin promoted cell proliferation and reduced its solubility and cytotoxicity. Lu et al. [65] fabricated gelatin nanofibers by electrospinning and compared the properties of the gelatin nanofibers cross-linked by vapor and liquid-phase GA. Their results indicated that the cross-linking by both vapor-phase and liquid-phase GA was able to improve the mechanical strength of the gelatin fibers. The cross-linking with liquid-phase GA resulted in more evenly cross-linked, close-packed, and higher tensile. The excessive cross-linking with vapor-phase GA led to membranes shrinking due to poorly cross-linked middle layers in their hierarchy structure. Therefore, a novel in situ cross-linking method was proposed by Slemming-Adamsen et al. [66] for the fabrication of electrospun gelatin nanofibers with 1-ethyl-3-(3-dimethylaminopropyl)-1-carbodiimide hydrochloride (EDC) and N-hydroxysuccinimide (NHS) as cross-linkers. Kwak et al. [67] prepared Phaeodactylum tricornutum (P. tricornutum) extract-loaded gelatin nanofibers via electrospinning. P. tricornutum is a diatom which exists in brackish and marine water worldwide. The water-soluble extracts of P. tricornutum have showed anti-inflammatory, analgesic, and free radical scavenging activities. The introduction of P. tricornutum extracts improved the conductivity of the gelatin solution and reduced the diameter of the nanofibers. Their results also confirmed the antimicrobial activity and noncytotoxicity of the P. tricornutum-loaded gelatin nanofibers.

Gelatin has also been used as a blend component to prepare nanofibrous membranes for tissue scaffolds, wound healing and health caring devices, and other biomedical applications. Shan et al. [68] prepared silk fibroin/gelatin (SF/GT) electrospun nanofibrous dressing loaded with astragaloside IV (AS), an effective component, to accelerate wound healing. They found that the nanofibers were able to significantly improve cell adhesion and proliferation, promote wound healing, induce proangiogenesis of partial thickness burn wound, and inhibit complications in wounds. In addition, SF significantly decreased the average diameter, degradation rate, and the facture strain under high break strength of the produced nanofibers.

In addition to the general properties of a condensed polyphenol, the grape seed polyphenol (GSP) has some unique properties, such as antimicrobial activity, anti-inflammatory activity, and reducing blood sugar and fat. Han et al. [69] fabricated GSP/gelatin composite nanofibers containing Ag-NPs by electrospinning as a novel biomaterial. The process is shown in Figure 11. GSP served as the reducing agent and the stabilizing agent for silver nanoparticles. The electrospun composite nanofibers had average diameters in the range of 150–230 nm and a uniform distribution of Ag-NPs with an average diameter of 11 nm that contributed to their good antibacterial performance. Further investigation also revealed that the composite nanofibers containing silver nanoparticles had no toxicity at such concentrations. Thus, they could be potentially utilized in biomedical field as a wound dressing.

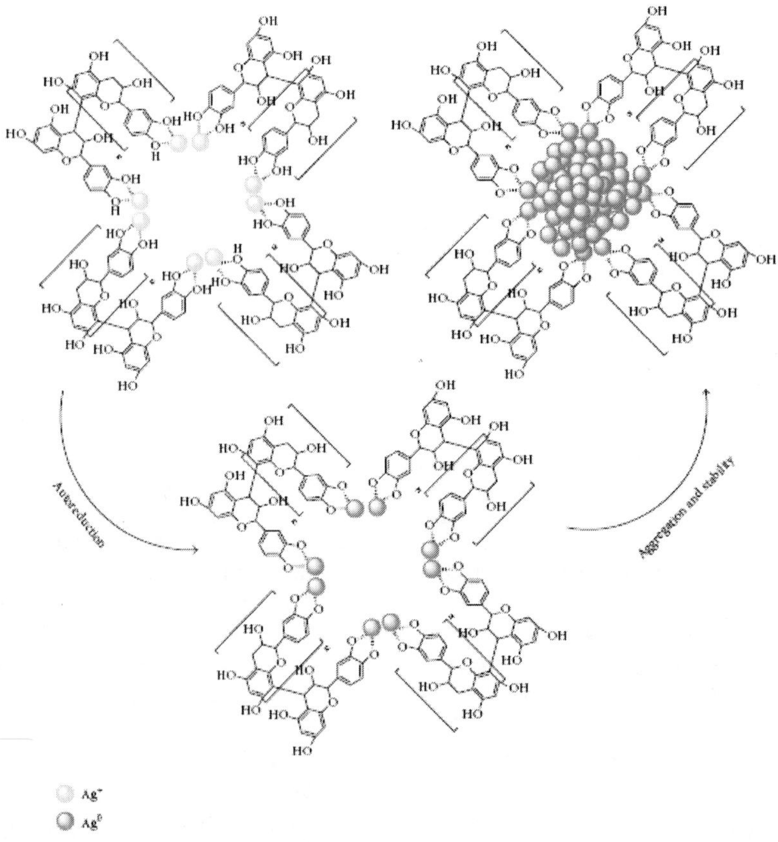

Figure 11. Schematic representation of the mechanism for preparing silver nanoparticles with GSP [69].

3.5. Hyaluronic Acid

Hyaluronic acid (HA) is one main glycosaminoglycan which is found in the ECM of many soft tissues of higher animals [70]. HA can decrease the formation of postsurgical adhesions due to its noncoagulant activity, modulation of fibrous tissue, barrier effect on fibrinogen, and inhibitory effect on the superoxide release from granulocytes [71]. Previous studies have demonstrated that HA nanofiber wound dressing showed better performance than the solid HA in wound healing due to its unique morphologies. However, HA solutions have high solution viscosity even at low concentrations, leading to issues in the production of consistent nanofibers [72].

Brenner et al. prepared pure HA nanofibers by electrospinning in a solvent system of aqueous ammonium hydroxide (25% NH_4OH) and DMF at pH of 11 to eliminate the solvent induced degradation effects on the biopolymer as reported in the previous studies. Continuous, cylindrical, and randomly oriented pure HA nanofibers with a diameter of 39 ± 12 nm and good biocompatibility were successfully electrospun [70].

Uppal et al. fabricated HA nanofibers by electrospinning with deionized water as a solvent and cocamidopropyl betaine as a surfactant. Based on the report of histopathologist, the sterilized HA nanofibers wound dressing showed better pathology performance than other sterilized wound dressings including solid HA, gauze with Vaseline dressing, adhesive bandage, and antibiotic wound dressing. The evaluation criteria included epithelial tissue gap, epithelial tissue detachment from the dermis, the presence of granulation tissue, and the assessment from clinic photographs based on the color of the wound and whether the wound was fully covered or not by the epithelial tissue. In addition, it showed higher air permeability than sterilized solid HA and gauze with Vaseline dressing, which contributed to the faster wound healing [72]. In all, the addition of antibiotic components could lead to better performance of HA nanofibers.

Chen et al. prepared dual functional core-sheath structured nanofibers with HA as the inner core and Ag-NPs-loaded PCL as extremal sheath by coaxial electrospinning. This design was used to control the slow release of HA from the core to mimic the biological function of HA in synovial fluid containing high concentration of HA. The sheath acted as a barrier with antimicrobial activity to prevent infection and alleviate adhesion. The results demonstrated that the core-sheath structure met the need of preventing peritendinous adhesion and postoperative infection through the slowly released HA and rapid release of Ag-NPs without exhibiting significant cytotoxicity [71].

3.6. Collagen

Collagen is the principal structural element of the native ECM in a three-dimensional network structure composed of multifibrils in a nanofiber scale. Due to a wealth of merits including biological origin, nonimmunogenicity, and excellent biocompatibility and biodegradability, collagen has been widely used as biomaterials in pharmaceutical and medical fields such as carriers for drug delivery, dressings for wound healing, and tissue engineering scaffold. Among all isoforms of collagen, the fibrillar structure of type I collagen has long been known for its important roles in cell attachment, proliferation, and differentiation [73, 74].

The adaptations of electrospinning to produce tissue engineering scaffolds composed of collagen nanofibers have shown that the structural properties of the electrospun collagen vary with the tissue of origin, the isoform, and the concentration of the collagen solution [75].

Collagen-composite fibrous mats have been fabricated with PEO [76], PCL [77, 78], polylactide-polyglycolide (PLGA) [75], poly(L-lactide-co-ε-caprolactone) (PLCL) [79], chitosan [73], hydroxyapatite [80, 81], and poly(L-lactic acid)-co-poly(ε-caprolactone) (P(LLA-CL)) [82]. Chen et al. [83] prepared the composite nanofibrous membrane (NFM) of type I collagen, chitosan, and PEO by electrospinning, which was further cross-linked with GA vapor. The cross-linking increased Young's modulus of the NFMs and decreased their ultimate tensile strength, tensile strain, and water sorption capability. The NFMs showed no cytotoxicity toward the growth of 3T3 fibroblasts and demonstrated good biocompatibility in vitro. In addition, the NFMs showed better wound healing performances than gauze and commercial collagen sponge wound dressings. This novel electrospun matrix can also be potentially used as a wound dressing for skin regeneration. Choi et al. [78] examined the feasibility of using PCL/collagen based nanofibers as a scaffold system for implantable engineered muscle organization. Their results showed that, in contrast to the randomly oriented nanofibers, unidirectionally oriented nanofibers could induce muscle cell alignment and myotube formation. The aligned composite nanofiber scaffolds seeded with skeletal muscle cells may be able to provide implantable functional muscle tissues. Liu et al. [84] explored the incorporation of neurotrophin (NT-3) and chondroitinase ABC (ChABC) into the electrospun collagen nanofiber for spinal cord injuries (SCI) treatment. The sustained release of NT-3 and ChABC for at least 28 days was achieved. Such biofunctional scaffolds may be found useful in SCI treatment by providing topographical and multiple biochemical cues to manipulate the growth inhibitory environment and promote axonal regeneration.

3.7. Silk Fibroin

Silk fibroin (SF) is a main component of silk. It is a type of fibrous proteins with remarkable mechanical properties produced in the form of fiber by silkworms and spiders. SF is also a promising wound dressing material due to its excellent mechanical properties, as well as its biocompatibility, controllable biodegradability, water-based processing, morphologic flexibility, and easily accessible chemical groups for functional modification. Despite its advantages, the low availability of SF limits the industrial scale production of its biomaterials. Generally, the electrospun SF fibers with desirable properties for bioengineering are prepared with SF at very high concentrations in the range of 20–40%. In addition, the highly crystalline β-sheet secondary structures result in the insolubility of SF, which limits its biomedical applications. Therefore, electrospun SF nanofibers are usually chemically treated with methanol [85], ethanol, propanol [86], and water vapor [87] to increase their stability. Strategies have been proposed to increase the viscoelasticity of SF systems by blending them with other polymers, which improves the mechanical properties while maintaining the biocompatibility of the obtained nanofibers [88–91].

Calamak et al. [92] fabricated silk fibroin nanofibers containing UV-reduced Ag-NPs by electrospinning method and posttreated the product with methanol and GA vapor. The posttreatments promoted the mechanical properties of the SF bionanotextiles. The results indicate that the methanol treatment was more effective in the crystallization transition of fibroin from random coil to β-sheet. In addition, Ag-NPs significantly decreased the diameter of the nanofibers and contributed to their antibacterial activity against Gram-positive bacteria S. aureus and Gram-negative bacteria P. aeruginosa.

Polyethylenimine (PEI) is a polycationic antimicrobial polymer with interesting ability to enter cells or permeabilize cell membranes. Çalamak et al. [90] prepared silk fibroin based antibacterial bionanotextiles and explored their application as wound dressing materials. The addition of PEI to the nanofibers decreased the average diameter of the nanofibers and increased their hydrophobicity, cell viability, and antimicrobial activity against both Gram-positive and Gram-negative bacteria. In contrast, the sulfated SF only showed antibacterial activity against P. aeruginosa.

Shahverdi et al. [89] fabricated silk fibroin blended with PLGA nanofibers via a dual source electrospinning setup, through which the electrospinning of SF and PLGA nanofibers could be separately optimized and achieved. They found that the concentrations of SF and PLGA predominantly affected the formation of the bead structure in the fibers. The addition of PLGA significantly improved the mechanical properties of the nanofibers. The incorporation of SF with PLGA displayed better cell attachment, proliferation, and viability performance than pure SF nanofibers and pure PLGA nanofibers.

A novel tissue engineering scaffold of electrospun silk fibroin/nanohydroxyapatite (nHA) biocomposite was prepared by an effective calcium and phosphate (Ca-P) alternate soaking method. The cell cultivation experiments demonstrated that the silk/nHA nanocomposite scaffold could support the early stage of osteoblast adhesion and had a significant effect on the differentiation stage. Their findings suggest that this composite scaffold can be a promising biomaterial for bone tissue engineering [93].

Ki et al. [94] determined the characteristics and performance of an electrospun membrane prepared with wool keratose (WK) and SF protein for metal ion absorption. The nanofibrous membranes exhibited higher metal ion absorption capacities than the conventional fibrous filters (wool sliver and filter paper). Particularly, the WK/SF blend nanofibrous membrane showed an extraordinary performance for the absorption of metal ions. In addition, its absorption capacity remained stable after several adsorption-desorption cycles. The electrospun WK/SF nanofibrous membrane is suitable for removing and recovering heavy metal ions in water.

4. APPLICATIONS OF RENEWABLE POLYMER BASED FIBERS IN WOUND DRESSING

Effective wound dressings are highly desired in various areas, such as promoting wound healing, hemostasis, skin regeneration, diabetic ulcers, and tissue engineering. A good wound dressing can maintain a moist environment at the interface, provide mechanical protection, prevent bacterial penetration, and allow gas-fluid exchange. In addition, it is nonadhesive to the wound and easily removed without trauma. However, the dressings used in specific fields may have their unique properties and requirements.

4.1. Promoting Wound Healing

A variety of fibers prepared with renewable materials including chitosan, fibrinogen, silk fibrin, bacterial cellulose, gelatin, and collagen and composites of these materials can promote wound healing and show antibacterial activity, good adhesion, good absorption, and nontoxicity and sufficient gas-exchange properties. The mats prepared with the fibers have unique porous structures and high surface areas, which are beneficial to cell regeneration and the wetness of the wound. In addition, the mats can protect the wound from the harmful external environments [95].

The bioactive molecules or drugs can be easily incorporated in the fibers by electrospinning to provide additional functions such as anti-inflammatory activity and tissue growth-promoting activity [96, 97]. Wu developed a composite dressing of alginate, hydroxyethyl cellulose, and growth factor and successfully applied it to the treatment of facial wounds [98].

Chitosan fibers blended with other polymers, such as PEO, PLA, PVA, and PCL, have been proved as promising candidates for wound healing dressing [99]. Huang et al. reported that chitosan/poly-L-lactic acid (PLLA) composite nanofiber prepared by electrospinning showed better performance in maintaining wettability than pure PLLA and thus accelerated wound healing [100]. Theisen et al. found that blending gelatin with PLLA led to a better performance for cell proliferation and ECM production and the cells were able to grow into tissue-like constructs [101]. Alginate might provide gelation and moist healing environment for chronic wounds, boosting their reepithelialization [102]. de Moraes and Beppu prepared a biocomposite membrane with sodium alginate and silk fibroin fibers. The incorporation of fibroin fibers enhanced the tensile strength of the fibers and resulted in the properties of a typical biomaterial for biomedical applications such as wound dressing. Its nontoxicity and flexibility also contributed to the fast wound healing [103].

4.2. Hemostasis

Timely and rapid hemostasis is very important to avoid further complications in surgical operation and emergency medical treatment. An effective hemostatic dressing should have properties such as excellent hemostasis, good biocompatibility, nontoxicity, no antigenicity, low cost, and easy processing [104].

Various renewable polymers, such as fibrinogen, fibrin, alginate, chitosan, cellulose, and gelatin, have been used in the preparation of hemostatic dressings. These materials in the form of fiber have shown excellent performances in hemostasis due to the large contact areas with blood which promotes the aggregation and adhesion of thrombocyte on the fibers. In addition, the soft and strong fiber can closely adhere to the wound and press the wound to promote hemostasis.

Zhao et al. prepared a nanofiber of fibrinogen fiber by electrospinning and applied it to hemostasis. The clinical study indicates that the dressing was able to significantly shorten bleeding time and decrease bleeding volume of splenic hemorrhage, hepatic hemorrhage, and femoral arteriovenous fistula, thus reducing mortality rate [105]. The fibrinogen forms a tough fibrin clots once it is activated by thrombin. The clot adheres to the wound and promotes hemostasis to accelerate wound healing. In addition, cross-linked fibrinogens have certain toughness and viscosity, which can prevent blood flushing and provide a physical hemostasis [106].

Fibrin is used in hemostasis due to its biocompatibility and excellent biological properties [107]. Fibrin is an initiator for the regulation of thrombosis and hemostasis. It is decomposed by fibrinogen, the major component of plasma proteins, when human body gets wounded and immediately covers the wound to accelerate the hemostasis [108]. During the healing process, fibrin can be completely degraded and absorbed by the tissue and causes no damage to the organism nearby.

Alginate is also an excellent hemostatic material. Pillai and Naieb prepared a ventral nonadhesive dressing with a soft net of pure calcium alginate, which showed an excellent hydrophilism. During the treatment, it swelled upon the absorption of exudate and formed a firm gel filled wound without adhesion, which maintained a moist environment for the wound and promoted its healing [109]. In addition, Ca^{2+} in the alginate fibers could be transferred into the blood by exchanging with Na^{+} and acted as a coagulation factor to accelerate the hemostasis.

Sibaja et al. synthesized a fiber by composing chitosan with alginate through wet extrusion and found that it was a biocompatible material with acceptable tensile properties for the inhibition of body fluids, such as blood, from flowing. The positive $-NH_3^+$ groups on the skeleton of chitosan can interact with the negative charges on the surface of the red blood cells and platelets, accelerate their aggregation, and promote the hemostasis. Studies have also shown that the incorporation of chitosan into alginate fibers can significantly enhance the tensile properties of the dressing [110].

Liu et al. prepared several novel nonwoven hemostatic dressings composed of chitosan derivatives including carboxymethyl chitosan fiber, hydroxypropyl chitosan fiber, hydroxyethyl chitosan fiber, and carboxymethyl hydroxyethyl chitosan fiber. The dressing showed good biocompatibility, excellent tensile strength, and easy processing properties and had no antigenicity. Their study also demonstrated that the dressings promoted hemostasis and wound healing by forming a colloidal solution during the swelling and dissolving. Therefore, the dressings are especially suitable in the field of in vivo hemostasis [111].

Dressings prepared with cellulose have been widely applied in the surgical hemostasis because they can quickly stop bleeding and have a strong affinity for water and normal saline. The dressing absorbs water from blood and is dissolved to form a gel that plugs the capillary end to increase the viscosity of the blood and decelerate blood and thus stops bleeding. The hemostasis of cellulose dressing is based on mechanical press, erythrocyte absorption, and thrombocyte aggregation. Wang et al. prepared a soluble hemostatic fiber of carboxymethyl cellulose, which showed an excellent hemostatic performance in the clinical experiments [112]. Compared with gelatin sponge, the dressing was able to more effectively control bleeding of small blood vessels during the surgery, especially the blood infiltration of wound and suture. It is believed that the carboxymethyl cellulose fiber can accelerate platelet adhesion and reduce postoperative hematoma.

Niu et al. blended gelatin with lidocaine and PVA to produce nonwoven hemostatic materials. The produced dressing had a thin and dense network structure with numerous loose gaps that increased the contact area for the blood and thus promoted thrombocytes aggregation and the formation of clot and accelerated hemostasis [113]. PVA is a high expansion material having a high degree of hydrophilicity, which makes the dressing expand quickly to partially press the wound and stanch the bleeding [114].

4.3. Skin Regeneration

Nowadays, burn healing is one of the most significant problems of modern surgery because of the high lethality and disability after the treatment for burns of large surface area. Patients with burn injuries undergo significant emotional and physical pain and trauma from the burn, dressing changes, and the related treatments during the healing process [115]. In addition, fluid loss, bacterial invasion, passive evaporative water, heat loss, and other complications can prevent wound from healing [116]. Therefore, the ideal burn dressing should have the following properties [117, 118]: (1) high exudate absorption capability, (2) ability to protect the wound from microorganisms, (3) being removable from the injured skin surface, (4) sufficient oxygen permeability, (5) high water vapor permeability, (6) nontoxicity, and (7) ability to accelerate scar formation. Covering large burned surfaces is still a challenge.

Fibrous wound dressing material is a promising candidate for skin regeneration. For example, Venugopal and Ramakrishna found that gelatin nanofibers could speed wound healing and could be used for cell transfer and skin regeneration [119]. Various blend fibers of renewable polymers, such as PEO/chitosan [120], PCL/collagen [119], PCL/gelatin [121], PLLA/chitosan [122], Zein/collagen [123], and PLLA/gelatin [124], are the dominate candidates.

The micro- and nanofibers of chitosan are an ideal candidate for burn dressing because of its high surface area to volume ratio [120]. The nanostructure of fibrous materials with small pore size functions as biological filters to effectively retain water and prevent infection from various bacteria. Kossovich et al. prepared a novel nanofiber dressing by blending chitosan with PEO through an electrospinning in glacial acetic acid. The clinical study indicates that the chitosan nanofiber dressing showed antibacterial properties and high absorption capacity and was able to effectively absorb exudate, ventilate wound, protect the wound from infection, and stimulate skin tissue regeneration. In addition, the chitosan-based dressing showed excellent biodegradation, which avoided the mechanical damage to the wound from the wound dressing removal, reduced pain from dressing removal and change, and fastened rehabilitation [120]. Figure 12 shows that the wound dressing applied to the donor wound of transplant skin removal promoted skin regeneration by keeping a moist environment and removing the surplus of exudate without dressing change.

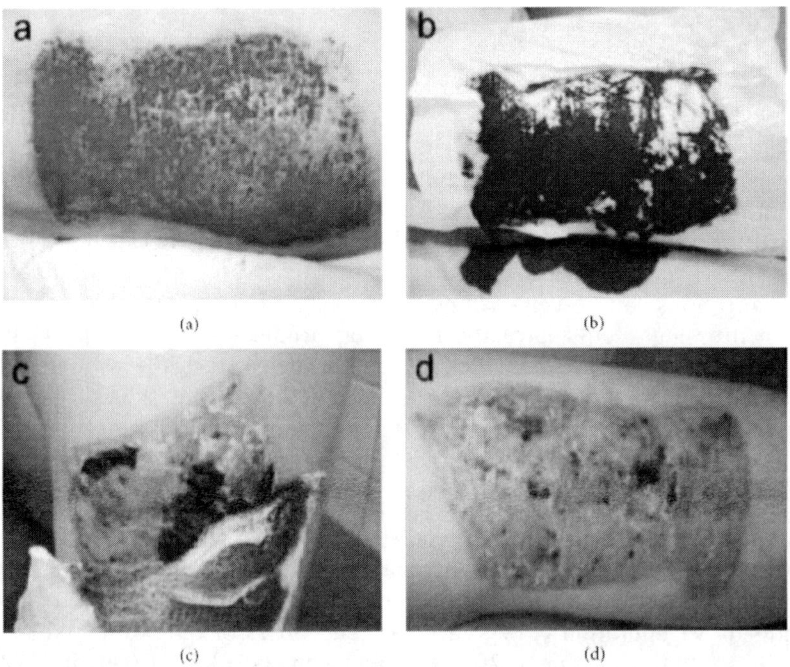

Figure 12. Donor wound: (a) before covering; (b) after covering; (c) 12 days after covering; and (d) 14 days after covering [121].

Venugopal et al. reported that composite fiber of collagen and PCL could support the growths of human dermal fibroblast and keratinocytes in skin engineering and the PCK film could not. They believe that cells migrated through pores in the fibers by amoeboid movement, pushing the surrounding fibers aside to expand. The dynamic architecture may provide cells with an opportunity to adjust the optimal pore diameter and grow into the scaffold. Chaisiri et al. prepared a skin tissue scaffold with composite fibers of gelatin and PCL to mimic the natural dermal ECM. The scaffold had a pore size of 40–70 micron and 98%-99% of porosity. In addition, it is biocompatible and can support cell migration and proliferation in its network [121].

4.4. Diabetic Ulcers

Diabetes mellitus is one of the most important public health problems worldwide because of its high prevalence and enormous social and economic consequences. It has been reported that there are more than 135 million diabetics in the world and the number is expected to increase to 300 million in the next 25 years [125].

Foot ulcers are a common, serious, and costly health issue for people with diabetes around the world. Two common chronic complications of diabetes mellitus including peripheral neuropathy and vascular insufficiency may lead to diabetic foot ulcers. Diabetic patients have a sharp decreased fibroblasts activity provoked by unavailability of glucose for aerobic metabolism, resulting in insufficient fibrous collagen tissue proliferation. Imbalance in these processes has been believed to be the main reason for the occurrence and persistence of this kind of disease. The wound healing process involves the interaction between various cellular groups, ECM, and growth factors, which is affected by the vascular insufficiency and severity of the lesion and infection.

Many renewable polymers, such as chitosan, alginate, cellulose, and silk, can be used to produce the dressings for the diabetic ulcers. Wang et al. developed a dressing composed of chitosan fibers, calcium alginate fibers, and chemically modified cellulose fibers and successfully used it to treat diabetic ulcers. In addition, epidermal growth factor (EGF) was incorporated into the renewable polymers to further promote the healing [126]. Schneider et al. prepared a silk mat with electrospun nanoscale silk fibers containing EGF, from which the EGF was slowly released in a time-dependent manner (25% EGF released in 170 h). Clinical study shows that the silk mat aided the healing by increasing time of wound closure by 90% by epidermal tongue, especially in chronic wounds including diabetic ulcers. Renewable polymers blended with metals can also show anti-inflammatory activity to accelerate the healing. Lázaro-Martínez reported a dressing of the blend of alginate fiber and hydrocolloid silver with excellent performances in decreasing inflammatory clinical signs and promoting healing on diabetic ulcers. The clinical study indicates that it reduced wound area with a median percentage of 47.7% and a mean percentage on the lesion of 58% at 2 weeks and 67.14% at 3 weeks. Wounds of all patients showed an obvious reduction in their size by 3-week treatment [127].

5. CONCLUSION

The nanofibers prepared from the renewable polymers can combine the excellent properties of the renewable polymers and nanofibers. They have become one of the most important materials used in wound dressing and attracted more and more attentions in recent years. The preparation methods, including electrospinning, bubbfil spinning, centrifugal spinning, and freeze-drying, have been widely used to prepare the nanofibers from the renewable polymers. Particularly the wound dressings prepared from chitosan, cellulose, alginate, gelatin, hyaluronic acid, collagen, and silk fibroin or their composites have been widely used as the wound dressing to promote wound healing, hemostasis, skin regeneration, diabetic ulcers, tissue engineering, and so on. Many of the dressings have been used in clinic. But for many renewable polymers which originated from natural components, there are many factors which influence the structure and property of the polymers and they are hard to be controlled. Although their properties are versatile, many of the properties are less prominent. When the renewable polymers are prepared to nanofibers and used for wound dressing, many of their properties, including antibacterial properties, hemostatic properties, biodegradability, cell adhesion and proliferation, and mechanical properties, need to be regulated by controlling the structure and properties of the nanofibers or compositing with other synthetic polymers to fit for the requirement of different wound. The above problems have become the future research keynotes for the nanofibers prepared from the renewable polymers.

ACKNOWLEDGMENTS

This work was supported by the Youth Talent Plan of Beijing City, the Combination Project of Guangdong Province, and the "YangFan" Innovative Research Team Project of Guangdong Province.

REFERENCES

1. P. Scarfato, L. Di Maio, and L. Incarnato, "Recent advances and migration issues in biodegradable polymers from renewable sources for food packaging," Journal of Applied Polymer Science, vol. 132, no. 48, Article ID 42597, 2015.
2. L. Yu, K. Dean, and L. Li, "Polymer blends and composites from renewable resources," Progress in Polymer Science, vol. 31, no. 6, pp. 576–602, 2006.
3. J. J. Blaker, K.-Y. Lee, and A. Bismarck, "Hierarchical composites made entirely from renewable resources," Journal of Biobased Materials and Bioenergy, vol. 5, no. 1, pp. 1–16, 2011.
4. V. Pillay, C. Dott, Y. E. Choonara et al., "A review of the effect of processing variables on the fabrication of electrospun nanofibers for drug delivery applications," Journal of Nanomaterials, vol. 2013, Article ID 789289, 22 pages, 2013.
5. R. Nayak, R. Padhye, I. L. Kyratzis, Y. B. Truong, and L. Arnold, "Recent advances in nanofibre fabrication techniques," Textile Research Journal, vol. 82, no. 2, pp. 129–147, 2012.
6. Y. J. Li, F. Chen, J. Nie, and D. Z. Yang, "Electrospun poly(lactic acid)/chitosan core-shell structure nanofibers from homogeneous solution," Carbohydrate Polymers, vol. 90, no. 4, pp. 1445–1451, 2012.
7. O. Ayaz, N. Ucar, E. Bahar, et al., "Properties of composite nanofiber produced by single and coaxial nozzle method used for electrospinning technique," International Journal of Chemical and Biological Engineering, vol. 6, pp. 52–56, 2012.
8. S. Agarwal, A. Greiner, and J. H. Wendorff, "Functional materials by electrospinning of polymers,"Progress in Polymer Science, vol. 38, no. 6, pp. 963–991, 2013.
9. Z. J. Yu, Y. H. Lin, W. W. Huang et al., "Multi spinnerets electrospinning with assistant sheath gas," inProceedings of the 9th IEEE International Conference on Nano/Micro Engineered and Molecular Systems, pp. 64–67, IEEE, Waikiki Beach, Hawaii, USA, April 2014.
10. I. G. Kim, J.-H. Lee, A. R. Unnithan, C.-H. Park, and C. S. Kim, "A comprehensive electric field analysis of cylinder-type multi-nozzle electrospinning system for mass production of nanofibers," Journal

of Industrial and Engineering Chemistry, vol. 31, pp. 251–256, 2015.

11. A. Y. Nuryantini, M. M. Munir, M. P. Ekaputra, T. Suciati, and Khairurrijal, "Electrospinning of poly(vinyl alcohol)/chitosan via multi-nozzle spinneret and drum collector," Advanced Materials Research, vol. 896, pp. 41–44, 2014.

12. N. Sasithorn and L. Martinová, "Fabrication of silk nanofibres with needle and roller electrospinning methods," Journal of Nanomaterials, vol. 2014, Article ID 947315, 9 pages, 2014.

13. X. Wang, H. Niu, X. Wang, and T. Lin, "Needleless electrospinning of uniform nanofibers using spiral coil spinnerets," Journal of Nanomaterials, vol. 2012, Article ID 785920, 9 pages, 2012.

14. R.-X. Chen, Y. Li, and J.-H. He, "Mini-review on Bubbfil spinning process for mass-production of nanofibers," Revista Materia, vol. 19, no. 4, Article ID 11560, pp. 325–344, 2014.

15. J.-H. He, H.-Y. Kong, R.-R. Yang et al., "Review on fiber morphology obtained by bubble electrospinning and blown bubble spinning," Thermal Science, vol. 16, no. 5, pp. 1263–1279, 2012.

16. Z.-B. Li, H.-Y. Liu, and H. Dou, "On air blowing direction in the blown bubble-spinning," Revista Materia, vol. 19, no. 4, pp. 345–349, 2014.

17. J.-H. He and Y. Liu, "Control of bubble size and bubble number in bubble electrospinning," Computers and Mathematics with Applications, vol. 64, no. 5, pp. 1033–1035, 2012.

18. G. J. Jiang, S. Zhang, Y. T. Wang, and X. H. Qin, "An improved free surface electrospinning with micro-bubble solution system for massive production of nanofibers," Materials Letters, vol. 144, pp. 22–25, 2015.

19. H. Dou, B.-Q. Zuo, and J.-H. He, "Blown bubble-spinning for fabrication of superfine fibers," Thermal Science, vol. 16, no. 5, pp. 1465–1466, 2012.

20. Y. Lu, Y. Li, S. Zhang et al., "Parameter study and characterization for polyacrylonitrile nanofibers fabricated via centrifugal spinning process," European Polymer Journal, vol. 49, no. 12, pp. 3834–3845, 2013.

21. M. A. Hammami, M. Krifa, and O. Harzallah, "Centrifugal force spinning of PA6 nanofibers—processability and morphology of

solution-spun fibers," The Journal of the Textile Institute, vol. 105, no. 6, pp. 637–647, 2014.

22. A. E. Erickson, D. Edmondson, F.-C. Chang et al., "High-throughput and high-yield fabrication of uniaxially-aligned chitosan-based nanofibers by centrifugal electrospinning," Carbohydrate Polymers, vol. 134, Article ID 10199, pp. 467–474, 2015.

23. J. Wu and J. C. Meredith, "Assembly of chitin nanofibers into porous biomimetic structures via freeze drying," ACS Macro Letters, vol. 3, no. 2, pp. 185–190, 2014.

24. N. Salamian, S. Irani, M. Zandi, S. M. Saeed, and S. M. Atyabi, "Cell attachment studies on electrospun nanofibrous PLGA and freeze-dried porous PLGA," Nano Bulletin, vol. 2, no. 1, Article ID 130103, 2013.

25. G. Ma, Z. Wang, J. Chen, R. Yin, B. Chen, and J. Nie, "Freeze-dried chitosan-sodium hyaluronate polyelectrolyte complex fibers as tissue engineering scaffolds," New Journal of Chemistry, vol. 38, no. 3, pp. 1211–1217, 2014.

26. M. Pakravan, M.-C. Heuzey, and A. Ajji, "A fundamental study of chitosan/PEO electrospinning,"Polymer, vol. 52, no. 21, pp. 4813–4824, 2011.

27. A. Sionkowska, "Current research on the blends of natural and synthetic polymers as new biomaterials: review," Progress in Polymer Science, vol. 36, no. 9, pp. 1254–1276, 2011.

28. M. Dash, F. Chiellini, R. M. Ottenbrite, and E. Chiellini, "Chitosan-a versatile semi-synthetic polymer in biomedical applications," Progress in Polymer Science, vol. 36, no. 8, pp. 981–1014, 2011.

29. K. Ohkawa, K.-I. Minato, G. Kumagai, S. Hayashi, and H. Yamamoto, "Chitosan nanofiber,"Biomacromolecules, vol. 7, no. 11, pp. 3291–3294, 2006.

30. V. Sencadas, D. M. Correia, A. Areias et al., "Determination of the parameters affecting electrospun chitosan fiber size distribution and morphology," Carbohydrate Polymers, vol. 87, no. 2, pp. 1295–1301, 2012.

31. M. Z. Elsabee, H. F. Naguib, and R. E. Morsi, "Chitosan based nanofibers, review," Materials Science and Engineering C, vol. 32, no. 7, pp. 1711–1726, 2012.

32. P. Su, C. J. Wang, X. Y. Yang et al., "Electrospinning of chitosan nanofibers: the favorable effect of metal ions," Carbohydrate Polymers, vol. 84, no. 1, pp. 239–246, 2011.

33. A. A. Nada, R. James, N. B. Shelke et al., "A smart methodology to fabricate electrospun chitosan nanofiber matrices for regenerative engineering applications," Polymers for Advanced Technologies, vol. 25, no. 5, pp. 507–515, 2014.

34. R. Zhao, X. Li, B. Sun et al., "Electrospun chitosan/sericin composite nanofibers with antibacterial property as potential wound dressings," International Journal of Biological Macromolecules, vol. 68, pp. 92–97, 2014.

35. Y. H. Zhao, Y. Zhou, X. M. Wu, L. Wang, L. Xu, and S. C. Wei, "A facile method for electrospinning of Ag nanoparticles/poly (vinyl alcohol)/carboxymethyl-chitosan nanofibers," Applied Surface Science, vol. 258, no. 22, pp. 8867–8873, 2012.

36. R. Liu, X. Xu, X. Zhuang, and B. Cheng, "Solution blowing of chitosan/PVA hydrogel nanofiber mats,"Carbohydrate Polymers, vol. 101, no. 1, pp. 1116–1121, 2014.

37. A. M. Abdelgawad, S. M. Hudson, and O. J. Rojas, "Antimicrobial wound dressing nanofiber mats from multicomponent (chitosan/silver-NPs/polyvinyl alcohol) systems," Carbohydrate Polymers, vol. 100, pp. 166–178, 2014.

38. H. Y. Ma, C. Burger, B. S. Hsiao, and B. Chu, "Fabrication and characterization of cellulose nanofiber based thin-film nanofibrous composite membranes," Journal of Membrane Science, vol. 454, pp. 272–282, 2014.

39. L. Fu, J. Zhang, and G. Yang, "Present status and applications of bacterial cellulose-based materials for skin tissue repair," Carbohydrate Polymers, vol. 92, no. 2, pp. 1432–1442, 2013.

40. L. M. M. Costa, G. M. Olyveira, P. Basmaji, and L. X. Filho, "Nanopores structure in electrospun bacterial cellulose," Journal of Biomaterials and Nanobiotechnology, vol. 3, pp. 92–96, 2012.

41. N. Shah, M. Ul-Islam, W. A. Khattak, and J. K. Park, "Overview of bacterial cellulose composites: a multipurpose advanced material," Carbohydrate Polymers, vol. 98, no. 2, pp. 1585–1598, 2013.

42. P. Chen, Y. S. Yun, H. Bak, S. Y. Cho, and H.-J. Jin, "Multiwalled carbon nanotubes-embedded electrospun bacterial cellulose nanofibers," Molecular Crystals and Liquid Crystals, vol. 519, pp. 169–178, 2010.

43. E. Omollo, C. Zhang, J. I. Mwasiagi, and S. Ncube, "Electrospinning cellulose acetate nanofibers and a study of their possible use in high-efficiency filtration," Journal of Industrial Textiles, 2014.

44. A. Khalf, K. Singarapu, and S. V. Madihally, "Cellulose acetate core-shell structured electrospun fiber: fabrication and characterization," Cellulose, vol. 22, no. 2, pp. 1389–1400, 2015.

45. K. T. Shalumon, K. H. Anulekha, S. V. Nair, S. V. Nair, K. P. Chennazhi, and R. Jayakumar, "Sodium alginate/poly(vinyl alcohol)/nano ZnO composite nanofibers for antibacterial wound dressings,"International Journal of Biological Macromolecules, vol. 49, no. 3, pp. 247–254, 2011.

46. N. Bhattarai and M. Zhang, "Controlled synthesis and structural stability of alginate-based nanofibers,"Nanotechnology, vol. 18, no. 45, Article ID 455601, 2007.

47. S. A. Stone, P. Gosavi, T. J. Athauda, and R. R. Ozer, "In situ citric acid crosslinking of alginate/polyvinyl alcohol electrospun nanofibers," Materials Letters, vol. 112, pp. 32–35, 2013.

48. V. Leung, R. Hartwell, S. S. Elizei, H. J. Yang, A. Ghahary, and F. Ko, "Postelectrospinning modifications for alginate nanofiber-based wound dressings," Journal of Biomedical Materials Research Part B Applied Biomaterials, vol. 102, no. 3, pp. 508–515, 2014.

49. K. Y. Lee, L. Jeong, Y. O. Kang, S. J. Lee, and W. H. Park, "Electrospinning of polysaccharides for regenerative medicine," Advanced Drug Delivery Reviews, vol. 61, no. 12, pp. 1020–1032, 2009.

50. D. W. Fang, Y. Liu, S. Jiang, J. Nie, and G. P. Ma, "Effect of intermolecular interaction on electrospinning of sodium alginate," Carbohydrate Polymers, vol. 85, no. 1, pp. 276–279, 2011.

51. G. P. Ma, D. W. Fang, Y. Liu, X. D. Zhu, and J. Nie, "Electrospun sodium alginate/poly(ethylene oxide) core-shell nanofibers scaffolds potential for tissue engineering applications," Carbohydrate Polymers, vol. 87, no. 1, pp. 737–743, 2012.

52. S. A. Park, K. E. Park, and W. D. Kim, "Preparation of sodium alginate/poly(ethylene oxide) blend nanofibers with lecithin," Macromolecular Research, vol. 18, no. 9, pp. 891–896, 2010.

53. S. Safi, M. Morshed, S. A. H. Ravandi, and M. Ghiaci, "Study of electrospinning of sodium alginate, blended solutions of sodium alginate/poly(vinyl alcohol) and sodium alginate/poly(ethylene oxide),"Journal of Applied Polymer Science, vol. 104, no. 5, pp. 3245–3255, 2007.

54. V. M. Merkle, P. L. Tran, M. Hutchinson et al., "Core-shell PVA/gelatin electrospun nanofibers promote human umbilical vein endothelial cell and smooth muscle cell proliferation and migration," Acta Biomaterialia, vol. 27, pp. 77–87, 2015.

55. R.-Y. Tsai, S.-C. Hung, J.-Y. Lai, D.-M. Wang, and H.-J. Hsieh, "Electrospun chitosan-gelatin-polyvinyl alcohol hybrid nanofibrous mats: production and characterization," Journal of the Taiwan Institute of Chemical Engineers, vol. 45, no. 4, pp. 1975–1981, 2014.

56. M. Fallah, S. H. Bahrami, and R. M. Marziyeh, "Fabrication and characterization of PCL/gelatin/curcumin nanofibers and their antibacterial properties," Journal of Industrial Textiles, pp. 1–16, 2015.

57. K. Jalaja, P. R. A. Kumar, T. Dey, S. C. Kundu, and N. R. James, "Modified dextran cross-linked electrospun gelatin nanofibres for biomedical applications," Carbohydrate Polymers, vol. 114, pp. 467–475, 2014.

58. T. H. Nguyen and B. T. Lee, "Fabrication and characterization of cross-linked gelatin electro-spun nano-fibers," Journal of Biomedical Science and Engineering, vol. 3, no. 12, pp. 1117–1124, 2010.

59. Y. Z. Zhang, J. Venugopal, Z.-M. Huang, C. T. Lim, and S. Ramakrishna, "Crosslinking of the electrospun gelatin nanofibers," Polymer, vol. 47, no. 8, pp. 2911–2917, 2006.

60. J. H. Ko, H. Yin, J. An et al., "Characterization of cross-linked gelatin nanofibers through electrospinning," Macromolecular Research, vol. 18, no. 2, pp. 137–143, 2010.

61. Z.-M. Huang, Y. Z. Zhang, S. Ramakrishna, and C. T. Lim, "Electrospinning and mechanical characterization of gelatin nanofibers," Polymer, vol. 45, no. 15, pp. 5361–5368, 2004.

62. C. S. Ki, D. H. Baek, K. D. Gang, K. H. Lee, I. C. Um, and Y. H. Park, "Characterization of gelatin nanofiber prepared from gelatin-formic acid solution," Polymer, vol. 46, no. 14, pp. 5094–5102, 2005.

63. J.-H. Song, H.-E. Kim, and H.-W. Kim, "Production of electrospun gelatin nanofiber by water-based co-solvent approach," Journal of Materials Science: Materials in Medicine, vol. 19, no. 1, pp. 95–102, 2008.

64. K. Sisson, C. Zhang, M. C. Farach-Carson, D. B. Chase, and J. F. Rabolt, "Evaluation of cross-linking methods for electrospun gelatin on cell growth and viability," Biomacromolecules, vol. 10, no. 7, pp. 1675–1680, 2009.

65. W. P. Lu, M. Ma, H. T. Xu, B. Zhang, X. F. Cao, and Y. C. Guo, "Gelatin nanofibers prepared by spiral-electrospinning and cross-linked by vapor and liquid-phase glutaraldehyde," Materials Letters, vol. 140, pp. 1–4, 2015.

66. P. Slemming-Adamsen, J. Song, M. Dong, F. Besenbacher, and M. Chen, "In situ cross-linked PNIPAM/gelatin nanofibers for thermo-responsive drug release," Macromolecular Materials and Engineering, vol. 300, no. 12, pp. 1226–1231, 2015.

67. H. W. Kwak, M. J. Kang, J. H. Bae et al., "Fabrication of Phaeodactylum tricornutum extract-loaded gelatin nanofibrous mats exhibiting antimicrobial activity," International Journal of Biological Macromolecules, vol. 63, pp. 198–204, 2014.

68. Y.-H. Shan, L.-H. Peng, X. Liu, X. Chen, J. Xiong, and J.-Q. Gao, "Silk fibroin/gelatin electrospun nanofibrous dressing functionalized with astragaloside IV induces healing and anti-scar effects on burn wound," International Journal of Pharmaceutics, vol. 479, no. 2, pp. 291–301, 2015.

69. X. T. Han, Z. G. Xing, S. X. Si, Y. Y. Yao, and Q. Y. Zhang, "Electrospun grape seed polyphenols/gelatin composite fibers contained silver nanoparticles as biomaterials," Fibers and Polymers, vol. 15, no. 12, pp. 2572–2580, 2014.

70. E. K. Brenner, J. D. Schiffman, E. A. Thompson, L. J. Toth, and C. L. Schauer, "Electrospinning of hyaluronic acid nanofibers from aqueous ammonium solutions," Carbohydrate Polymers, vol. 87, no. 1, pp. 926–929, 2012.

71. C.-H. Chen, S.-H. Chen, K. T. Shalumon, and J.-P. Chen, "Dual functional core-sheath electrospun hyaluronic acid/polycaprolactone nanofibrous membranes embedded with

silver nanoparticles for prevention of peritendinous adhesion," Acta Biomaterialia, vol. 26, pp. 225–235, 2015.

72. R. Uppal, G. N. Ramaswamy, C. Arnold, R. Goodband, and Y. Wang, "Hyaluronic acid nanofiber wound dressing—production, characterization, and in vivo behavior," Journal of Biomedical Materials Research Part B: Applied Biomaterials, vol. 97, no. 1, pp. 20–29, 2011.

73. Z. G. Chen, P. W. Wang, B. Wei, X. M. Mo, and F. Z. Cui, "Electrospun collagen-chitosan nanofiber: a biomimetic extracellular matrix for endothelial cell and smooth muscle cell," Acta Biomaterialia, vol. 6, no. 2, pp. 372–382, 2010.

74. K. S. Rho, L. Jeong, G. Lee et al., "Electrospinning of collagen nanofibers: effects on the behavior of normal human keratinocytes and early-stage wound healing," Biomaterials, vol. 27, no. 8, pp. 1452–1461, 2006.

75. S.-J. Liu, Y.-C. Kau, C.-Y. Chou, J.-K. Chen, R.-C. Wu, and W.-L. Yeh, "Electrospun PLGA/collagen nanofibrous membrane as early-stage wound dressing," Journal of Membrane Science, vol. 355, no. 1-2, pp. 53–59, 2010.

76. L. Huang, K. Nagapudi, P. R. Apkarian, and E. L. Chaikof, "Engineered collagen-PEO nanofibers and fabrics," Journal of Biomaterials Science—Polymer Edition, vol. 12, no. 9, pp. 979–993, 2001.

77. E. Schnell, K. Klinkhammer, S. Balzer et al., "Guidance of glial cell migration and axonal growth on electrospun nanofibers of poly-ε-caprolactone and a collagen/poly-ε-caprolactone blend," Biomaterials, vol. 28, no. 19, pp. 3012–3025, 2007.

78. J. S. Choi, S. J. Lee, G. J. Christ, A. Atala, and J. J. Yoo, "The influence of electrospun aligned poly(ε-caprolactone)/collagen nanofiber meshes on the formation of self-aligned skeletal muscle myotubes,"Biomaterials, vol. 29, no. 19, pp. 2899–2906, 2008.

79. I. K. Kwon and T. Matsuda, "Co-electrospun nanofiber fabrics of poly (L-lactide-co-ϵ-caprolactone) with type I collagen or heparin," Biomacromolecules, vol. 6, no. 4, pp. 2096–2105, 2005.

80. J. Venugopal, S. Low, A. T. Choon, T. S. S. Kumar, and S. Ramakrishna, "Mineralization of osteoblasts with electrospun collagen/hydroxyapatite nanofibers," Journal of Materials Science: Materials in Medicine, vol. 19, no. 5, pp. 2039–2046, 2008.

81. J.-H. Song, H.-E. Kim, and H.-W. Kim, "Electrospun fibrous web of collagen-apatite precipitated nanocomposite for bone regeneration," Journal of Materials Science: Materials in Medicine, vol. 19, no. 8, pp. 2925–2932, 2008.

82. W. He, T. Yong, W. E. Teo, Z. Ma, and S. Ramakrishna, "Fabrication and endothelialization of collagen-blended biodegradable polymer nanofibers: potential vascular graft for blood vessel tissue engineering,"Tissue Engineering, vol. 11, no. 9-10, pp. 1574–1588, 2005.

83. J.-P. Chen, G.-Y. Chang, and J.-K. Chen, "Electrospun collagen/chitosan nanofibrous membrane as wound dressing," Colloids and Surfaces A: Physicochemical and Engineering Aspects, vol. 313-314, pp. 183–188, 2008.

84. T. Liu, J. Xu, B. P. Chan, and S. Y. Chew, "Sustained release of neurotrophin-3 and chondroitinase ABC from electrospun collagen nanofiber scaffold for spinal cord injury repair," Journal of Biomedical Materials Research Part A, vol. 100, no. 1, pp. 236–242, 2011.

85. S. H. Kim, Y. S. Nam, T. S. Lee, and W. H. Park, "Silk fibroin nanofiber. Electrospinning, properties, and structure," Polymer Journal, vol. 35, no. 2, pp. 185–190, 2003.

86. L. Jeong, K. Y. Lee, J. W. Liu, and W. H. Park, "Time-resolved structural investigation of regenerated silk fibroin nanofibers treated with solvent vapor," International Journal of Biological Macromolecules, vol. 38, no. 2, pp. 140–144, 2006.

87. B.-M. Min, L. Jeong, K. Y. Lee, and W. H. Park, "Regenerated silk fibroin nanofibers: water vapor-induced structural changes and their effects on the behavior of normal human cells," Macromolecular Bioscience, vol. 6, no. 4, pp. 285–292, 2006.

88. L. Jeong, D. Cho, O. H. Kwon, B.-M. Min, and W. H. Park, "Cellular response of silk fibroin nanofibers containing silver nanoparticles in vitro," Macromolecular Research, vol. 22, no. 7, pp. 796–803, 2014.

89. S. Shahverdi, M. Hajimiri, M. A. Esfandiari et al., "Fabrication and structure analysis of poly(lactide-co-glycolic acid)/silk fibroin hybrid scaffold for wound dressing applications," International Journal of Pharmaceutics, vol. 473, no. 1-2, pp. 345–355, 2014.

90. S. Çalamak, C. Erdoğdu, M. Özalp, and K. Ulubayram, "Silk fibroin based antibacterial bionanotextiles as wound dressing

materials," Materials Science and Engineering C, vol. 43, pp. 11–20, 2014.

91. J. Chutipakdeevong, U. Ruktanonchai, and P. Supaphol, "Hybrid biomimetic electrospun fibrous mats derived from poly(ε-caprolactone) and silk fibroin protein for wound dressing application," Journal of Applied Polymer Science, vol. 132, no. 11, Article ID 41653, 2015.

92. S. Calamak, E. A. Aksoy, C. Erdogdu, M. Sagıroglu, and K. Ulubayram, "Silver nanoparticle containing silk fibroin bionanotextiles," Journal of Nanoparticle Research, vol. 17, no. 2, pp. 1–9, 2015.

93. K. Wei, Y. Li, K.-O. Kim et al., "Fabrication of nano-hydroxyapatite on electrospun silk fibroin nanofiber and their effects in osteoblastic behavior," Journal of Biomedical Materials Research Part A, vol. 97, no. 3, pp. 272–280, 2011.

94. C. S. Ki, E. H. Gang, I. C. Um, and Y. H. Park, "Nanofibrous membrane of wool keratose/silk fibroin blend for heavy metal ion adsorption," Journal of Membrane Science, vol. 302, no. 1-2, pp. 20–26, 2007.

95. C. S. Jiang and X. B. Chen, "Research process in the preparation of biomedical dressings by electrospinning," Synthetic Fiber Industry, vol. 35, no. 3, pp. 44–46, 2012.

96. T. T. T. Nguyen, C. Ghosh, S.-G. Hwang, L. D. Tran, and J. S. Park, "Characteristics of curcumin-loaded poly (lactic acid) nanofibers for wound healing," Journal of Materials Science, vol. 48, no. 20, pp. 7125–7133, 2013.

97. Y.-F. Goh, I. Shakir, and R. Hussain, "Electrospun fibers for tissue engineering, drug delivery, and wound dressing," Journal of Materials Science, vol. 48, no. 8, pp. 3027–3054, 2013.

98. T. Wu, "Absorbable composite dressing," China, CN102008740-A, CN102008740-B.

99. J. Fang, H. Niu, T. Lin, and X. Wang, "Applications of electrospun nanofibers," Chinese Science Bulletin, vol. 53, no. 15, pp. 2265–2286, 2008.

100. X.-F. Huang, J.-W. Jia, Z.-K. Wang, and Q.-L. Hu, "A novel chitosan-based sponge coated with self-assembled thrombin/tannic acid multilayer films as a hemostatic dressing," Chinese Journal of Polymer Science, vol. 33, no. 2, pp. 284–290, 2015.

101. C. Theisen, S. F. Winkelmann, K. Knappstein et al., "Influence of nanofibers on growth and gene expression of human tendon derived fibroblast," BioMedical Engineering Online, vol. 9, article 9, 2010.

102. M. Dubský, Š. Kubinová, J. Širc et al., "Nanofibers prepared by needleless electrospinning technology as scaffolds for wound healing," Journal of Materials Science: Materials in Medicine, vol. 23, no. 4, pp. 931–941, 2012.

103. M. A. de Moraes and M. M. Beppu, "Biocomposite membranes of sodium alginate and silk fibroin fibers for biomedical applications," Journal of Applied Polymer Science, vol. 130, no. 5, pp. 3451–3457, 2013.

104. Z. X. Zhou, X. L. Wei, J. R. Yao, et al., "Evaluation of biological hemostatic dressings and gauzes in surgical wounds," Journal of Clinical Rehabilitative Tissue Engineering Research, vol. 14, no. 51, pp. 9635–9638, 2010.

105. P. Jiang, "Preparation of the PLLA from the fibrinogen and applications in the wound bleeding pig model," Journal of Trauma and Emergency, vol. 1, no. 4, p. 54, 2013.

106. X. Zhao, X. H. Cao, Y. Y. Ma et al., "Hemostatic effect of fibrin hemostatic dressing," China Blood Transfusion, vol. 23, no. 4, pp. 250–252, 2010.

107. K. Y. Baik, S. Y. Park, S. Namgung et al., "Synthetic nanowire/nanotube-based solid substrates for controlled cell growth," Nano Convergence, vol. 1, article 28, 2014.

108. T. Rajangam, H.-J. Paik, and S. S. A. An, "Development of fibrinogen microspheres as a biodegradable carrier for tissue engineering," Biochip Journal, vol. 5, no. 2, pp. 175–183, 2011.

109. R. G. Pillai and Z. A. Naieb, "Plastibell circumcision supported by a calcium-alginate fibre dressing to reduce bleeding," Arab Journal of Urology, vol. 13, no. 3, pp. 179–181, 2015.

110. B. Sibaja, E. Culbertson, P. Marshall et al., "Preparation of alginate-chitosan fibers with potential biomedical applications," Carbohydrate Polymers, vol. 134, pp. 598–608, 2015.

111. W. S. Liu, B. Q. Han, and Q. S. Gu, "Preparation method and application of the hemostasis and healing materials from the water soluble chitosan fibres," CN 200710015743.6, 2007.

112. H. M. Wang, J. Y. Pan, B. Yang, and K. X. Liu, "Animal test and clinical application of total soluble hemostatic fiber," Journal of Clinical Pharmacology and Therapeutics in China, vol. 11, no. 9, pp. s1017–s1020, 2006.

113. J. Z. Niu, B. P. Li, H. T. Lan, H. Wang, and K. Li, "Effect of lidocaine gelatin fiber net on the hemostatic and analgesic effect of endoscopic sinus surgery," Chinese Journal of Primary Medicine and Pharmacy, vol. 20, no. 12, pp. 1776–1777, 2013.

114. S. Liu, J. Z. Niu, H. J. Ge, et al., "Clinical observation of the hemostatic effect of gelatin fiber net," Journal of Chinese Trauma, vol. 16, no. 3, p. 181, 2000.

115. D. A. Das, K. A. Grimmer, A. L. Sparnon, S. E. McRae, and B. H. Thomas, "The efficacy of playing a virtual game in modulating pain for children with a cuteburn injuries: a randomized controlled trial,"BMC Pediatrics, vol. 5, article 1, 2005.

116. S. C. Sharma, M. M. Bagree, A. L. Bhat, B. B. Banga, and M. P. Singh, "Amniotic membrane is an effective burn dressing material," The Japanese Journal of Surgery, vol. 15, no. 2, pp. 140–143, 1985.

117. G. D. Mogoşanu and A. M. Grumezescu, "Natural and synthetic polymers for wounds and burns dressing," International Journal of Pharmaceutics, vol. 463, no. 2, pp. 127–136, 2014.

118. E. G. Kee, R. M. Kimble, L. Cuttle, and K. Stockton, "Comparison of three different dressings for partial thickness burns in children: study protocol for a randomised controlled trial," Trials, vol. 14, article 403, 8 pages, 2013.

119. J. Venugopal and S. Ramakrishna, "Applications of polymer nanofibers in biomedicine and biotechnology," Applied Biochemistry and Biotechnology, vol. 125, no. 3, pp. 147–157, 2005.

120. L. Y. Kossovich, Y. Salkovskiy, and I. V. Kirillova, "Electrospun chitosan nanofiber materials as burn dressing," in Proceedings of the 6th World Congress of Biomechanics (WCB '10), C. T. Lim and J. C. H. Goh, Eds., vol. 31 of IFMBE Proceedings, pp. 1212–1214, Singapore, August 2010.

121. O. Chaisiri, N. Chanunpanich, and B. S. Hanpanich, "Polycarpolactone fibers gellation with gelatin ground substance: Engineered skin extracellular matrix aims for using as tissue engineering skin," in The 15th International Conference on Biomedical Engineering: ICBME 2013, 4th to 7th December 2013,

Singapore, vol. 43 of IFMBE Proceedings, pp. 267–270, Springer, Berlin, Germany, 2014.

122. W. Hu, Z.-M. Huang, S.-Y. Meng, and C.-L. He, "Fabrication and characterization of chitosan coated braided PLLA wire using aligned electrospun fibers," Journal of Materials Science: Materials in Medicine, vol. 20, no. 11, pp. 2275–2284, 2009.

123. N. Alhusein, I. S. Blagbrough, M. L. Beeton, A. Bolhuis, and P. A. D. Bank, "Electrospun zein/PCL fibrous matrices release tetracycline in a controlled manner, killing Staphylococcus aureus both in biofilms and ex vivo on pig skin, and are compatible with human skin cells," Pharmaceutical Research, vol. 33, no. 1, pp. 237–246, 2016.

124. J.-B. Li, J. Han, and J. Ren, "Interaction of human fibroblasts with electrospun composites gelatin/PLLA, chitosan/PLLA and PLLA fibrous scaffolds," Journal of Shanghai Jiaotong University (Science), vol. 17, no. 5, pp. 559–566, 2012.

125. J. C. Dumville, M. O. Soares, S. O'meara, and N. Cullum, "Systematic review and mixed treatment comparison: dressings to heal diabetic Foot ulcers," Diabetologia, vol. 55, no. 7, pp. 1902–1910, 2012.

126. X. Wang, D. Zhang, and F. Shi, "Wound dressing useful for treating venous stasis ulcers and etc," China, WO2012092812-A1 CN102580135-A.

127. J. L. Lázaro-Martínez, A. Cecilia-Matilla, J. Aragón-Sánchez, E. García-Morales, Y. Garcia-Alvarez, and F. Alvaro-Afonso, "Treatment of infected diabetic foot ulcers clinical effectiveness of a dressing of alginate and hydrocolloid, with silver fiber. Analysis of results of a series of cases," Revista de Enfermería, vol. 36, no. 11, pp. 29–34, 2013.

7
CHAPTER

Fabrication of Polyimide Porous Nanostructures for Low-k Materials

Takayuki Ishizaka[1] and Hitoshi Kasai[1]

[1] *Research Center for Compact System, AIST (National Institute of Advanced Industrial Science and Technology), Sendai, Japan*

[2] *Institute of Multidisciplinary Research for Advanced Materials, Tohoku University, Sendai, Japan*

1. INTRODUCTION

The processing speeds of microchips continue to increase as integration densities increase in the semiconductor industry. When the device size is decreased, the so-called resistance–capacitance (RC) delay and the crosstalk noise between metal interconnects offset any gain in chip performance. To avoid these undesirable phenomena, lower dielectric insulating layers must be employed. In the near future, it will be need to develop dielectric materials with ultralow dielectric constants (ultralow-k: $k < 2.0$), and a replacement dielectric for carbon-doped silicon dioxide (SiOC) will be required [1]. In addition, high-thermal stability, good adhesion to metals and chemical stability are also important requirements for the interlayer dielectrics. Polyimides (PIs) are among the most promising candidates for use as next-generation interlayer dielectrics because of satisfying all above-mentioned requirements. Indeed, PIs have been widely employed in the fields of microelectronics applications such as substrate of flexible printed circuitry boards, insulating layers in multilevel very-large-scale integrated (VLSI) circuits and buffer coatings in electronic packages. However, the dielectric constant (k) is about 2.4~3.0 even in fluorinated PIs [2-7], which are insufficient for the requirement of ultralow-*k* materials (*k* < 2.0).

According to the Maxwell-Garnett model [8], it is well known that porous structures in PI films could substantially reduce dielectric constants, because the dielectric constant of air is unity. Therefore, many studies on porous PI films have been reported. The porous PI films have been generally prepared by pyrolysis of thermally labile polymer units in phase-separated PI composite films [9-12]. Namely, so-called nanofoams were produced in a PI film, resulting in providing a low dielectric constant (*k* = 2.56). Their pore sizes are also easily controlled by a component ratio of the copolymer. On the other hand, we have proposed the deposition of PI nanoparticles (NPs) onto a substrate as a novel alternative approach toward the preparation of low-*k* (*k* < 2.5) PI films, that is, by introducing air voids between the PI NPs (Fig. 1a). To obtain ultralow-*k* PI films, we have focused on incorporating pores into the individual PI NPs and the subsequent assembly of multilayered films of porous PI NPs (Fig. 1b). This strategy must be a relatively simple and effective means for introducing air voids uniformly into ultralow-*k* PI films.

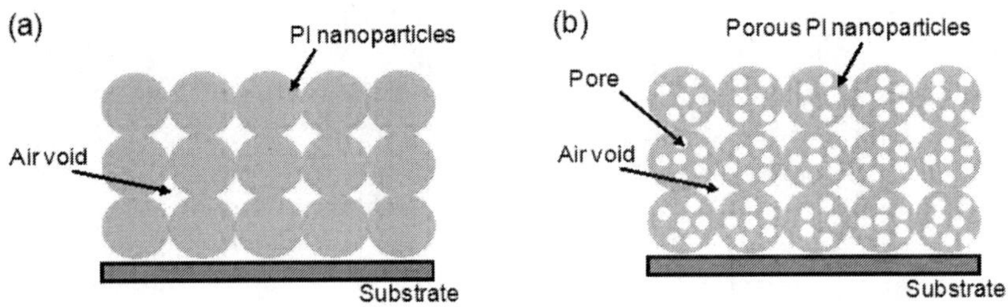

Figure 1. Schematic illustration of our strategy to reduce the dielectric constant (*k*).

In this chapter, we represent fabrication of PI NPs and morphological controlled PI NPs, *i.e.*, variously-sized NPs [13], soccer-ball-like NPs [14], cage-like microparticles [14], golf-ball-like NPs [15-17] and hollow NPs [18], using our technique, the *reprecipitation method* [19]. Furthermore, fabrication of multilayered films of porous PI NPs and their dielectric property are described [18,20].

2. POLYIMIDE NANOPARTICLES

Polymer fine particles are very useful for chromatographic packings; functional coatings; inks and toners; additives for foods, medicines and cosmetics; supports for catalysts; drug delivery systems; optics; and so forth. Polymer fine particles are usually manufactured during polymer synthesis from a monomer, for example, in suspension, emulsion and dispersion polymerizations. The preparations of fine particles of vinyl polymers, *e.g.* polystyrene (PS) and poly(methyl methacrylate) (PMMA), have been well-established by above-mentioned methods, and circumstantially described elsewhere. However, there are a few reports on fine particles of non-vinyl polymers including PI. Lin *et al.* [21] performed the fabrication of submicrometer PI spheres by a cooling a *N*-methyl-2-pyrrolidone (NMP) solution of PI (not PAA), which were given in high temperatures (90°C). Chai *et al.* [22] and Xiong *et al.* [23] obtained PI spheres by the dropwise addition of the precipitant (water or ethanol) into PI solutions. These methods are useful for only soluble PIs because PIs are generally insoluble in common solvents. On the other hand, precursor polymer of PI, poly(amic acid) (PAA), which is generally synthesized by polymerization of tetracarboxylic acid dianhydride and diamine, dissolves in some organic solvents such as NMP, and is easily converted into PI by the imidization treatment, namely, heat treatment or addition of cyclodehydration reagents (Fig. 2). Therefore, PI particles are often fabricated from PAA solutions or PAA particles. Nagata *et al.* [24], Asao *et al.* [25] and Basset *et al.* [26] have reported on poor-solubility PI microparticles prepared by the precipitation polymerization method, *i.e.*, the thermal imidization of PAA dissolved in NMP, which acts as a good solvent for PAA but a poor solvent for PI. Okamura *et al.* [27] precipitated uniform PAA particles of submicron scale by the precipitation polymerization between pyromeritic acid dianhydride and 4,4'-diaminodiphenyl ether in acetone, followed by imidizing these PAA particles. Recently, a few studies on fabrication of polymeric particles composed of PI and other materials (*e.g.* PS, silica) have been also reported [28-31]. On the other hand, we have reported on the reprecipitation method, which is a simple and facile fabrication technique for polymer particles and organic nanocrystals [19,32-34]. We have successfully fabricated PI NPs via the reprecipitation method [13-18]. Furthermore, novel composite PI NPs have been also fabricated by this method [35,36].

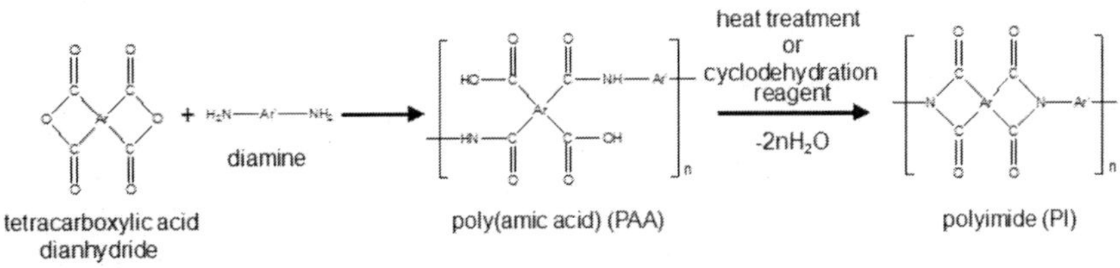

Figure 2. Synthesis route to polyimide.

2.1. Fabrication of Polyimide Nanoparticles Via the Reprecipitation Method

We have successfully fabricated NPs of various PIs shown in Fig. 3. The fabrication procedure is described as follows. Because PIs are generally insoluble in common organic solvents, we used PAA solutions as starting materials for fabrication of PI NPs. Figure 4 shows the schematic representation of PI NPs via the reprecipitation method. PAA NPs were firstly fabricated by the reprecipitation method and then were converted to PI NPs through the "two-steps imidization", *i.e.* chemical imidization of PAA NPs in the dispersion medium with cyclodehydration reagent, followed by thermal imidization. In a typical experiment, a dispersion liquid of PAA NPs was obtained by injecting NMP solution of PAA (100 μl) into a vigorously stirred poor solvent (10 ml, containing 0.1 wt% of dispersing agent, Acrydic A-1381 (Dainippon ink and chemicals, Inc.)). In the present case, carbon disulfide (CS_2), cyclohexane and their mixture were used as the poor solvent. Chemical imidization of PAA NPs was performed by adding 100 μl of pyridine/acetic anhydride mixture (1:1) into the PAA NPs dispersion liquid and stirring for 2 h. After the chemical imidization, the PI NPs were centrifuged and dried *in vacuo*. Finally, PI NPs were cured at 270°C for 1 h.

PAA NPs and PI NPs obtained through the two-steps imidization were all spherical shape as shown inFig. 5. Moreover, DLS measurement showed that the mean particle sizes and their distributions almost consisted before and after the chemical imidization [13]. These results indicate that no aggregation occurred during the imidization and we can control size of PI NPs by size-controlling PAA NPs.

A conversion to PI was estimated from IR spectra shown in Fig. 6. After the two-steps imidization (c), the absorption band around 1550 cm^{-1} (amide II: CNH vibration) and 1690 cm^{-1} (Amide I: C=O stretching vibration) disappeared, while the absorption band of imide IV (bending vibration of cyclic C=O), imide II (C-N stretching vibration), and imide I (stretching vibration of cyclic C=O) newly appeared at 720 cm^{-1}, 1380 cm^{-1} and 1720 cm^{-1}, respectively. A conversion to PI was determined by the following equation [25,37].

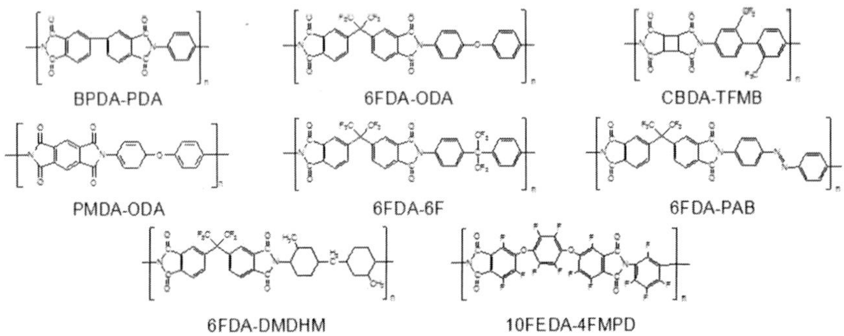

Figure 3. Structure of polyimides used in our study.

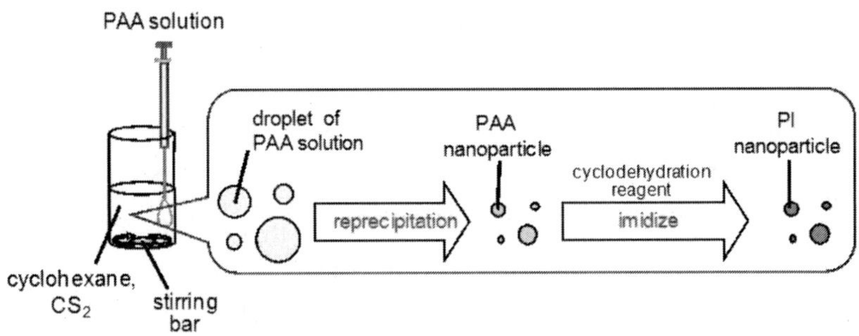

Figure 4. Schematic representation of PI nanoparticles via the reprecipitation method.

$$Conv.(\%) = \frac{\left(D_{1380\mathrm{cm}^{-1}} / D_{1500\mathrm{cm}^{-1}}\right)_{\mathrm{NPs}}}{\left(D_{1380\mathrm{cm}^{-1}} / D_{1500\mathrm{cm}^{-1}}\right)_{\mathrm{Bulk}}} \times 100, \quad (1)$$

where D is the optical density of each absorption bands. The absorption band at 1500 cm^{-1} (C-C stretching vibration of *p*-substituted benzene) was selected as an internal standard because its intensity remains unchanged before and after imidization. A completely-imidized sample, bulk PI, was prepared by curing a PAA cast film at 200°C for 1 h and successively at 350°C for 1 h in nitrogen atmosphere. According to the above equation, the conversion of the PI NPs was 73 % after the chemical imidization, and then the conversion of almost 100 % was found to be achieved by the subsequent curing at 270°C for 1 h. These results indicate that the imidization have been carried out quantitatively without changing their morphology.

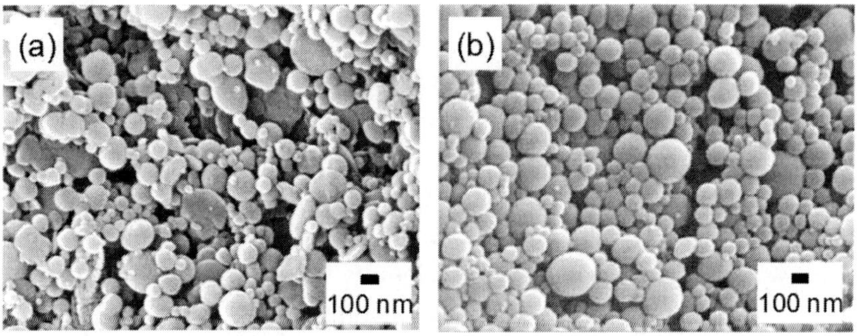

Figure 5. SEM images of 6FDA-ODA-type (a) PAA and (b) PI nanoparticles [13].

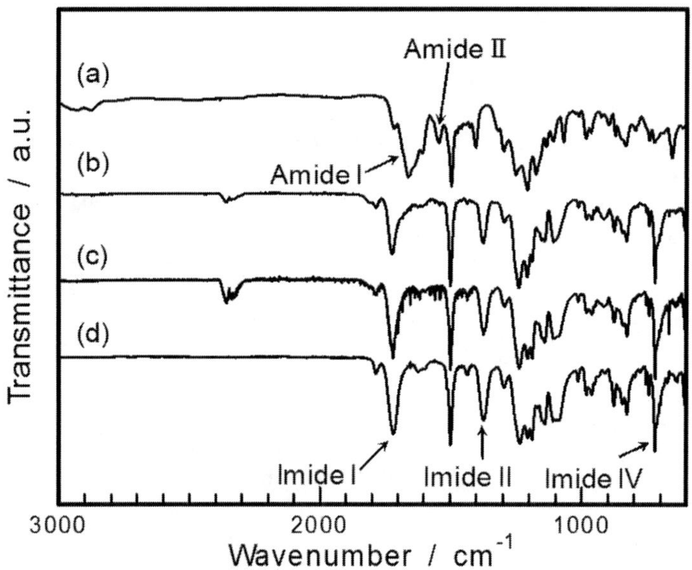

Figure 6. IR specta of (a) PAA nanoparticles, (b) nanoparticles after chemical imidization, (c) nanoparticles after two-steps imidization and (d) bulk polyimide.

2.2. Size-Control of Polyimide Nanoparticles

As mentioned above, size-control of PI NPs will be performed by controlling size of PAA NPs. Size of PAA NPs can be decided by reprecipitation conditions such as concentration of a PAA solution (C), temperature of a poor solvent (T_{poor}) and species of a poor solvent (*e.g.* CS_2 and cyclohexane). Relationships between sizes of final PI NPs and reprecipitation conditions are shown in Fig. 7. As shown in Fig. 7a, a mean particle size increased with increase in C. This tendency is same to that in other organic compound systems previously reported [33]. To understand this phenomenon, we have to consider on the fabrication mechanism of PAA NPs using the reprecipitation method. In the reprecipitation process, first, fine droplets are dispersed in cyclohexane after injection of PAA solution. Next, NMP is removed from droplets by rapid dissolution into cyclohexane, and then PAA NPs form because PAA is insoluble in cyclohexane-rich solvent. When the concentration of the injected solution is higher, higher-concentrated droplets are produced without changing the droplet size, providing lager-sized NPs. On the other hand, a mean particle size decreased with increase in T_{poor} up to 20°C, and same size was given above 20°C (Fig. 7b). This result is explained as follows. NMP diffuses from droplets into cyclohexane slowly because of its high viscosity and poor miscibility with cyclohexane. Hence, droplets grow by agglutination between them before completion of the reprecipitation process, resulting in lager-sized particles are obtained. Under the higher temperature condition, however, the reprecipitation process is performed in a shorter time because the diffusion rate becomes drastically higher. Thus, decrease in the particle size caused by a shorter time period for growth of droplets at higher poor solvent temperature. In addition, a mean particle size also decreased with increasing a volume rate (Φ_{CS2}) of CS_2 in a cyclohexane/CS_2 mixture (Fig. 7c). PAA NPs with a size of *ca.* 50 nm were obtained under the condition using pure CS_2 as a poor solvent. This result is explained by considering miscibility between NMP and poor solvents (cyclohexane and CS_2). A difference in solubility parameters between two kinds of solvents is often used for evaluating their miscibility, and the smaller difference means higher miscibility. The solubility parameters (δ) of NMP, CS_2, and cyclohexane were calculated by the Hansen method [38] and the values of 22.9, 20.5 and 16.8 $MPa^{1/2}$were obtained, respectively. The differences in solubility parameters indicate that NMP/CS_2 system is higher miscibility than NMP/cyclohexane system. In the case of use of CS_2/cyclohexane mixture as a poor solvent (Fig. 7c), increase in Φ_{CS2} suggests improvement of the miscibility between NMP and poor solvents due to increase in high-miscible CS_2 for NMP. The improvement results in same effect to the temperature dependence, *i.e.*, a shorter time period for droplets growth and a decrease in a size of PAA NP.

Thus, we could successfully obtain size-controlled PI NPs via the two-step imidization process. Figure 8 shows SEM images of PI NPs, which were size-controlled in the range of 20 ~ 500 nm by fabricating under various reprecipitation conditions. All PI NPs were almost spherical, free of aggregations, and quantitatively imidized.

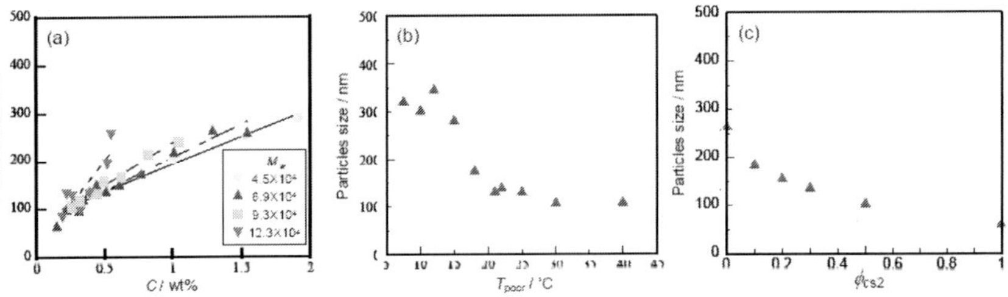

Figure 7. Relationships between size of PI nanoparticles and reprecipitation conditions [13]. Dependence of (a) concentration of PAA solution under T_{poor} = 22°C, Φ_{CS2}=0; (b) temperature of poor solvent under C=0.5 wt%, Φ_{CS2}=0, Mw=69,000; (c) volume fraction of CS_2 in poor solvent (mixture of $CS_2$and cyclohexane) under C=1.5 wt% and T_{poor}=22°C, Mw=69,000.

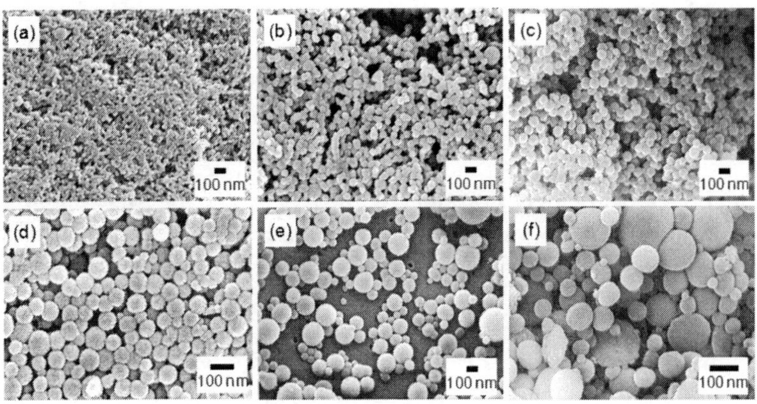

Figure 8. SEM images of 6FDA-ODA PI nanoparticles fabricated under various conditions [13]. Mean particles size: (a) 20 nm, (b) 50 nm, (c) 80 nm, (d) 100 nm, (e) 250 nm, (f) 500 nm.

3. POROUS POLYIMIDE NANOPARTICLES

Porous NPs are promising materials for low dielectric fillers, adsorbents, drug delivery systems, catalyst carriers and optics. Some studies on fabrication of porous polymer particles have been reported [39-44]. Okubo *et al.* [39] have reported fabrication of micron-sized PMMA/PS composite particles having 50~150 nm-sized dents on their surface. They achieved preparation of these particles by the seeded dispersion polymerization of styrene with PMMA seed particles in a methanol/water medium in the presence of decalin droplets. Griffiths *et al.* [40] have reported porous PS particles fabricated by the surfactant-free polymerization of styrene under the presence of water-soluble natural polymers. Kao *et al.* [41] fabricated porous poly(styrene-divinylbenzen) (PS-DVB) micoparticles by the method of multistep swelling and polymerization involving the use of polymeric porogens as follows. Monodispersed polystyrene seed particles were first prepared by the dispersion polymerization. These seeds were swollen with styrene monomer and then polymerized. Subsequently, styrene and DVB were both absorbed into the particles obtained in the previous step, and then copolymerized within the enlarged particles. In the final step, polystyrene in the particles was eluted by toluene. Unsal *et al.* [42] have also reported poly(glycidyl methacrylate-ethylene dimethacrylate) porous particles prepared by the method of swelling and seeded polymerization using PS seed particles. Koushik *et al.* [43] obtained porous deslorelin-poly(lactide-co-glycolide) microparticles by a supercritical CO_2 treatment of these sphere particles. Albrecht *et al.* [44] have reported microporous particles of poly(ether imide) prepared by a spraying/coagulation process as follows. A NMP solution of the polymer was sprayed through a hollow needle to form droplets. The droplets fell through an air gap in a water coagulation bath, which induced a phase inversion and initiated a fixation of a particle shape, and finally, the microparticles owning micoropores inside were produced. As above-described, most of porous particles sized in the micrometer range, and few studies had been reported on porous polyimide particles [24,29]. Recently, we have successfully fabricated various types of porous polyimide NPs using the reprecipitation method [14-18].

3.1. Fabrication Procedure of Porous Pi Nanoparticles Via The Reprecipitation Method

We have focused on incorporating pores into every single PI NPs by using microphase separation of porogen within the particles, and found that LiCl [14], poly(acrylic acid) (PA) [15,16], poly(sodium 4-styrenesulfonate) (PSS)

[17], poly(methyl methacrylate) (PMMA) [18] and polyvinylpyrrolidone (PVP) [18], act as useful porogens for fabrication of porous PI NPs. The reprecipitation method was also employed in order to obtain porous PI NPs as follows. A NMP solution of PAA was prepared, and NMP solution of porogen was added subsequently. And then, PAA NPs were fabricated by injecting the polymer blend solution into cyclohexane, and two-step-imidized subsequently.

3.2. Cage-Like Polyimide Microparticles

Interestingly, as shown in Fig. 9a and b, we could obtain some doughnut-like and hollow polyimide NPs without porogens under specific reprecipitation conditions (use of high concentrated PAA solution (e.g. 1.5 wt%) and low temperature cyclohexane (e.g. 7°C)), under which particles are mainly fabricated into the sizes larger than 1000 nm. Furthermore, we sometimes produced intriguing and uniquely shaped objects under otherwise identical conditions. For example, Figure 9c displays 300-nm-diameter polymeric soccer ball-shaped particles that exhibit both pentagonal and hexagonal facets. The line widths of the facet-frames, which are composed of polymer, were ca. 50 nm. We are unaware of any such organic and/or polymeric nanomaterials having been reported previously. We suspect that these strangely shaped NPs were fabricated through phase separation of the polymer in droplets of the solution, as indicated in Fig. 10. First, immediately after injecting the solution, the droplets were produced in cyclohexane (Fig. 10a). As N-methyl-2-pyrrolidinone (NMP) molecules in the droplets gradually dissolved in cyclohexane, the droplet size reduced. As the polymer began to precipitate at the surface of the droplet, capsule structures having sizes greater than 1 μm were formed (Fig. 10b). At this point, if all of the NMP molecules completely diffused from these capsule structures, then hollow structured nanomaterials, such as those displayed in Fig. 10c, were obtained. If, however, some of the NMP molecules remained at the surface layer, so that the polymer molecules had the flexibility to move, shrinkage of the polymer structure in Fig.10b would lead to particles having their most stable structure as that presented in Fig. 10d. Unfortunately, further research will be necessary to improve the reproducibility of our results; indeed, we successfully obtained soccer ball-shaped particles such as those in Fig. 10d less than 5% of the time. Nevertheless, we believe that the ability to prepare these new types of nanosized polymer particles should be of interest to many researchers in this field.

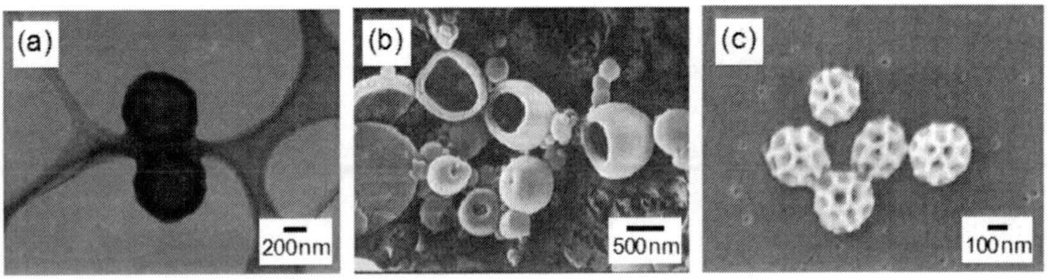

Figure 9. a) TEM and (b), (c) SEM images of PI nanostructures fabricated by using high concentrated PAA solution and low temperature cyclohexane [14].

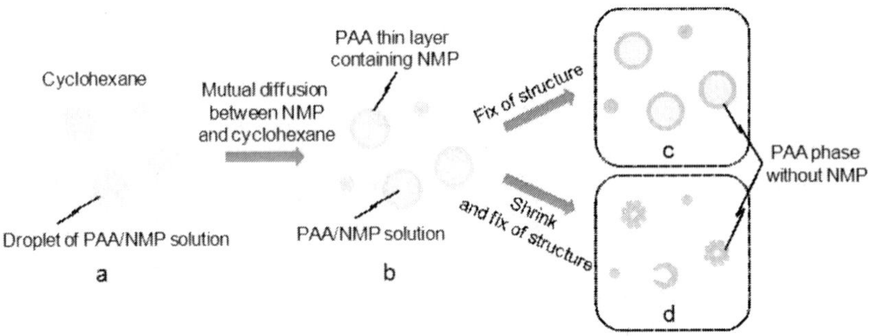

Figure 10. Schematic illustration of the process of fabricating the polyimide nanostructures.

Here, we fine-tuned the reprecipitation method to improve the reproducibility of forming porous polyimide nanostructured materials by using LiCl as a porogen. In short, a solution of poly(amic acid) (PAA) including LiCl was rapidly injected into vigorously stirred cyclohexane at room temperature.Figure 11 displays SEM images of the porous PI particles obtained after imidization and removal of the salt through washing with water. Increasing the weight ratio of LiCl to PAA in the injected solution led to an increase in the number of holes in the porous particles. The diameters of the holes on the surfaces of the particles were almost all ca. 100 nm. Figure 12 presents TEM images of the porous PI particles. We observed that spherical holes existed within the particles, which could be divided roughly into two types of porous materials. When the particle size was 700 nm or more, we observed hollow-type particles, each having a single huge hole in their center (Fig. 12a); for particle sizes below 700 nm, we observed a number of holes of almost equal size within the particles (Fig. 12b).

We have suggested a mechanism for generating these porous particles as follows. Initially, NMP droplets containing PAA and LiCl were generated in the poor solvent immediately after injection.

Next, reprecipitation of the PAA began as the NMP molecules gradually dissolved in the cyclohexane. At this stage, porous particles featuring some holes were formed in the droplets through phase separation of the polymer units. At the same time, LiCl and NMP species became concentrated within the holes. And then, NMP and LiCl eluted perfectly from the polymer particle and into the cyclohexane. Indeed, we did not detect any LiCl in the final products. Finally, the porous PI particles were then obtained through subsequent chemical imidization.

Figure 11. SEM images of porous PI particles fabricated using LiCl as a porogen [14]. Content of LiCl (relative to PAA) in the NMP solution: (a) 5, (b) 10, (c) 20 wt%.

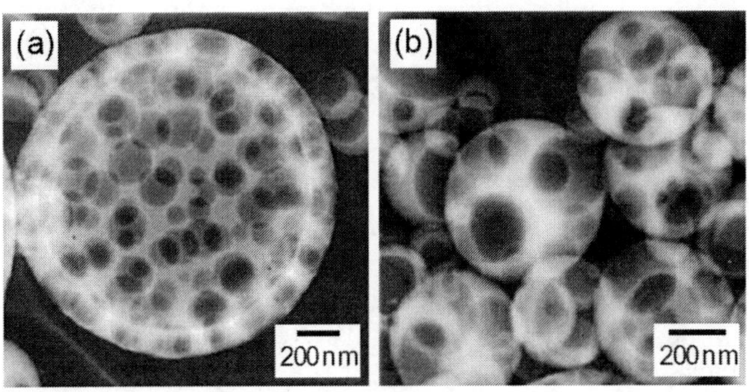

Figure 12. TEM images of the porous PI particles fabricated using LiCl as a porogen [14]. (a) Large (>700 nm) hollow-type porous particles containing a single huge hole; (b) small (<700 nm) porous particles containing multiple and equally sized holes.

3.3. Golf-Ball-Like Polyimide Nanoparticles

To fabricate porous PI NPs, the first polymer evaluated as a porogen was poly(vinyl alcohol) (PVA,*Mw*=500). Although both PVA and PAA (6FDA-ODA, *Mw*=69,000) dissolved in NMP, only some of the resulting PI NPs possessed nanopores (Fig. 13a and b). A fraction of porous PI NPs was not, however, affected by a content of PVA added. The second polymer evaluated as a porogen was poly(acrylic acid) (PA, *Mw*=2,000). Porous PI NPs were successfully fabricated as indicated inFig.13c–e. The diameters of pores were in the range of 20~70 nm and the surface morphology remained almost constant when the PA content was greater than 40 wt %. Thus, PA acted as a progen for fabrication of porous PI NPs, and we have also obtained those consisted of other types of PI (10FEDA-4FMPD, BPDA-PDA and CBDA-TFMB) (Fig. 13f–h)

To consider the mechanism of formation for the porous structure, we observed SEM images of the porous PAA and/or PI NPs at each stages of their fabrication (Fig. 14). It is apparent that nanopores were already formed on the surface of the PAA NPs prior to their imidization. In addition, the surface morphology was similar before and after performing the two-step imidization process. These findings have led us to propose a plausible mechanism of formation for the porous NPs (Fig. 15). At first, fine droplets consisting of mixture of NMP, PAA, and PA were generated in cyclohexane immediately after injection. According to the solubility parameters, cyclohexane is a much poorer solvent for PA than for PAA. As the cyclohexane and NMP diffused into one another, PAA and PA began to precipitate and microphase-separate in the surface layer of the fine droplets. At this stage, the content of cyclohexane increased gradually within each fine droplet, with NMP remaining as the main component, but with a reduced solvent power. Thus, PA-rich microdomains, which consist mainly of PA and NMP, formed within the fine droplets, because PA is more miscible with NMP. Because these internal PA-rich microdomains did not possess such a clear interface, they might have then diffused toward the surface layer. Finally, the discontinuous and isolated PA-rich microdomains that laid in the surface layer were eliminated, leading to the formation of micropores. Undoubtedly, the microphase separation giving these porous surface nanostructure is a much more complicated process than this mechanism described above; it must also depend on the molar mass of PA, the interfacial tension, the viscosity of the solvents, the mutual diffusion, and the compatibility between PAA and the porogen. With regard to this mechanism, the selection of the porogen is very important aspect of the successful formation of the porous NPs. A suitable porogen must be compatible with PAA to some extent; a suitable choice can be determined qualitatively by comparing the difference between the solubility parameters (δ) of PAA and the porogen [45]. The solubility parameter of PA ($\delta_{PA=}$ 23.6 MPa$^{1/2}$) is closer to that of PAA ($\delta_{6FDA\text{-}ODA=}$20.8 MPa$^{1/2}$) than it is to that of PVA (δ_{PA}=30.5 MPa$^{1/2}$), which implies that PA is more compatible with PAA. Therefore, we obtained better results when using PA as the porogen.

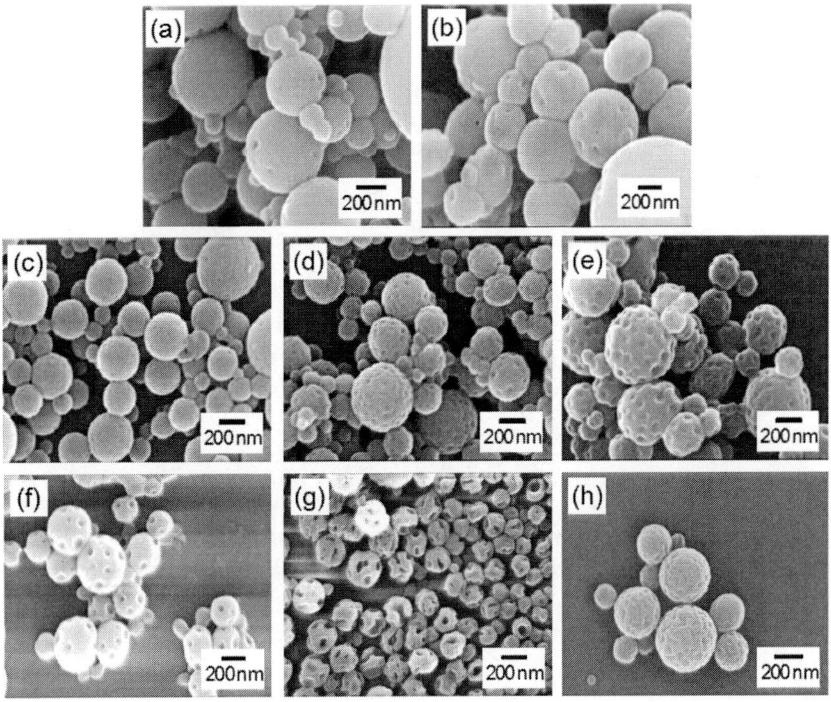

Figure 13. SEM images of porous PI nanoparticles fabricated by using various blend polymer solutions [15,16]. (a) PVA (20 wt% to PAA)/6FDA-ODA, (b) PVA (50 wt%)/6FDA-ODA, (c) PA (20 wt%)/6FDA-ODA, (d) PA (40 wt%)/6FDA-ODA, (e) PA (60 wt%)/6FDA-ODA, (f) PA (40 wt%)/10FEDA-4FMPD, (g) PA (20 wt%)/BPDA-PDA, (h) PA (40 wt%)/CBDA-TFMB.

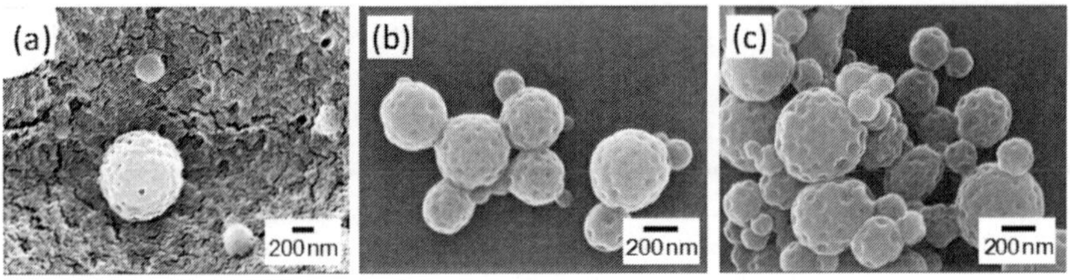

Figure 14. SEM images of porous PAA and/or PI nanoparticles at each stage of fabrication [15]. (a) immediately after injection, (b) after chemical imidization, (c) after thermal imidization.

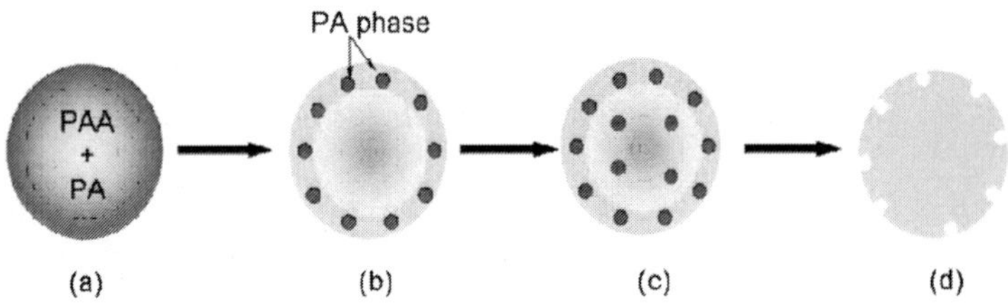

Figure 15. Schematic representation of the mechanism of formation of porous nanoparticles [15]. (a) A fine droplet of NMP, PAA, and PA formed immediately after injection; (b, c) possible intermediate states, in which the yellow regions represent zones in which NMP and cyclohexane exchange through mutual diffusion processes; (d) the resulting porous PAA NP.

It would be necessary to eliminate residual PA from the porous PI NPs if they were to be used in device applications. Figure 16 displays the results of thermogravimetric analysis (TGA) measurements. TGA curve i is that of ordinary PI NPs that lack porous structures; they decomposed thermally at a temperature above 550°C. On the other hand, PA began to decompose at ca. 200°C (TGA curve iv). The TGA curve ii is that of the chemically imidized porous PI NPs, which also decomposed gradually at ca. 200°C as a result of the removal of its residual PA and solvents. The thermally imidized porous PI NPs (TGA curve iii) yielded their 5% weight-loss temperature at 400°C; they were thermally stable up to 300°C, almost identical to the behavior of the ordinary PI NPs (TGA curve i). This result suggests that almost no PA remained within the thermally imidized porous PI NPs. Similar conclusions were drawn from the IR spectra of the various NPs. The spectrum of the thermally imidized porous PI NPs was virtually identical to that of the nonporous PI NPs. This finding suggests that PA was almost completely eliminated from the porous PI NPs during thermal imidization, consistent with the results of TGA.

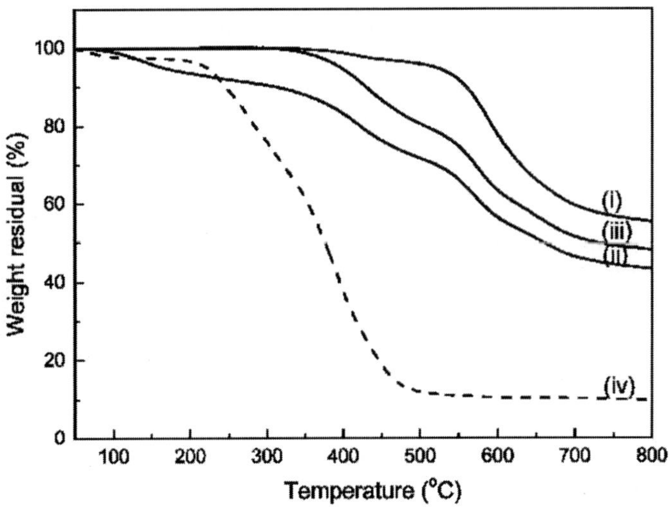

Figure 16. TGA (weight loss) curves recorded under N_2 [15]: (i) PI nanoparticles, (ii) chemically imidized porous PI nanoparticles, (iii) thermally imidized porous PI nanoparticles, (iv) PA powder. PA (40 wt %) was added to samples (ii) and (iii).

3.4. Hollow Polyimide Nanoparticles

By using other porogen (poly(methyl methacrylate) (PMMA) or polyvinylpyrrolidone (PVP)), hollow/multi-hollow PI NPs could be fabricated via the reprecipitation method. Figure 17 reveals that multi-hollow PI NPs were formed when we used PMMA or PVP as a porogen. Only a few small hollow cores (diameters: < 20 nm) appeared inside the PI NPs after thermal imidization when the PMMA content was 20 wt% (Fig. 17a). Upon increasing the PMMA content, larger hollow cores (diameters: ca. 100 nm), surrounded by several smaller cores, were generated; *i.e.*, the volume of the hollow structures increased accordingly (Fig. 17a–c). In contrast, we obtained only larger hollow pores ranging in a diameter from 100 to 200 nm when using PVP as a porogen (Fig. 17d, e). The number of hollow cores increased upon increasing the PVP content up to 40 wt%, but, due to high viscosity, agglomeration of the PAA NPs occurred at higher PVP contents. Clearly, the choice of a suitable porogen is an important feature affecting the generation of hollow structures. Suitable porogens can be determined qualitatively by comparing their solubility parameter with that of PAA. The solubility parameter of PAA (6FDA-ODA) ($\delta_{6FDA\text{-}ODA}$ = 20.8 MPa$^{1/2}$) is closer to that of PMMA (δ_{PMMA} = 19.7 MPa$^{1/2}$) than that of PA (δ_{PA} = 23.6 MPa$^{1/2}$), which acts as the porogen for fabrication of golf-ball-like particles (see section 3.3), indicating that PMMA is more compatible with PAA. Such a relatively high compatibility makes the PMMA phase penetrate deeper and remain within the PAA NPs, resulting in the formation of multi-hollow structures after the removal of the PMMA phase, rather than merely superficial pores. When we employed PVP (δ_{PVP} = 19.9) as an even more compatible porogen for PAA (6FDA-ODA), the PVP phases penetrated further to generate several large cores within the individual PI NPs. These findings indicate quite dramatically that the compatibility of the porogen and the PAA influences the number of hollow cores within the resulting PI NPs.

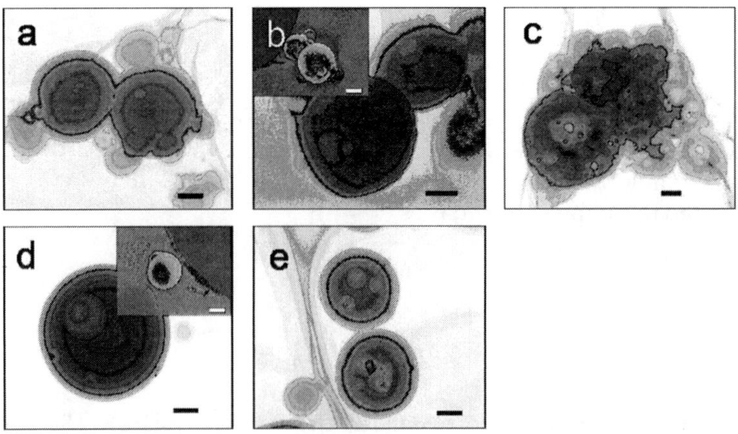

Figure 17. TEM images (scale bars: 100 nm) of multi-hollow PI(6FDA-ODA) NPs prepared using (a) PMMA 20 wt %, (b) PMMA 40 wt %, (c) PMMA 80 wt %, (d) PVP 20 wt %, and (e) PVP 40 wt % [18]. The insets display corresponding SEM images (scale bars: 200 nm).

The selection of a suitable porogen for a specific PAA allows the formation of single hollow core within each PI NP. Our approach is also applicable to other PAAs—as long as they can dissolve in the solvent. The second PI that we evaluated was PI(CBDA-TFMB) ($\delta_{CBDA\text{-}TFMB}$ = 20.0). Interestingly, the TEM images in Fig. 18 reveal only one type of hollow morphology formed—namely, a single hollow core—when using either PMMA or PVP as the porogen. Because these porogens are more compatible with PAA(CBDA-TFMB) than with PAA(6FDA-ODA), the porogen phase penetrated deeper into the center of each PI NP and integrated into a single domain, which ultimately formed the single hollow core. Moreover, the volume fraction of the hollow cores increased upon increasing the content of added PVP/PMMA. At 80 wt%, we prepared hollow PI NPs possessing core diameters in the range of 200~400 nm—the highest degree of porosity obtained in this present study.

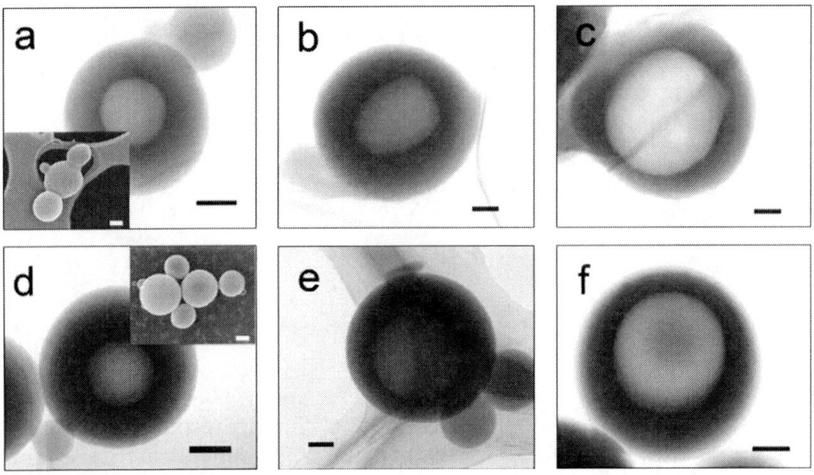

Figure 18. TEM images (scale bars: 100 nm) of hollow PI(CBDA-TFMB) NPs prepared using PMMA or PVP as the porogen at contents of (a) PMMA 20 wt %, (b) PMMA 40 wt %, (c) PMMA 80 wt %, (d) PVP 20 wt %, (e) PVP 40 wt %, and (f) PVP 80 wt % [18]. The insets display corresponding SEM images (scale bars: 200 nm).

Table 1. Solubility Parameters calculated using the Hoftyzer and van Krevelen method [46].

polymer	Solubility parameter ($MPa^{1/2}$)
PAA(6FDA-ODA)	20.8
PAA(CBDA-TFMB)	20.0
poly(methyl methacrylate) (PMMA)	19.7
poly(acrylic acid) (PA)	23.6
polyvinylpyrrolidone (PVP)	19.9

According to TGA and IR spectra of the resulting hollow PI NPs, little PVP remained within the thermally imidized hollow PI NPs, *i.e.*, the hollow PI NPs formed through thermal decomposition of the microphase-separated porogen within the composite PI/porogen NPs. This result is caused by high pyrolysis temperature of PVP (drastic weight loss from 400 to 470°C). On the other hand, PMMA was dramatically decomposed from 350 to 420°C. Therefore, no PMMA remained within the hollow PI NPs derived from the composite PI/PMMA NPs due to its lower pyrolysis temperature compared with that of PVP.

4. LOW-*K* POROUS POLYIMIDE FILMS

Porous PI films have been generally prepared by pyrolysis of thermally labile polymer units in phase-separated PI composite films, for example, derived from poly(propylene oxide)-, PMMA- and poly(α-methyl-styrene)-PI block copolymer [9,10]; blend solution containing polyurethane and PAA [11]; and polyacrylamide- and PMMA-grafted PI [12]. They have achieved 10~30% decrease in a dielectric constant of polyimide films, and low-*k* porous films ($k \approx 2.2$) have been provided. Also, other methods, such as using water droplets and LiCl crystal as templates, and extracting a porogen polymer from composite PI film by supercritical CO_2, have been known [47-49]. On the other hand, we have obtained ultralow-*k* porous films by depositing porous PI NPs onto substrates [18,20].

4.1. Preparation of Multilayered Films of Polyimide Nanoparticles

We employed the electrophoretic deposition method [50] to assemble the obtained hollow PI NPs into multilayered films. The typical experiment procedure is described as follows. DC voltages were applied to indium tin oxide (ITO) electrodes immersed in a face-to-face arrangement in a suspension of PI NPs. The PI NPs, which were minus-charged in cyclohexane, were electrophoretically deposited on the anodes, and multilayered films were quickly prepared within 1 minute. The films were dried and then cured at 270 °C for 1 h to perform pyrolysis of a porogen. The electrophoretic deposition process was repeated to fill some cracks formed through coalescence of the NPs during evaporation of dispersion medium. And then, PAA solution was spin-coated onto the resulting film. After thermal imidization, a multilayer PI NPs film having a smooth surface was obtained.

Uniform films of PI NPs were easily obtained by this method as shown in Fig. 19. Furthermore, multilayered films with different thickness were prepared by changing dispersion concentration of PI NPs. As shown in Fig. 20a, film thickness increased with increase in the dispersion concentration. On the other hand, applied voltage also influenced the film thickness dramatically (Fig. 20b). Under using the dispersion liquid with solid concentration of 0.2 wt%, films became thicker as the applied voltage increased from 50 to 2000 V/cm. However, no further increase of the film thickness was observed when the applied voltage was higher than 2000 V/cm. After the electrophoretic deposition process under the applied voltage above 2000 V/cm, a transparent liquid left in sample tubes. This result indicates 100% consumption of PI NPs. Thus, the film thickness was controlled in the range of several hundred nm ~ several μm by varing dispersion concentration and applied voltage.

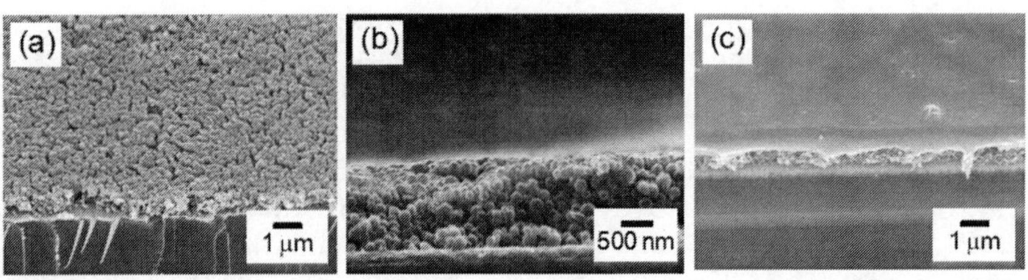

Figure 19. SEM images of multilayered films of PI nanoparticles on ITO substrates. (a) after 1st electrophoretic deposition, (b) cross section, (c) after further spin coat.

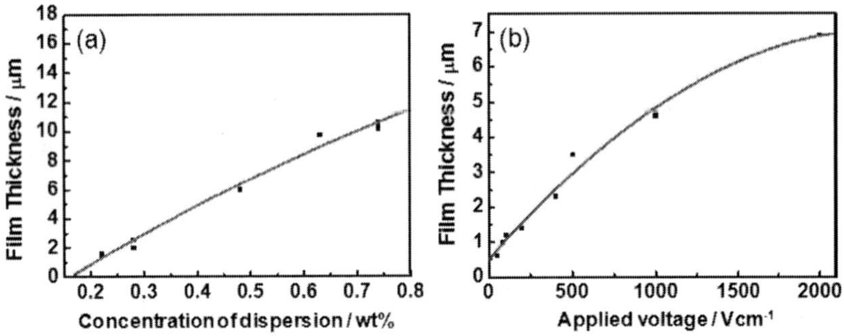

Figure 20. Dependence of (a) concentration of PI nanoparticles dispersion, (b) applied voltage, on thickness of layered film of PI nanoparlticles.

4.2. Dielectric Constant of Multilayered Films Assembled By Various Type of Pi Nanoparticles

We evaluated a dielectric constant of layered films of PI NPs prepared. Capacitance, C_p, of films was measured by a LCR meter in air and at room temperature, and a dielectric constant, k, was calculated by the equation (2).

$$k = \frac{tC_p}{\varepsilon_0 S}, \tag{2}$$

Here, C_p is a capacitance measured, t is a thickness of a film, ε_0 is the electric constant and S is an electrode area. The dielectric constants of various films, *i.e.*, cast films (non particles films), layered films of nonporous NPs and layered films of porous NPs, were summarized in Table 2. The dielectric constants of the cast films consisted of BPDA-PDA, 6FDA-ODA and CBDA-TFMB were 3.17, 2.76 and 2.52, respectively. The dielectric constants of multilayer films of nonporous PI NPs were somewhat lower, resulting from generation of air voids between NPs. The porosity of the porous films was 15~22%, calculated by using the Maxwell-Garnett model, the equation (3) [8].

$$\frac{k_{\text{porous}} - 1}{k_{\text{porous}} + 2} = \left(1 - \frac{P}{100}\right)\frac{k_{\text{solid}} - 1}{k_{\text{solid}} + 2}, \tag{3}$$

where k_{solid} is a dielectric constant of a nonporous film, and P is a porosity of a porous film. It was even lower than that of rhombohedral packing of spherical particles (porosity=26% [51]). Polydispersity of PI NPs may result in a denser packing arrangement. Moreover, the dielectric constants decreased significantly upon increasing the degree of porosity of the PI films (porosity =31~33%), *i.e.*, using hollow PI NPs instead of nonporous PI NPs. Amazingly, the dielectric constant was ca. 1.9 for the multilayer films formed from the hollow PI(CBDA-TFMB) NPs. This dielectric constant is even lower than that of a cast film of poly(tetrafluoroethylene), which has the lowest dielectric constant among polymeric insulators. Thus, the introduction of air voids between and within the PI NPs reduced the dielectric constant significantly. In addition, the electrophoretic deposition appears to be an effective means of generating multilayer ultralow-*k* films from hollow PI NPs.

Table 2. Dielectric constants and porosities of various type of PI films.

PI structures	Film types	Film thickness (μm)	Dielectric constant (*k*) @1MHz	Porosity (%)
BPDA-PDA	cast	2.14	3.17	—
	nonporous NPs	2.62	2.67	15
	golf-ball-like NPs	5.81	2.27	31
6FDA-ODA	cast	1.39	2.76	—
	nonporous NPs	3.18	2.36	17
	hollow NPs	1.90	2.00	32
CBDA-TFMB	cast	1.39	2.52	—
	nonporous NPs	4.40	2.07	22
	hollow NPs	2.50	1.87	33

5. CONCLUSION

We proposed the depositing of porous polyimide nanoparticles (PI NPs) onto a substrate as the novel strategy, that is, introducing air voids within and between the PI NPs, toward the preparation of ultralow-k (k < 2.0) PI films. We presented fabrication of porous PI NPs and porous films, which are assembled by the porous NPs, for low-*k* materials on the basis of this strategy. In particular, Size-controlled NPs were successfully fabricated by using our technique, the *reprecipitation method,*followed by the two-step imidization. Moreover, addition of various progens, such as LiCl, poly(acrylic acid) (PA), poly(methyl methacrylate) (PMMA) and polyvinylpyrrolidone (PVP), to starting solutions of poly(amic acid) (PAA) gave cage-like, golf-ball-like, multi-hollow and hollow NPs, respectively. Also, multilayered films of porous PI NPs were successfully prepared by the electrophoretic deposition method within 1 minute. We obtained a particularly low dielectric constant (k= ca. 1.9) for a multilayer film assembled by one such type of hollow PI NPs. Thus, our strategy is an effective means for preparing ultralow-k PI films featuring unique porous structures, which are alternative candidates for use in dielectric interlayer applications.

ACKNOWLEDGEMENT

Authors would like to thank Dr. M. Suzuki, Mr. H. Mitsui, Dr. G. Zhao, Prof. H. Oikawa, Prof. H. Nakanishi, Tohoku University; Prof. M. Hasegawa, Toho University and Prof. T. Furukawa, Tokyo University of Science for useful assistances and supports in this study.

REFERENCES

1. International Technology Roadmap for Semiconductors (ITRS) 2010 Update, Interconnect. Available:http://www.itrs.net/Links/2010ITRS/ Home 2010. htm.Accessed 2012 Sep 27.

2. M R Anderson, R M Davis, C D Taylor, M Parker, S Clark, D Marciu, M Miller, 2001Thin Polyimide Films Prepared by Ionic Self-Assembly. Langmuir 1783808385

3. A E Eichstadt, T C Ward, M D Bagwell, I V Farr, D L Dunson, J E Mcgrath, 2002Synthesis and Characterization of Amorphous Partially Aliphatic Polyimide Copolymers Based on Bisphenol-A Dianhydride. Macromolecules 3575617568

4. S Ando, T Matsuura, S Sasaki, 1992Perfluorinated Polyimide SynthesisMacromolecules2558585860

5. A E Feiring, B C Auman, E R Wonchoba, 1993Synthesis and Properties of Fluorinated Polyimides from Novel 2,2'-bis(fluoroalkoxy benzidines. Macromolecules2627792784

6. M Hasegawa, M Horiuchi, Y Wada, 2007Polyimides Containing Trans-1,4-cyclohexane Unit (II). Low-K and Low-CTE Semi- and Wholly Cycloaliphatic PolyimidesHigh Perform. Polym. 19175193

7. M Hasegawa, 2001Semi-Aromatic Polyimides with Low Dielectric Constant and Low CTEHigh Perform. Polym. 13: S93S106.

8. J Garnett, C M (1904Colours in Metal Glasses and in Metallic FilmsPhilos. Trans. R. Soc. London Ser. A 203385420

9. J L Hedrick, R D Miller, C J Hawker, K R Carther, W Volksen, D Y Yoon, M Trollsas, 1998Templating Nanoporosity in Thin-film Dielectric Insulators. Adv. Mater. 1010491053

10. K R Carter, DiPetro R A, Sanchez M I, Swanson S A (2001Nanoporous Polyimides Derived from Highly Fluorinated Polyimide/Poly(propylene Oxide) Copolymers. Chem. Mater. 13213221

11. P Krishnan, S G Cheng, C Z Cheng, Y S Cheng, J W C (2003Preparation of Nanoporous Polyimide Films from Poly(urethane-imide) by Thermal Treatment. Macromol. Mater. Eng. 288730736

12. W-C Wang, R H Vora, E-T Kang, K-G Neoh, C-K Ong, L-F Chen, 2004Nanoporous Ultra-low-κ Films Prepared from Fluorinated Polyimide with Grafted Poly(acrylic acid) Side Chains. Adv. Mater. 165457

13. M Suzuki, H Kasai, T Ishizaka, H Miura, S Okada, H Oikawa, T Nihira, H Fukuro, H Nakanishi, 2007Fabrication of Size-controlled Polyimide Nanoparticles. J. Nanosci. Nanotechnol. 727482752

14. H Kasai, H Mitsui, G Zhao, T Ishizaka, M Suzuki, H Oikawa, H Nakanishi, 2008Fabrication of Porous Nanoscale Polyimide Structures. Chem. Lett. 3710561057

15. G Zhao, T Ishizaka, H Kasai, H Oikawa, H Nakanishi, 2007Fabrication of Unique Porous Polyimide Nanoparticles Using Reprecipitation Method.Chem. Mater. 1919011905

16. G Zhao, T Ishizaka, H Kasai, H Nakanishi, H Oikawa, 2007Introducing Porosity into Polyimide NanoparticlesJ. Nanosci. Nanotechnol. 815

17. G Zhao, T Ishizaka, H Kasai, M Hasegawa, H Nakanishi, H Oikawa, 2009Using a Polyelectrolyte to Fabricate Porous Polyimide Nanoparticles with Crater-like PoresPolym. Adv. Technol. 204347

18. G Zhao, T Ishizaka, H Kasai, M Hasegawa, T Furukawa, H Nakanishi, H Oikawa, 2009Ultralow-dielectric-constant Films Prepared from Hollow Polyimide Nanoparticles Possessing Controllable Core SizesChem. Mater. 21419424

19. H Kasai, H S Nalwa, H Oikawa, S Okada, H Matsuda, N Minami, A Kakuta, K Ono, A Mukoh, H Naknishi, 1992A Novel Preparation Method of Organic Microcrystals. Jpn. J. Appl. Phys. 31: L1132L1134.

20. G Zhao, T Ishizaka, H Kasai, H Oikawa, H Nakanishi, 2007Preparation of Multilayered Film of Polyimide Nanoparticles for Low-k ApplicationsMol. Cryst. Liq. Cryst. 4643138

21. T Lin, K W Stickney, M Rogers, J S Riffle, J E Mcgrath, H Marand, T H Yu, R M Davis, 1993Preparation of Sub-micron Polyimide Particles by Precipitation from Solution. Polymer 34772777

22. Z Chai, X Zheng, X Sun, 2003Preparation of Polymer Microspheres from SolutionsJ. Polym. Sci. Part B: Polym. Phys. 41159165

23. J Y Xiong, X Y Liu, S B Chen, T S Chung, 2004Surfactant Free Fabrication of Polyimide Nanoparticles. Appl. Phys. Lett. 8557335735

24. Y Nagata, Y Ohnishi, T Kajiyama, 1996Highly Crystalline Polyimide ParticlesPolym. J. 28980985

25. K Asao, H Ohnishi, H Morita, 2000Preparation of Polyimide Particles by Precipitation Polymerization.Kobunshi Ronbunshuin Japanese 57271276

26. F Basset, A Lefrant, T Pascal, B Gallot, B Sillion, 1998Crystalline Polyimide Particles Generated via Thermal Imidization in a Heterogeneous MediumPolym. Adv. Technol. 9202209

27. A Okamura, K Fujimoto, H Kawaguchi, H Nishizawa, O Hirai, 1996Preprints of 9th Polymeric Microspheres Symp.: 167 p.

28. S Omi, A Matsuda, K Imamura, M Nagai, G H Ma, 1999Synthesis of Monodisperse Polymeric Microspheres Including Polyimide Prepolymer by Using SPG Emulsification TechniqueColloids surf. A 153373381

29. S Watanabe, K Ueno, M Murata, Y Masuda, 2006Preparation of Polystyrene-Polyimide Particles by Dispersion Polymerization of Styrene Using Poly(amic acid) as a Stabilizer.Polym. J. 38471476

30. S Watanabe, K Ueno, K Kudoh, M Murata, Y Masuda, 2000Preparation of Core-shell Polystyrene-polyimide Particles by Dispersion Polymerization of Styrene Using Poly(amic acid) as a Stabilizer. Macromol. Rapid Commun. 2113231326

31. T H Kim, C D Ki, H Cho, T Chang, J Y Chang, 2005Facile Preparation of Core-Shell Type Molecularly Imprinted Particles: Molecular Imprinting into Aromatic Polyimide Coated on Silica SpheresMacromolecules3864236428

32. H Kasai, H Kamatani, Y Yoshikawa, S Okada, H Oikawa, A Watanabe, O Ito, H Nakanishi, 1997Crystal Size Dependence of Emission from Perylene Microcrystals. Chem. Lett. 1111811182

33. H Kasai, H S Nalwa, S Okada, H Oikawa, H Nakanishi, 2000Fabrication and Spectroscopic Characterization of Organic Nanocrystals. In: Nalwa H S, editor. Handbook of Nanostructured Materials and Nanotechnology, San Diego:Academic Press. Chap. 8.

34. H Kasai, S Okazaki, T Hanada, S Okada, H Oikawa, T Adschiri, K Arai, K Yase, H Nakanishi, 2000Preparation of C60 Microcrystals Using High-Temperature and High-Pressure Liquid Crystallization Method. Chem. Lett. 1213921393

35. T Ishizaka, H Kasai, H Nakanishi, 2004Intensity Controllable Luminescence of Eu3+-doped Polyimide Nanoparticles by UV-irradiation and Thermal Treatment. Jpn. J. Appl. Phys. 43: L516518

36. T Ishizaka, H Kasai, H Nakanishi, 2008Fabrication of Eu-complex/Polyimide Compostie Nanoparticles. J. Mater. Sci. 44166169

37. T Nishino, M Kotera, N Inayoshi, N Miki, K Nakamae, 2000Residual Stress and Microstructures of Aromatic Polyimide with Different Imidization ProcessesPolymer4169136918

38. A Barton, F M (1975Solubility Parameters. Chem. Rev. 75731753

39. M Okubo, R Takekoh, A Suzuki, 2002Preparation of Micron-sized, Monodisperse Poly(methyl methacrylate)/Polystyrene Composite Particles Having a Large Number of Dents on Their Surfaces by Seeded Dispersion Polymerization in the Presence of Decalin. Colloid Polym. Sci. 28010571061

40. P C Griffiths, C Wellappili, A R Hemsley, R Stephens, 2004Ultra-porous Hollow Particles. Colloid Poly. Sci. 28211551159

41. C-Y Kao, T-C Lo, W-C Lee, 2003Influence of Polyvinylpyrrolidone on the Hydrophobic Properties of Partially Porous Poly(Styrene-Divinylbenzene) Particles for Biological Applications.J. Appl. Polym. Sci. 8718181824

42. E Unsal, T Irmak, E Durusoy, M Tuncel, A Tuncel, 2006Monodisperse Porous Polymer Particles with Polyionic Ligands for Ion Exchange Separation of ProteinsAnal. Chim. Acta 570240248

43. K Koushik, U B Kompella, 2004Preparation of Large Porous Deslorelin-PLGA Microparticles with Reduced Residual Solvent and Cellular Uptake Using a Supercritical CO2 Process. Pharm. Res. 21524535

44. W Albrecht, K Lützow, T Weigel, T Groth, M Schossig, A Lendlein, 2006Development of Highly Porous Microparticles from Poly(ether imide) Prepared by a Spraying/Coagulation Process. J. Membr. Sci. 273106115

45. M M Coleman, C J Serman, D E Bhagwagar, P C Painter, 1990A practical Guide to Polymer Miscibility. Polymer 3111871203

46. D Krevelen, W V Nijenhuis, K T (2009Cohesive Properties and Solubility. In: Properties of Polymers, 4th Edition. Amsterdam: Elsevier Science. 189228

47. H Yabu, M Tanaka, K Ijiro, M Shimomura, 2003Preparation of Honeycomb-patterned Polyimide Films by Self-organizationLangmuir1962976300

48. S Niyogi, B Adhikari, 2002Preparation and Characterization of a Polyimide MembraneEur. Polym. J. 3812371243

49. A Mochizuki, T Fukuoka, M Kanada, N Kinjou, T Yamamoto, 2002Development of Photosensitive Porous Polyimide with Low Dielectric ConstantJ. Photopolym. Sci. Technol. 15159166

50. M Trau, D A Saville, I A Aksay, 1996Field-induced Layering of Colloidal CrystalsScience272706709

51. L C Graton, H J Fraser, 1935Systematic Packing of Spheres with Particular Reference to Porosity and Permeability. J. Geol. 43: 785.

8

CHAPTER

Electrochemical Properties of LLTO/Fluoropolymer-Shell Cellulose-Core Fibrous Membrane for Separator of High Performance Lithium-Ion Battery

Fenglin Huang †, Wenting Liu †, Peiying Li, Jinxia Ning and Qufu Wei *

Key Laboratory of Eco-Textiles, Ministry of Education, Jiangnan University, Wuxi 214122, China

ABSTRACT

A superfine $Li_{0.33}La_{0.557}TiO_3$ (LLTO, 69.4 nm) was successfully synthesized by a facile solvent-thermal method to enhance the electrochemical properties of the lithium-ion battery separator. Co-axial nanofiber of cellulose and Poly(vinylidene fluoride-*co*-hexafluoropropylene) (PVDF-HFP) was prepared by a co-axial electrospinning technique, in which the shell material was PVDF-HFP and the core was cellulose. LLTO superfine nanoparticles were incorporated into the shell of the PVDF-HFP. The core–shell composite nanofibrous membrane showed good wettability (16.5°, contact angle), high porosity (69.77%), and super electrolyte compatibility (497%, electrolyte uptake). It had a higher ionic conductivity (13.897 $mS{\cdot}cm^{-1}$) than those of pure polymer fibrous membrane and commercial separator. In addition, the rate capability (155.56 $mAh{\cdot}g^{-1}$) was also superior to the compared separator. These excellent performances endowed LLTO composite nanofibrous membrane as a promising separator for high-performance lithium-ion batteries.

Keywords: LLTO; electrospinning; nanofiber; separator

1. INTRODUCTION

High energy demands have accelerated emergent efforts to develop high-performance lithium-ion batteries (LIBs) due to their high specific energy and long cycle lifetime [1,2,3]. A separator in LIB is considered as a key component to prevent safety issues, because it can isolate the cathode and anode to prevent electrical short circuits and at the same time allow rapid transport of ionic charge carriers [4,5,6]. Polyolefin separators such as polyethylene (PE) and polypropylene (PP) have been widely used in LIBs [7]. However, their shrinkage at high working temperature leads to poor thermal property, which would cause a short circuit between electrodes [8]. Furthermore, the large difference in polarity between the polyolefin and electrolyte leads to high resistance and low electrolyte retention, which will influence the cycling performance of LIBs. To overcome these problems, some new materials and functional methods have been attempted for application in separators.

Semi-crystalline polyvinylidene fluoride (PVDF) and its copolymer Poly(vinylidene fluoride-*co*-hexafluoropropylene) (PVDF-HFP) have received special attentions as promising host polymers for separators of LIBs due to higher polarity and ionic conductivity [9,10]. Electrospinning is reported as one of the efficient methods to improve the wettability performance of the fibrous membrane due to its low diameter and high porosity [11,12]. In our previous study, a cellulose/PVDF-HFP nanofibrous membrane was successfully prepared by co-axial electrospinning. The composite membrane demonstrated good mechanical property, superior flame retardancy, excellent thermal stability and good electrolyte wettability [13].

Inorganic ceramics always attract great interest since they may present several advantages, *i.e.*, high electrochemical stability window, high thermal stability, high mechanical resistance [14,15,16]. Li-conducting ceramics based on $Li_{3x}La_{2/3-x}TiO_3$ (named hereafter LLTO) hold an important place among them because of their high ionic conductivity at room temperature, *i.e.*, $\sigma = 10^{-3}$ S·cm^{-1} for $x = 0.10$, and low electronic conductivity, *i.e.*, $\sigma = 10^{-8}$ S·cm^{-1} [17,18]. Inspired by its excellent conductivity in Li-ion battery applications, various methods, such as the solid-state reaction method, sol–gel method and solutions precipitation, are applied to synthesize perovskite-type lithium ion-conducting oxides [19,20]. However, it was considered that the LLTOs reported in literature might not be suitable for mingling in nanofibers because of their large diameters (larger than 100 nm). In this work, a superfine $Li_{0.33}La_{0.557}TiO_3$ was facilely synthesized by the solvent thermal method, and then was added into the shell of nanofibers which were co-axially electrospun with cellulose acetate and PVDF-HFP. The electrochemical properties and cycling performance of the lithium-ion batteries were evaluated and compared to those of commercial separator.

2. EXPERIMENTAL SECTION

2.1. Materials

Poly(vinylidene fluoride-*co*-hexafluoropropylene) (PVDF-HFP) (molecular weight 50,000 $g \cdot mol^{-1}$) was purchased from Shanghai 3F New Materials Co., Ltd (Shanghai, China). Cellulose acetate, lithium nitrate, lanthanum nitrate, butyltitanate, dimethylacetamide (DMAc) and acetone were supplied by Sinopharm Chemical Reagent Co., Ltd (Shanghai, China) and used without further purification. The electrolyte was composed of 1 M LiPF6 dissolved in ethylene carbonate, dimethyl carbonate, and ethylene methyl carbonate (1:1:1, *v*/*v*/*v*).

2.2. Preparation of LLTO Nanoparticles and Co-Axial Electrospinning of LLTO Composite Nanofibrous Membrane

$Li_{0.33}La_{0.557}TiO_3$ nanoparticles were synthesized by lithium nitrate, lanthanum nitrate and butyl titanate with the starting stoichiometry amount 0.33:0.55:1.00, using a solution thermal method. Citric acid as the complexing agent was mixed in an ethanol solution. Provide quantitative of lithium nitrate and butyl titanate was added in the solution with stirring at 80 °C for 30 min. Lanthanum nitrate solution was then added to the reaction system in drops, followed by stirring for 80 min at ambient temperature. This solution was then transferred into a 200 mL Teflon-lined stainless steel autoclave and heated at 180 °C for 10 h. After slowly cooling it to room temperature, the dried mixture was paralyzed at 350 °C for 4 h and then calcined at 900 °C for 2 h with the heating rate of 5 °C/min. The calcined product was ground in an agate mortar and then sieved through #325 mesh for further use.

The two polymer solutions were independently fed through concentrically conFigured needles with outer and inner diameters of 1.3 and 0.3 mm, respectively. The shell solution was prepared by dissolving

PVDF-HFP in a mixture of DMAc and acetone (3/7, *w*/*w*) to 10 wt% containing different amounts of LLTO particles (0, 2, 5 and 8 wt%). The core solution was prepared by dissolving cellulose acetate into DMAc and acetone (2/1, *w*/*w*) to 15 wt%. Coaxial electrospinning was conducted according to our previous work [13]. The as-prepared electrospun nanofibers were then dried in a vacuum oven at 60 °C for 12 h. Hydrolysis of cellulose acetate/PVDF-HFP membrane was performed in 0.05 M LiOH aqueous solution at ambient temperature for 10 h to produce composite nanofibrous membrane. Then the obtained membrane was rinsed in distilled water and dried under vacuum at 60 °C for 12 h.

2.3. Characterizations and Electrochemical Evaluation

The morphologies of LLTO particles and nanofibers were examined using a field emission scanning electron microscope (FE-SEM, Hitachi, Tokyo, Japan) and transmission electron microscope (TEM, JEM-2100HR, JEOL, Tokyo, Japan). The structure of LLTO particles was investigated using wide-angle X-ray diffraction (WXRD, D8, Bruker, Karlsruhe, Germany).

The thermal stability was analyzed by using a thermogravimetric analyzer (TGA, Q500, TA Instruments, New Castle, DE, USA) under a flowing nitrogen atmosphere at a heating rate of 10 $C \cdot min^{-1}$ with the temperature range of 30–900 °C.

The wettability was examined by a sessile drop method and the porosity of membranes was measured by immersing the membrane in *n*-butanol for 1 h [13].

The testing cells had a typical coin-type construction using lithium foils as both counter electrode and reference electrode. The cells were assembled in an argon-filled glove box.

Charge-discharge tests were carried out at a current density from 0.2 C to 5.0 C in a range of 2.8–4.2 V *vs.* Li/Li^+. All the tests were performed at 20 °C. The ionic conductivity and interfacial resistances of LLTO/PVDF-HFP/cellulose nanofibrous membrane were evaluated using electrochemical impedance spectroscopy (EIS) measurement in combination with an Electrochemical Workstation (CHENHUA, CHI710b, Shanghai, China). The electrochemical stability of the separators was measured by a linear sweep voltammograms (LSV) on a working electrode of stainless-steel and a counter electrode of lithium metal at the potential range between 2.5 V and 6.0 V under the scan rate of 1.0 $mV \cdot s^{-1}$ at 20 °C.

3. RESULTS AND DISCUSSION

3.1. Structure of LLTO and Composite Nanofibrous Membrane

Figure 1 shows the uniformly nano-sized morphologies of the LLTO, after calcination at the temperature at which crystallization finishes (900 °C). It is observed from the images that the size of the particle ranges from 42 nm to 99 nm, with an average size of 69.4 nm, which is significantly finer than those of references listed in Table 1. The smaller diameter of the particles endows them as very suitable functional particles for doping in polymer nanofibers.

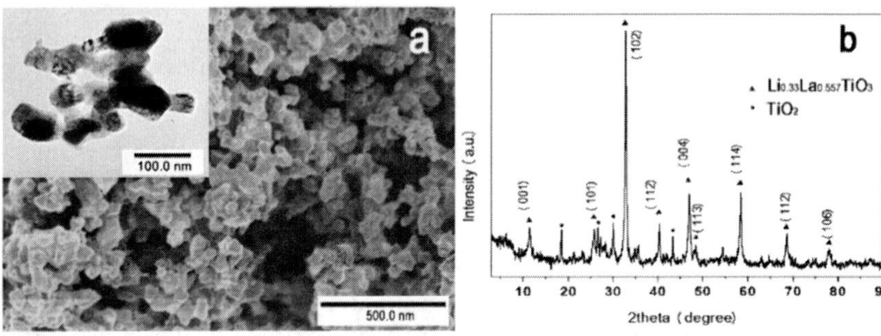

Figure 1. (**a**) SEM and TEM (inset) images and (**b**) XRD pattern of $Li_{0.33}La_{0.557}TiO_3$particles.

Figure 1b shows the powder X-ray diffraction pattern of LLTO particles sintered at 900 °C. The main phase is crystallized into an orthorhombic perovskite which depends on the preparation condition. The small amount of rutile TiO_2 and unknown phases are observed from the XRD pattern, which might be produced during the sintering process. The average metal composition of the sintered sample is determined to be La:Li:Ti = 0.55:0.33:1 according to the standard atlas and references [22], which is consistent with the nominal composition.

Table 1. LLTO particles prepared by various methods.

Authors	Forms	Methods	Diameters
Abhilash *et al.* [21]	$Li_{0.5}La_{0.5}TiO_3$	solid-state reaction	80 nm
Tang *et al.* [22]	$Li_{0.34}La_{0.51}TiO_{2.94}$	solid-state reaction	100 nm
Hua *et al.* [23]	$Li_{0.27}La_{0.54}TiO_{2.945}$	sol–gel	1.5 μm
Yoon *et al.* [24]	$Li_{0.35}La_{0.55}TiO_3$	solid-state reaction	1.5–1.6 μm
Geng *et al.* [25]	$Li_{0.5}La_{0.5}TiO_3$	sol–gel	200 nm
Murugesan *et al.* [26]	$Li_{0.3}La_{0.566}TiO_3$	Pechini-type	100 nm
Liang *et al.* [19]	$Li_{0.125}La_{0.625}TiO_3$	sol–gel	200 nm
Ionela *et al.* [27]	$La_{0.66}Li_{0.33}TiO_3$	sol-gel	123.01 nm
Anatolii Belous *et al.* [28]	$Li_{0.5}La_{0.5}TiO_3$	Solution precipitations	100 nm
Our work	$Li_{0.35}La_{0.55}TiO_3$	Thermal-solution	69.4 nm

The morphological features of the composite nanofibrous membranes observed in SEM and TEM are shown in Figure 2. The diameter of core–shell nanofibers fabricated with a 3:3 flow rate ratio ranges from 652 nm to 779 nm, averaging at 712 ± 65 nm (Figure 2a). The average diameters of composite nanofibers containing 2%, 5% and 8% LLTO nanoparticles are 795 ± 89 nm (Figure 2b), 822 ± 76 nm (Figure 2c), and 888 ± 112 nm (Figure 2d), respectively, showing a little variation among all four samples. The cross-sectional structures of the three composite membranes show clear core–shell structures with apparently different core and shell thicknesses (inset SEM a images in Figure 2b–d), consistent with the biphasic fibers with a darker cellulose core and lighter PVDF-HFP shell in the TEM images (inset TEM images in Figure 2b–d). The addition of 2% LLTO leads to the formation of a small number of irregularities and uneven surface morphology, and the resultant LLTO composite nanofibers display a wider fiber diameter distribution. With further increase in LLTO content, the surface roughness of the composite nanofiber increases, and some LLTO particles begin to agglomerate and form clusters.

3.2. Thermal Stability

The evaluation of the thermal properties of the composite nanofibrous membranes is very important for the determination of the performance of the separator in the range of temperatures in which the battery must be stable. Figure 3 shows the TGA results of the composite nanofibrous membranes. The LLTO composite nanofibrous membranes experience thermal degradation all around 300 °C (the temperature at which the weight loss of the sample is 1%), suggesting the excellent thermal properties originate from the intrinsic thermal stability of pure cellulose nanofibers (at 300 °C the weight loss of the sample is 10.5%). The separator containing 8% LLTO has the maximum decomposing temperature of 477 °C, and 11.8% of residues at 900 °C, which are slightly higher than 5% and 2% LLTO composite separator having 9.07% and 6.18% of residues, respectively. The enhanced decomposing temperature and reduced weight loss may be ascribed to the doping of LLTO nanoparticles and its efficiency to retard the thermal degradation of cellulose.

Figure 2. (**a**) SEM image of cellulose/PVDF-HFP nanofibers; (**b**) composite nanofibers containing 2% LLTO; (**c**) 5% LLTO and (**d**) 8% LLTO; inset images are TEM and SEM images of cross-sectional.

3.3. Liquid Electrolyte Uptake

A separator with good wettability can retain the electrolyte effectively and facilitate electrolyte diffusion smoothly. The wettability of the LLTO composite nanofibrous membranes was evaluated by contact angle measurements. Our previous study showed that the contact angle of the commercial separator was 69.29° [13], indicating its poor wettability.

For the cellulose/PVDF-HFP core-shell nanofibrous membrane, the initiating contact angle is 27.3°, and the droplet of the electrolyte is instantly (2.1 s, Figure 4) infiltrated into the membrane. Initiating contact angles of membranes with LLTO doping is finally decreased to 16.5° (8% LLTO) due to the hydrophilic nature of LLTO, suggesting the better interfacial compatibility between the separator and electrolyte. Separator porosity is another critical factor for electrolyte uptake. The porosity and electrolyte uptake of composite nanofibrous membranes are shown in Table 2. The porosity of the nanofibrous membrane is fairly higher than that of the Celgard 2300 separator (47.88%). Both higher interfacial compatibility and porosity of the LLTO composite nanofibrous membrane lead to the higher electrolyte uptake of 497% (8% LLTO), *ca.* 255% and 40% higher than those of the Celgard 2300 separator (140%) and PVDF-HFP/cellulose nanofibrous membrane (355%).

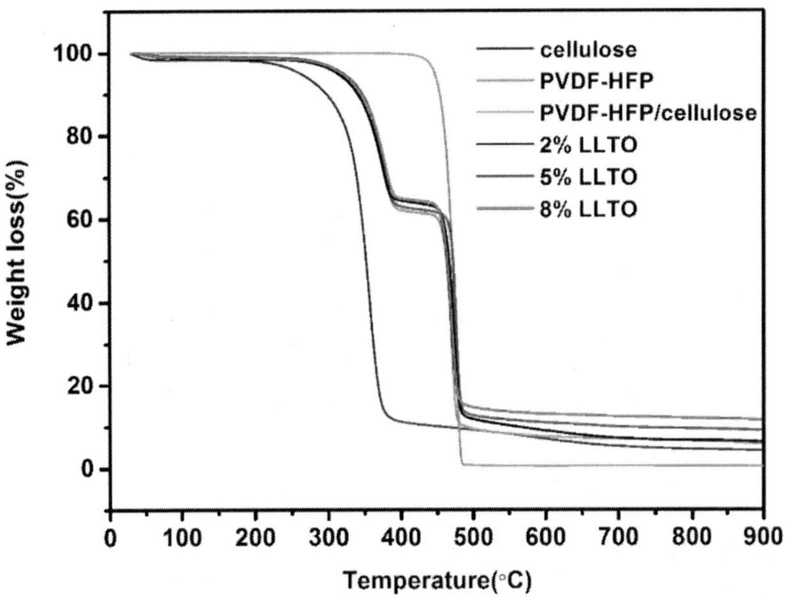

Figure 3. TGA curves of PVDF-HFP/cellulose and LLTO composite nanofibrous membrane.

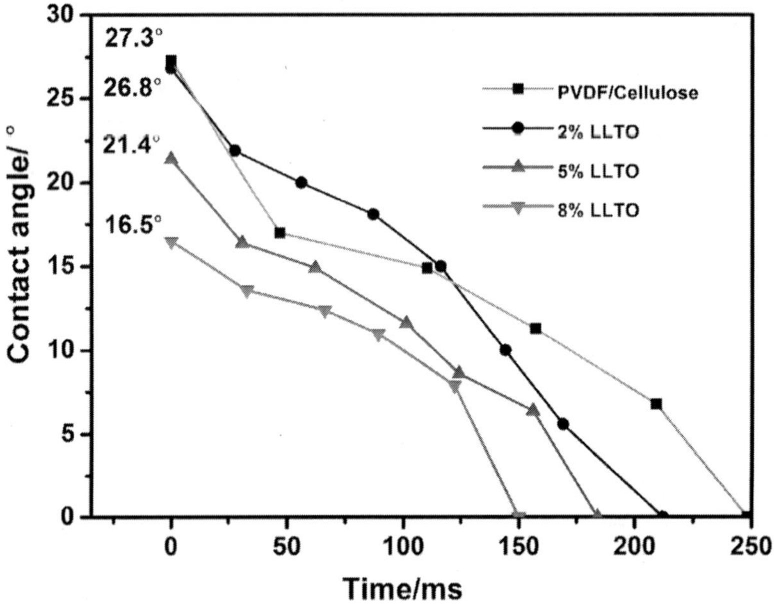

Figure 4. Surface contact angle of PVDF-HFP/cellulose and LLTO composite nanofibrous membrane.

Table 2. Porosity, electrolyte uptake, ionic conductivity of PVDF-HFP/cellulose and LLTO composite nanofibrous separators.

Separator	Thickness (μm)	Porosity (%)	Electrolyte Uptake (%)	Resistance (Ω)	Ionic Conductivity ($mS\cdot cm^{-1}$)
PVDF-HFP/cellulose	59	47.88	140	1.107	2.082
2% LLTO	48	66.36	355	0.415	4.518
5% LLTO	47	69.77	384	0.258	7.116
8% LLTO	37	67.45	487	0.104	13.897

3.4. Electrochemical and Battery Performance

The ionic conductivity is the most important factor that directly affects the electrochemical performance of an electrolyte.Figure 5 shows the ionic conductivity dependence of the PVDF-HFP/cellulose membrane and its composites, the values of which are calculated from the Nyquist plots by intercepts of the plots shown in the inset of Figure 4. The ionic conductivity of the cellulose/PVDF-HFP nanofibrous separator is calculated to be 2.082 $mS\cdot cm^{-1}$, while the value of the commercial separator (Celgard® 2300, Celgard, NC, USA) is 0.88 mS·cm^{-1} [13]. The value of the as-prepared composite membrane containing 8% LLTO is up to 13.897 mS·cm^{-1}, which is much higher than that of the separator containing 2% LLTO (4.518 mS·cm^{-1}) and 5% LLTO (7.116 mS·cm^{-1}). The specifications and conductivity values of as-prepared separators are listed inTable 2. The higher ionic conductivity of the composite membranes might be attributed to the higher electrolyte uptake and the conductivity nature of LLTO particles.

The interfacial properties of the separator with the Li electrode are another important aspect that affects the electrochemical performance of cell. Figure 6 shows the AC impedance spectra of the PVDF-HFP/cellulose membrane and its composites, the diameters of semi-circles represent the interfacial resistances between the Li electrode and the electrolyte-soaked separator. It can be seen from Figure 6 that interfacial resistances are 45, 55, 65 Ω for the composite membrane containing 8%, 5%, 2% LLTO, which are much higher than that of the PVDF-HFP/cellulose membrane (95 Ω). The smaller interfacial resistance means better compatibility between the separator and Li electrode.

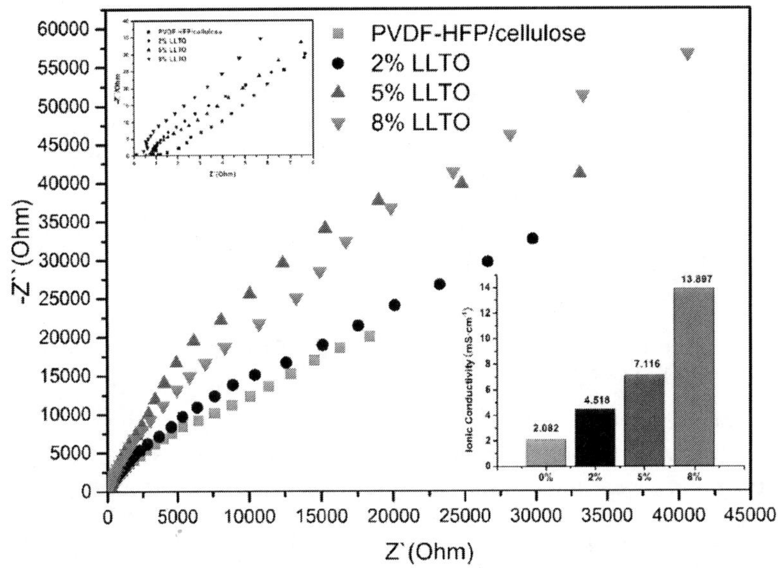

Figure 5. Nyquist plots of stainless steel/electrolyte-soaked separator/stainless steel cells at 20 °C, insets are magnified Nyquist plot and ionic conductivity of separators.

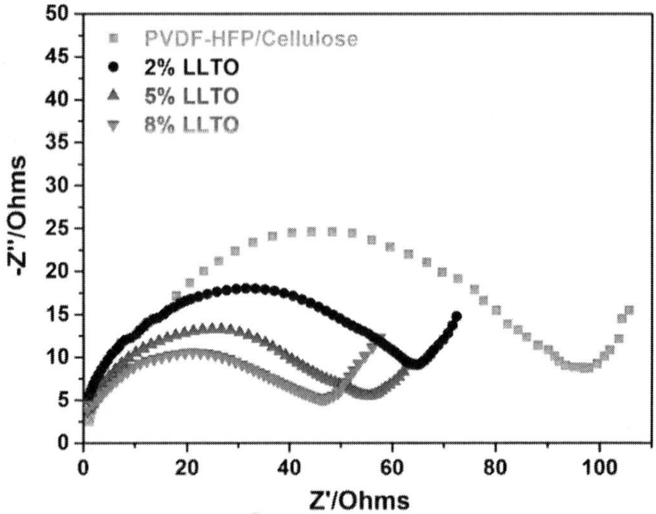

Figure 6. AC impedance spectra of Li/electrolyte-soaked separator/Li cells at 20 °C.

The electrochemical stability of the LLTO composite separator is conducted by linear sweep voltammograms testing to verify whether the composite separator is still stable in the electrochemical stability window. The results are shown inFigure 7 and it is clearly revealed that no appreciable decomposition of cellulose/PVDF-HFP and its composite separators takes place below 5.0 V *vs.* Li/Li^+, which is almost comparable to that of the Celgard® 2300 separator [13]. This result demonstrates the good electrochemical stability of the PVDF-HFP/cellulose and LLTO composite nanofibrous separator, revealing that it could be used as a promising alternative to the commercial separator for high-voltage lithium-ion batteries.

The electrochemical properties of PVDF-HFP/cellulose and its composite separators are investigated by charge and discharge testing and cycling performance characterization (Figure 8). The discharge capacity of batteries assembled with PVDF-HFP/cellulose is around 135.91 $mAh{\cdot}g^{-1}$ at 0.2 C, approximately 20% higher than the 114.8 $mAh{\cdot}g^{-1}$ discharge capacity for Celgard® 2300 separators [13]. The capacities of the LLTO composite separator are found to be 148.00 $mAh{\cdot}g^{-1}$(2% LLTO), 152.01 $mAh{\cdot}g^{-1}$ (5% LLTO) and 155.56 $mAh{\cdot}g^{-1}$ (8% LLTO) and are also superior to those in the references [13]. The utilization efficiency charge-discharge characteristics in the first cycle confirm that the composite membrane containing LLTO exhibits superior kinetic properties to those of commercial separators.

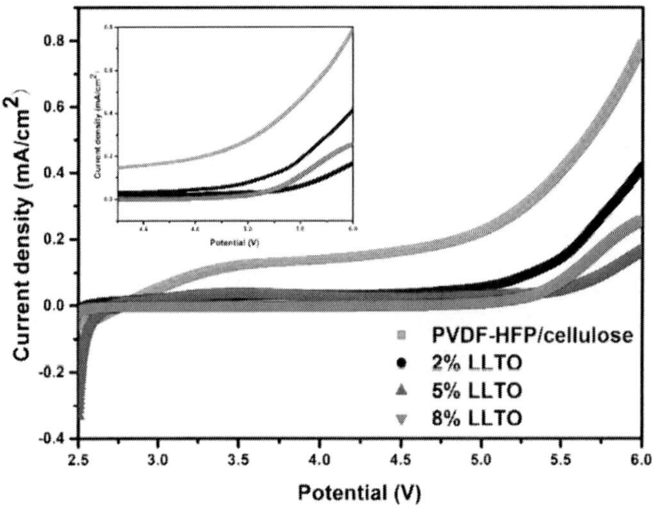

Figure 7. LSV curve of PVDF-HFP/cellulose and LLTO composite separators at a scan rate of 1.0 $mV \cdot s^{-1}$.

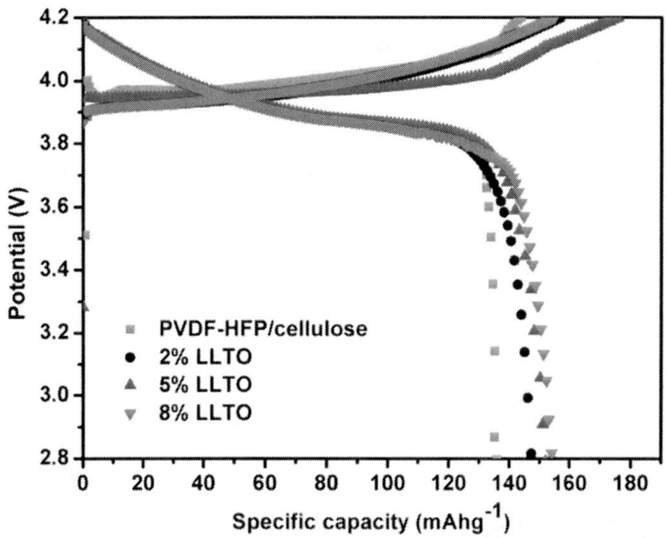

Figure 8. Charge-discharge characteristics of cells containing various separators at first cycle.

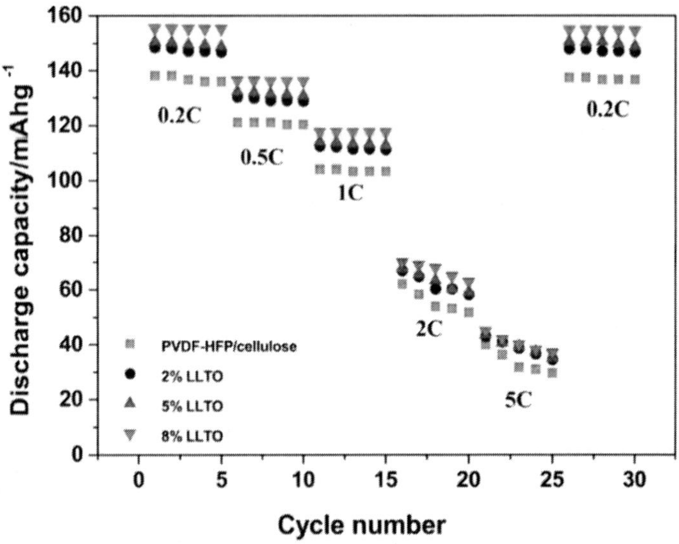

Figure 9. Rate performance of cells using PVDF-HFP/cellulose and LLTO composite nanofibrous membrane.

The cycle performances of the cells at different C-rates are shown in Figure 9. Similar rate capabilities are observed for those kinds of cells with various separators. With increased current densities of 0.2, 0.5, 1, 2 and 5 C, the discharge capacities decrease from 155.30 mAh·g^{-1} to 136.17, 117.65, 62.95 and 37.01 mAh·g^{-1} for cells containing 8% LLTO particles. The discharge capacity for the cell with composite separators fades with LLTO reducing from 8% to 2%. After being cycled at 5 C several times, the cycle rate is returned to 0.2 C. The discharge capacities are regained at 153.8 mAh·g^{-1} for the LLTO composite separator, which maintains about 98.9% of the initial discharge capacity. The superior discharge capacity and cycling performance could be ascribed to the higher ionic conductivity and better interfacial compatibility of the electrolyte-soaked composite nanofibrous membrane.

4. CONCLUSIONS

The superfine $Li_{0.33}La_{0.557}TiO_3$ (LLTO, 69.4 nm) was successfully synthesized and characterized. It was found that the as-prepared membranes exhibited enhanced ionic conductivities due to the conductivity of LLTO particles. The composite LLTO nanofibrous membrane containing 8% LLTO was found to exhibit the highest ionic conductivity (13.897 mS·cm^{-1}). Our LSV results also showed that the electrochemical stability of the composite nanofibrous membranes improved with an increase in the LLTO content. Moreover, cells assembled with the composite separator exhibited high storage and enhanced rate performances compared to the battery using the pure polymer nanofibrous separator. All these results suggest that this composite nanofiber membrane would be a promising separator for high-performance lithium-ion batteries.

ACKNOWLEDGMENTS

Authors are grateful to the financial support from the National Natural Science Foundation of China (51203064), Industry-Academia-Research Joint Innovation Fund of Jiangsu Province (BY2014023-04), and Fundamental Research Funds for the Central Universities (JUSRP2016010A).

AUTHOR CONTRIBUTIONS

Fenglin Huang and Qufu Wei conceived and designed the experiments; Fenglin Huang, Wenting Liu, Peiying Li, Jinxia Ning and Qufu Wei performed the experiments; all authors analyzed the data; Peiying Li and Jinxia Ning contributed materials and tools; and all authors wrote the paper.

REFERENCES

1. Fang, M.-D.; Ho, T.-S.; Yen, J.-P.; Lin, Y.-R.; Hong, J.-L.; Wu, S.-H.; Jow, J.-J. Preparation of advanced carbon anode materials from mesocarbon microbeads for use in high C-rate lithium ion batteries. *Materials* **2015**, *8*, 3550–3561.

2. Huang, X.S. Evaluation of a polymethylpentene fiber mat formed directly on an anode as a battery separator. *J. Membr. Sci.* **2014**, *466*, 331–337.

3. Liang, X.X.; Yang, Y.; Jin, X.; Huang, Z.L.; Kang, F.Y. The high performances of SiO_2/Al_2O_3-coated electrospun polyimide

fibrous separator for lithium-ion battery. *J. Membr. Sci.* **2015**, *493*, 1–7.

4. Shi, C.; Zhang, P.; Huang, S.H.; He, X.Y.; Yang, P.T.; Wu, D.Z.; Sun, D.H.; Zhao, J.B. Functional separator consisted of polyimide nonwoven fabrics and polyethylene coating layer for lithium-ion batteries. *J. Power Sour.* **2015**, *298*, 158–165.

5. Cai, C.; Wang, Y. Novel nanocomposite materials for advanced Li-ion rechargeable batteries. *Materials* **2009**, *2*, 1205–1238.

6. Liu, J.S.; Li, W.S.; Zuo, X.X.; Liu, S.Q.; Li, Z. Polyethylene-supported polyvinylidene fluoride–cellulose acetate butyrate blended polymer electrolyte for lithium ion battery. *J. Power Sour.* **2013**, *226*, 101–106.

7. Xu, Q.; Kong, Q.S.; Liu, Z.H.; Wang, X.J.; Liu, R.Z.; Zhang, J.J.; Yue, L.P.; Duan, Y.L.; Cui, G.L. Cellulose/Polysulfonamide Composite Membrane as a High Performance Lithium-Ion Battery Separator. *ACS Sustain. Chem. Eng.* **2014**, *2*, 194–199.

8. Xiao, S.Y.; Yang, Y.Q.; Li, M.X.; Wang, F.X.; Chang, Z.; Wu, Y.P.; Liu, X. A composite membrane based on a biocompatible cellulose as a host of gel polymer electrolyte for lithium ion batteries. *J. Power Sour.* **2014**, *270*, 53–58.

9. Göeren, A.; Costa, C.M.; Tamaño Machiavello, M.N.; Cíntora-Juárez, D.; Nunes-Pereira, J.; Tirado, J.L.; Silva, M.M.; Gomez Ribelles, J.L.; Lanceros-Méndez, S. Effect of the degree of porosity on the performance of poly (vinylidene fluoride-trifluoroethylene)/poly (ethylene oxide) blend membranes for lithium-ion battery separators. *Solid State Ion.***2015**, *280*, 1–9.

10. Seidel, S.M.; Jeschke, S.; Vettikuzha, P.; Wiemhoefer, H.D. PVDF-HFP/ether-modified polysiloxane membranes obtained via airbrush spraying as active separators for application in lithium ion batteries. *Chem. Commun.* **2015**, *51*, 12048–12051.

11. Shayapat, J.; Chung, O.H.; Park, J.S. Electrospun polyimide-composite separator for lithium-ion batteries. *Electrochim. Acta* **2015**, *170*, 110–121.

12. Yanilmaz, M.; Zhang, X.W. Polymethylmethacrylate/Polyacrylonitrile Membranes via Centrifugal Spinning as Separator in Li-Ion Batteries. *Polymers* **2015**, *7*, 629–643.

13. Huang, F.L.; Xu, Y.F.; Peng, B.; Su, Y.F.; Jiang, F.; Hsieh, Y.L.; Wei, Q.F. Coaxial Electrospun Cellulose-Core Fluoropolymer-Shell Fibrous Membrane from Recycled Cigarette Filter as Separator for High Performance Lithium-Ion Battery. *ACS Sustain. Chem. Eng.* **2015**, *3*, 932–940.

14. Zhu, X.M.; Jiang, X.Y.; Ai, X.P.; Yang, H.X.; Cao, Y.L. A Highly Thermostable Ceramic-Grafted Microporous Polyethylene Separator for Safer Lithium-Ion Batteries. *ACS Appl. Mater. Inter.* **2015**, *7*, 24119–24126.

15. Xu, W.X.; Wang, Z.Y.; Shi, L.Y.; Ma, Y.; Yuan, S.; Sun, L.N.; Zhao, Y.; Zhang, M.H.; Zhu, J.F. Layer-by-Layer Deposition of Organic-Inorganic Hybrid Multilayer on Microporous Polyethylene Separator to Enhance the Electrochemical Performance of Lithium-Ion Battery. *ACS Appl. Mater. Inter.* **2015**, *37*, 20678–20686.

16. Liang, F.; Hayashi, K. A High-Energy-Density Mixed-Aprotic-Aqueous Sodium-Air Cell with a Ceramic Separator and a Porous Carbon Electrode. *J. Electrochem. Soc.* **2015**, *162*, A1215–A1219.

17. Inaguma, Y.; Nakashima, M. A rechargeable lithium-air battery using a lithium ion-conducting lanthanum lithium titanate ceramics as an electrolyte separator. *J. Power Sour.* **2013**, *228*, 53–58.

18. Piana, M.; Wandt, J.; Meini, S.; Buchberger, I.; Tsiouvaras, N.; Gasteiger, H.A. Stability of a Pyrrolidinium-Based Ionic Liquid in Li-O_2 Cells. *J. Electrochem. Soc.* **2014**, *161*, A1992–A2001.

19. Liang, Y.Z.; Ji, L.W.; Guo, B.K.; Lin, Z.; Yao, Y.F.; Li, Y.; Alcoutlabi, M.; Qiu, Y.P.; Zhang, X.W. Preparation and electrochemical characterization of ionic-conducting lithium lanthanum titanate oxide/polyacrylonitrile submicron composite fiber-based lithium-ion battery separators. *J. Power Sour.* **2011**, *196*, 436–441.

20. Zhang, X.W.; Ji, L.W.; Toprakci, O.; Liang, Y.Z.; Alcoutlabi, M. Electrospun Nanofiber-Based Anodes, Cathodes, and Separators for Advanced Lithium-Ion Batteries. *Polym. Rev.* **2011**, *51*, 239–264.

21. Abhilasha, K.P.; Christopher Selvina, P.; Nalinib, B.; Nithyadharsenic, P.; Pillai, B.C. Investigations on pure and Ag doped lithium lanthanum titanate (LLTO) nanocrystalline ceramic electrolytes for rechargeable lithium-ion batteries.*Ceram. Int.* **2013**, *39*, 947–952.

22. Tang, H.; Xu, J. Enhanced electrochemical performance of $LiFePO_4$ coated with $Li_{0.34}La_{0.51}TiO_{2.94}$ by rheological phase reaction method. *Mater. Sci. Eng. B* **2013**, *178*, 1503–1508.

23. Hua, C.X.; Fang, X.P.; Wang, Z.X.; Chen, L.Q. Lithium storage in perovskite lithium lanthanum titanate. *Electrochem. Commun.* **2013**, *32*, 5–8.

24. Yoon, J.; Hunter, G.; Akbar, S.; Dutta, P.K. Interface reaction and its effect on the performance of a CO_2 gas sensor based on $Li_{0.35}La_{0.55}TiO_3$ electrolyte and Li_2CO_3 sensing electrode. *Sens. Actuators B Chem.* **2013**, *182*, 95–103.

25. Geng, H.X.; Lan, J.L.; Mei, A.; Lin, Y.H.; Nan, C.W. Effect of sintering temperature on microstructure and transport properties of $Li_{3x}La_{2/3-x}TiO_3$ with different lithium contents. *Electrochim. Acta* **2011**, *56*, 3406–3414.

26. Vijayakumar, M.; Inaguma, Y.; Mashiko, W.; Crosnier-Lopez, M.P.; Bohnke, C. Synthesis of Fine Powders of $Li_{3x}La_{2/3-x}TiO_3$ Perovskite by a Polymerizable Precursor Method. *Chem. Mater.* **2004**, *16*, 2719–2724.

27. Popovici, I.C.; Chirila, E.; Popescu, V.; Ciupina, V.; Pordan, G. Sol–gel preparation and characterization of perovskite lanthanum lithium titanate. *J. Mater. Sci.* **2007**, *42*, 3373–3377.

28. Belous, A.; Yanchevskiy, O.; V'yunov, O.; Bohnke, O.; Bohnke, C.; Berre, F.L.; Fourquet, J.L. Peculiarities of $Li_{0.5}La_{0.5}TiO_3$ Formation During the Synthesis by Solid-State Reaction or Precipitation from Solutions. *Chem. Mater.***2004**, *16*, 407–417.

9
CHAPTER

Auto-Reparation of Polyimide Film Coatings for Aerospace Applications Challenges and Perspectives

A.A. Périchaud[1], R.M. Iskakov[2, 3], Andrey Kurbatov[4], T. Z. Akhmetov[4], O.Y. Prokohdko[5], Irina V. Razumovskaya[6], Sergey L. Bazhenov[6], P.Y. Apel[7], V. Yu. Voytekunas[8] and M.J.M. Abadie[8,9]

[1] *Catalyse Ltd., Marseille, France*

[2] *Institute of Chemical Sciences, Almaty, Kazakhstan*

[3] *Kazakh British Technical University, Almaty, Kazakhstan*

[4] *Centre of Physico-Chemical Analyses of al-Farabi Kazakh National University, Almaty, Kazakhstan*

[5] *Physics Department of al-Farabi Kazakh National University, Almaty, Kazakhstan*

[6] *Solid State Physics Department of Federal Educational Establishment of Higher Professional Education "Moscow State Pedagogical University", Russia*

[7] *Joint Institute of Nuclear Researches, Dubna, Moscow, Russia*

[8] *Laboratory of Polymer Science & Advanced Organic Materials LEMP/MAO, CC 021, Université Montpellier II, Sciences et Techniques du Languedoc, Place Eugène Bataillon, France*

[9] *School of Materials Science & Engineering, Block N4.1, College of Engineering, Nanyang Technological University, Singapore*

1. INTRODUCTION

In the aerospace industry metal parts are substantially replaced by polymer matrix composites presenting many advantages compared with the metal parts they are replacing. However composite damage is very difficult to detect because it often forms a sub-laminate locations, invisible to the naked eye. To overcome these problems, research in the last two decades has led to the development of self healing polymeric materials that mimic some features found in biological systems. The development of self-healing polymeric materials, those that practically imitate self-healing process of wounds is a new and challenging problem for space applications to protect different types of devices made with composite materials and represents a new paradigm in materials design [1]. The pioneer works in auto-reparation of composites were developed by White S. R et al [2-6]. Their originality was to suggest the encapsulation of monomer – dicyclopentadiene and Grubbs catalysor disseminated uniformly into the composites. Under loading cracks are formed and microcapsules open up releasing monomer and initiator which fulfill the cracks by polymerization and crosslinking reactions and allow the composite to retrieve his original structure without losing its initial mechanical properties - Figure 1.

The concept is strongly dependant on the reactivity of the monomers used as well as the initiator systems [7].

A sub-orbital spaceflight (above 100 km) or a low Earth orbit (LEO) impose on the composites the following conditions : absence of oxygen, temperature variable from -120°C (dark side) to +300°C (solar side), UV and heavy ions radiation, low pressure 10^{-4} Pa. For example, the International Space Station is in a LEO that varies from 320 km (199 miles) to 400 km (249 miles) above the Earth's surface.

The methodology developed was to test the feasibility of self-healing by using an encapsulated multifunctional acrylates, i.e. trimethylol propane triacrylate (TMPTA) in presence of a radical photoinitiator (Darocur® 1173).

The novelty of the work was to develop new polymeric film composites, based on Polyimide containing in the same reservoir microencapsulated active monomer and generator of active species, to prevent destruction of the damaged sample.

The new concept tested was to crosslink the monomer in presence of a radical photoinitiator which, upon exposure to UV, produces active centres, permitting the self-healing of the damaged composites.[8]. A schematic of our system is briefly described in Figure 2.

Three major problems have been studied :

1. Develop a suitable microcapsule which could, under loading, liberate the active principle in presence of initiator.

2. Distribute homogeneously the microcapsules into the polymeric film used as coatings

3. Test the new super-composite film and its auto-reparation once micro cracks have been formed, to prevent crack growth and macro damage of the composite.

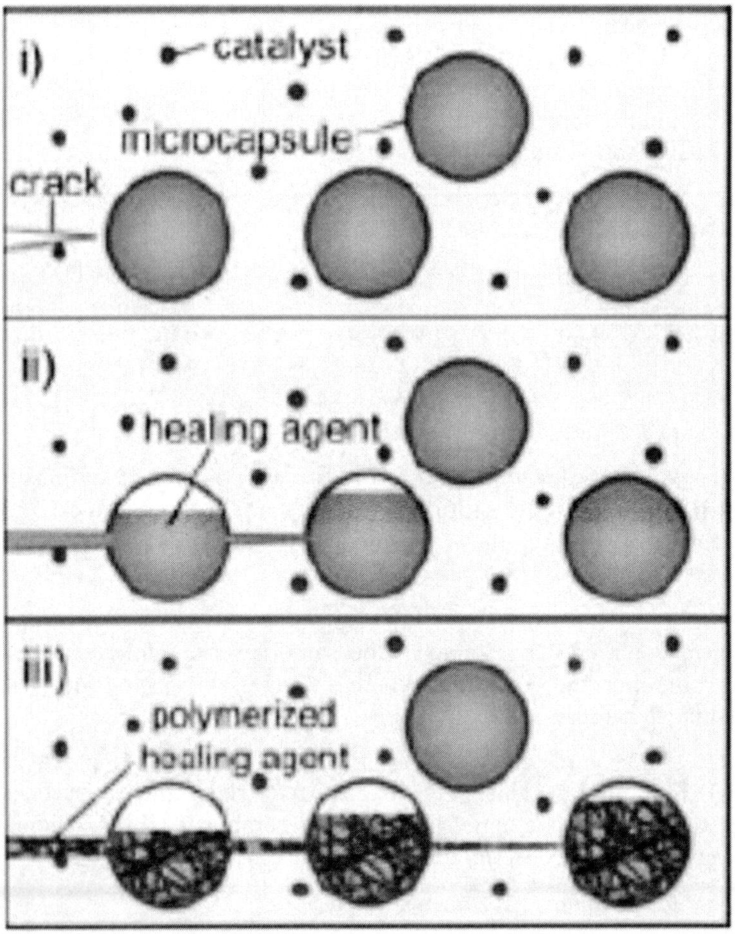

Figure 1. Concept of reservoirs developed by White S. R et al

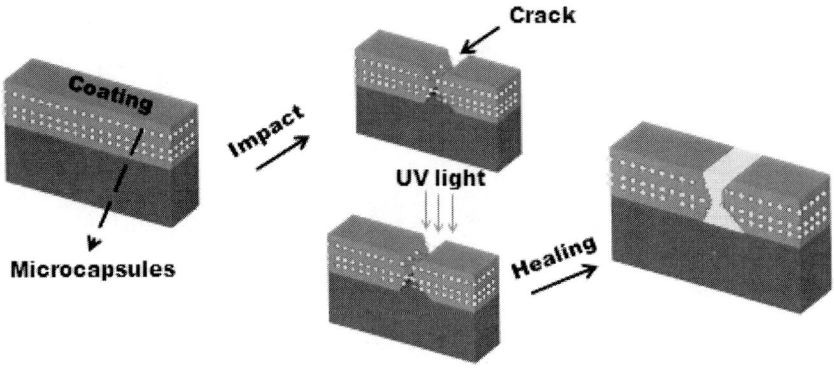

Figure 2. Self-healing of film coatings

2. MICROENCAPSULATION

Microcapsules shell should be resistant to high temperatures (> 300 °C) and have average diameters ranging 1-20 μm in order to put them in the host material.

Different models of microcapsules have been tested: Poly(urea/formaldehyde), Polyurethane and Silica gel

2.1. Synthesis Of Poly(Urea-Formaldehyde) Microcapsules

The healing agent (trimethylol propane triacrylate TMPTA monomer) and the photoinitiator (Darocur®1173) presented in – Figure 3, are encapsulated in poly(urea-formaldehyde) by *in situ* polymerization –Figure 4.

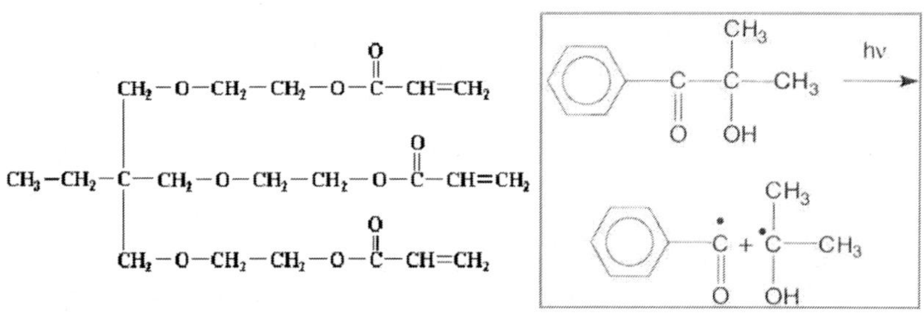

Figure 3. TMPTA monomer and Darocur® 1173 photoinitiator

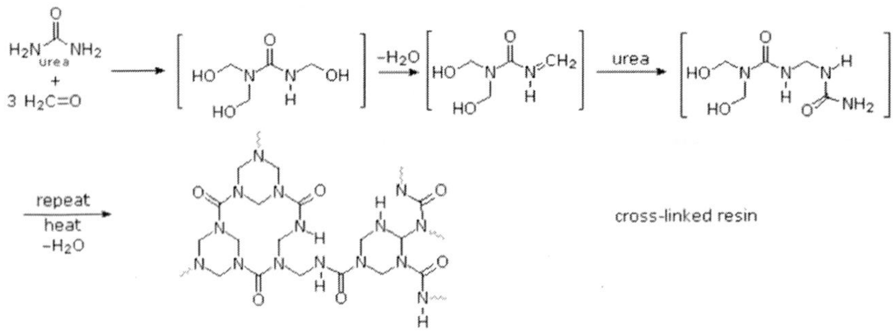

Figure 4. Poly(urea-formaldehyde) resin

These microcapsules are carried out by *in situ* polymerization in direct emulsion (Oil/Water). At room temperature, deionized water and aqueous solution of ethylene maleic anhydride (EMA) are mixed in a reactor. The reactor is placed in a temperature-controlled oil bath on a programmable hot plate with external temperature probe and the solution is agitated with a mechanical stirrer. Under agitation, urea, ammonium chloride and resorcinol were dissolved in the solution. The pH is raised from ~2.6 to 3.5 by drop-wise addition of sodium hydroxide (NaOH) and hydrochloric acid (HCl). One or two drops of 1-octanol are added to eliminate surface bubbles. Then, the healing agent (TMPTA monomer) and the photoinitiator (Darocur® 1173) are added slowly in the solution to lead to a stable emulsion. After 10 minutes of stabilization, an aqueous solution of formaldehyde (37 wt. %) is incorporated in the emulsion to obtain 1:1:9 molar ratio of formaldehyde to urea. The emulsion is covered and heated at 55 °C for 5 hours. After 5 hours of continuous agitation, the mixer and hot plate are switched off. Once cooled to ambient temperature, particles are filtered on Büchner and are washed with a water/ethanol solution (50 wt. %/50 wt. %). Then, particles are dried in air at room temperature for 24-48 h – Figure 5.

Scanning Electron Microscopy - The morphology of particles (surface and form) is investigated by Environmental Scanning Electron Microscopy (ESEM). The particles are spherical with average diameter in the range of 2-30 µm – Figures 6 (a) & 6 (b). The shell of microparticles presents two type of surface: one has a smooth surface – 6 (a), or a rough surface – Figure 6 (b). The Figure 6 (c)illustrates a poly(urea-formaldehyde) microcapsule which has been ruptured under physical constraint (pressure).

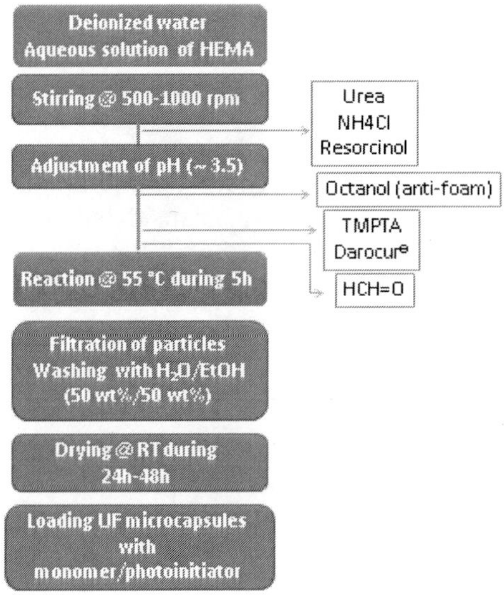

Figure 5. Schematic of the encapsulation process with poly(urea-formaldehyde)

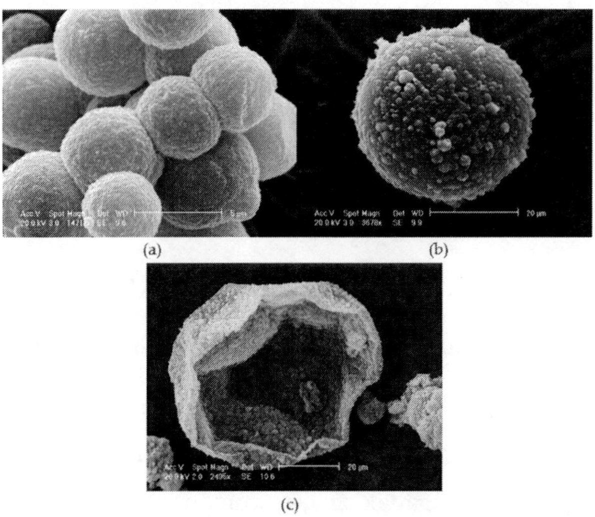

Figure 6. ESEM image of (a) UF microcapsules loading in TMPTA monomer and Darocur® 1173 photoinitiator with a smooth surface, (b) UF microcapsules loading in TMPTA monomer and Darocur® 1173 photoinitiator with a rough surface and (c) ESEM image of ruptured Urea-Formaldehyde microcapsule

Infrared Analysis - The comparison between infrared spectra of UF empty microparticles – Figure 7 (a) and UF microparticles loaded in TMPTA – Figure 3 (b), shows the presence of new bands, in particular at 1722 cm^{-1} (C=O stretching vibration), 983 cm^{-1} (wagging of the $=CH_2$ group), and 808 cm^{-1} (twisting of the $=CH_2$ group), characteristic bands of TMPTA which indicates that this compound has been encapsulated in UF microcapsules.

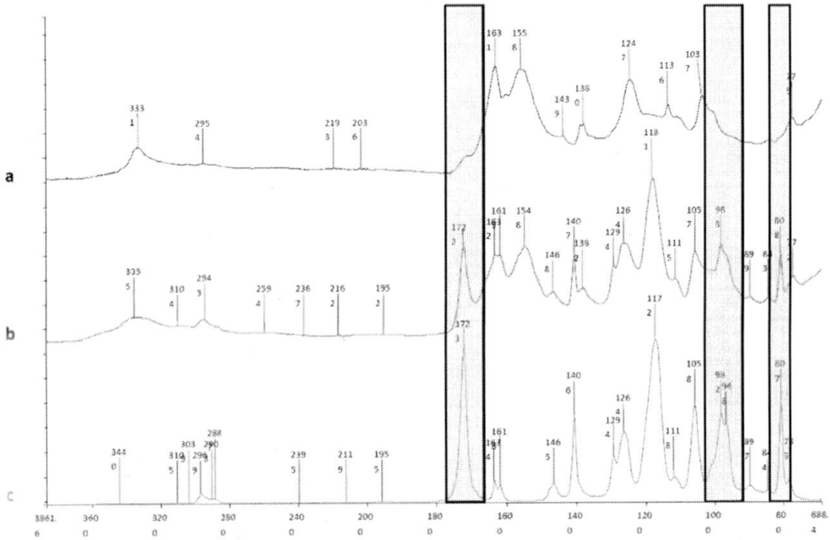

Figure 7. Infrared spectrum of (a) Empty UF microcapsules, (b) broken UF microcapsules loaded with TMPTA and (c) TMPTA monomer

Thermogravimetric Analysis - Empty UF microcapsule shows – Figure 8, a weight loss (around 10 %) until 110 °C corresponding to the evaporation of free water in microcapsules, and an important decreasing in weight (around 80 %) starting from 180 °C due to the decomposition of urea-formaldehyde microparticles. Therefore the urea formaldehyde could not be used to self-heal spatial device composites because of polymer decomposition below 300 °C (maximum temperature in space).

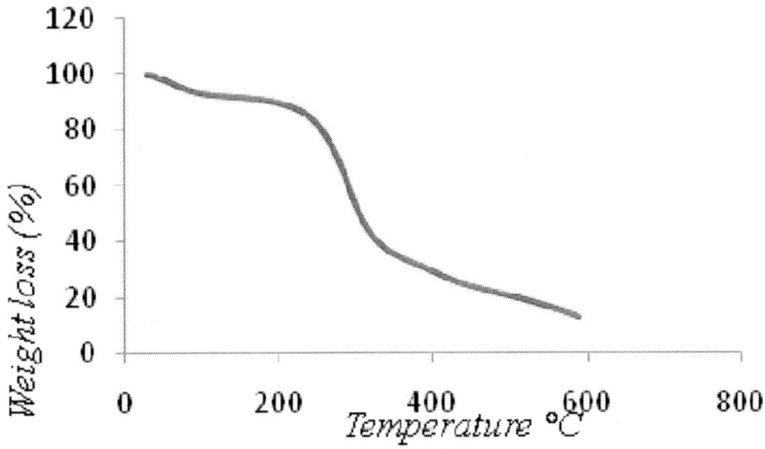

Figure 8. TGA of empty UF microcapsules

2.2. Synthesis of Poly(Urethane) Microcapsules By Interfacial Polycondensation

The TMPTA monomer and the Darocur® 1173 photoinitiator are encapsulated in polyurethane microcapsules by *in situ* polymerization. Microcapsules are obtained by *in situ* polymerization in direct emulsion (Oil/Water). At room temperature, deionized water and polyvinyl alcohol PVOH (3 wt. %) are mixed in a reactor equipped with a mechanical stirrer. Then, an organic solution of monomer (Hexamethylene diisocyanate HMDI/chloroform) containing the monomer (TMPTA) and photoinitiator (Darocur® 1173) to be encapsulated is added to the aqueous solution leading to a stable Oil/Water emulsion. After 5 minutes of stabilization, a hydrophilic monomer (Ethylene Diamine EDA in excess) is added to the emulsion – Figure 9. The reaction is continued until stabilization of the pH (around 5 hours).

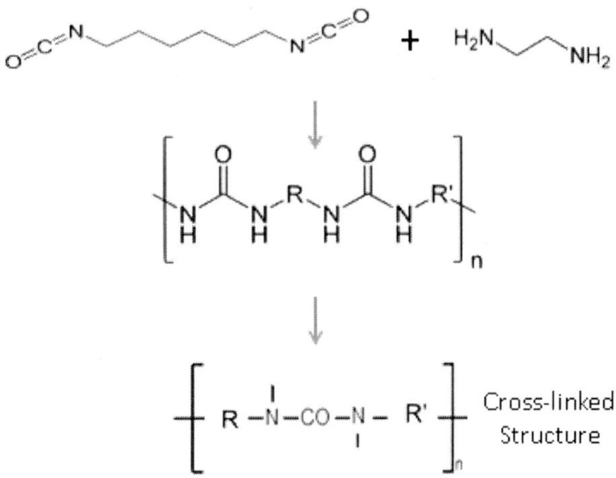

Figure 9. Cross-linked structure of polyurethane with R = $-(CH_2)_6-$ and R' = $-(CH_2)_2-$

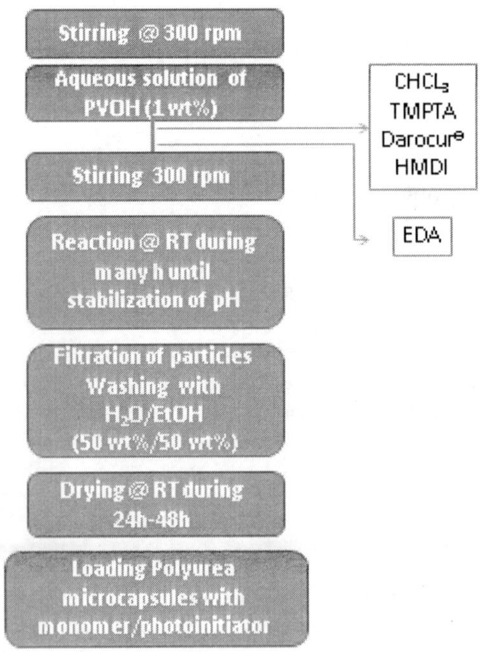

Figure 10. Schematic of the encapsulation process with polyurethane

Then, particles are separated under vacuum with a Büchner system and are rinsed with an ethanol/deionized water mixture (50 wt. %/50 wt. %). Microparticles are collected and are air dried during 24-48 h – Figure 10.

Scanning electron microscopy - Particles were examined by ESEM. The particles are spherical with average diameter in the range of 200-300 μm and the membrane of microparticles is smooth, as shown in – Figure 11.

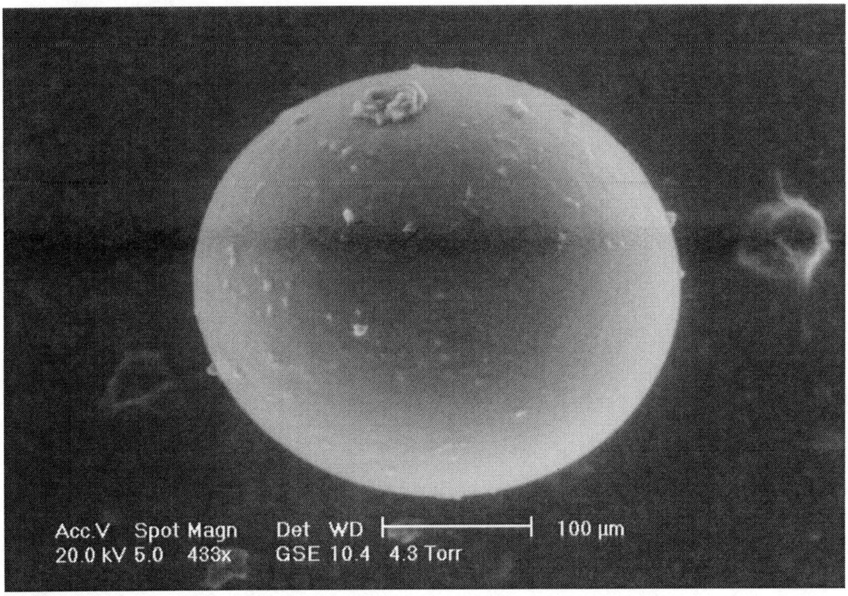

Figure 11. ESEM image of Poly(urethane) microcapsules loaded with TMPTA monomer and Darocur® 1173 photoinitiator

Infrared Analysis - The comparison between Infrared spectrums of the empty polyurethane microparticles - Figure 12 (b), and polyurethane microparticles loaded in TMPTA – Figure 12(c) shows the presence of new bands, in particular at 1721 and 983 cm^{-1} and confirm that TMPTA monomer has been encapsulated in polyurethane microcapsules.

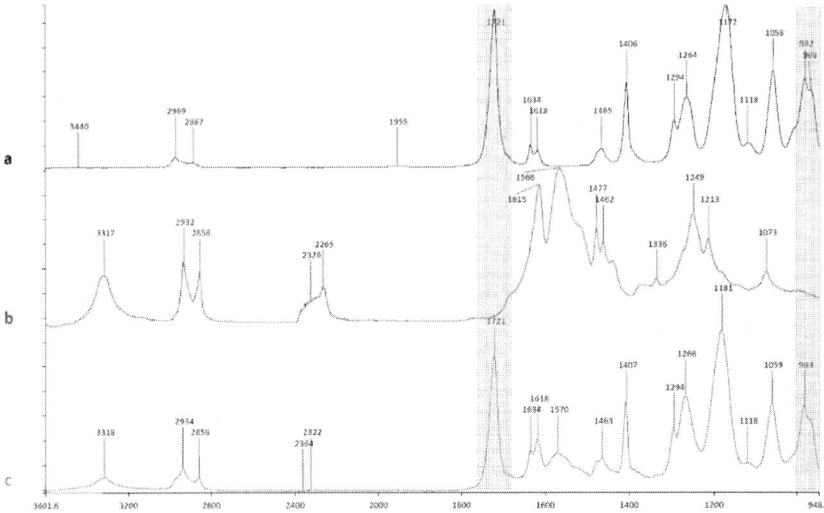

Figure 12. Infrared spectrum of (a) TMPTA, (b) empty polyurethane microparticles and (c) polyurethane microcapsules loaded with TMPTA monomer

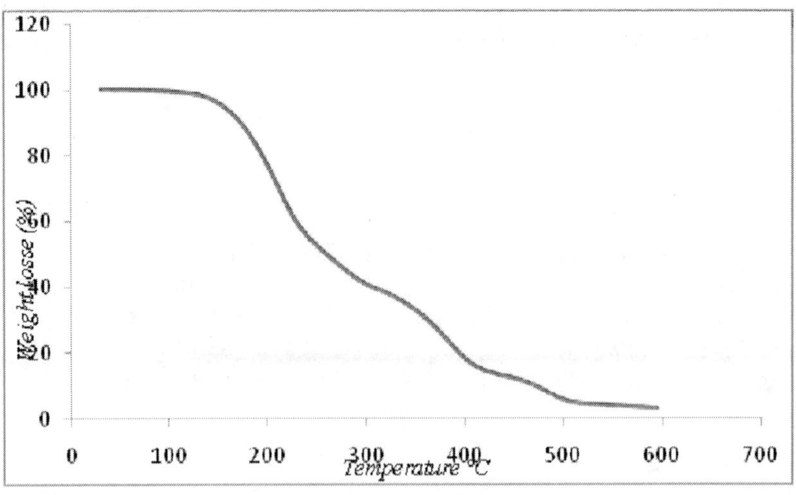

Figure 13. TGA of empty polyurethane microcapsules

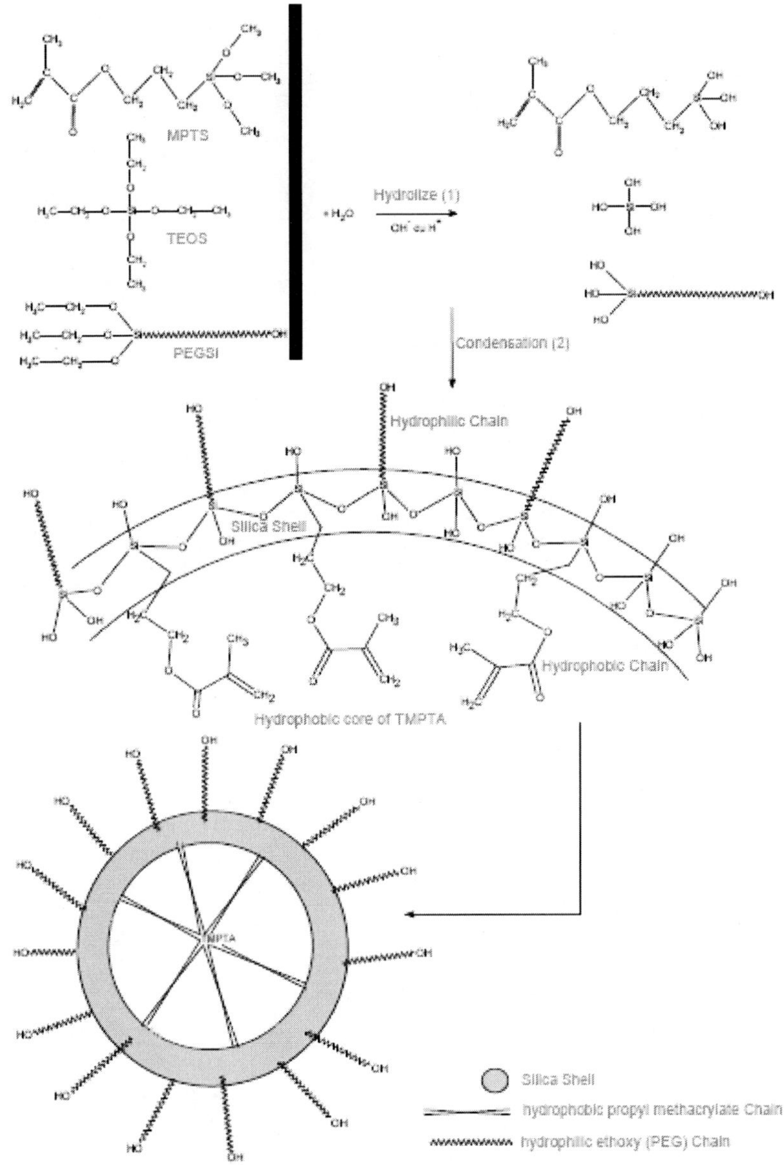

Figure 14. Schematic Road for Synthesis of Silica-Gel Microcapsules Loaded with TMPTA

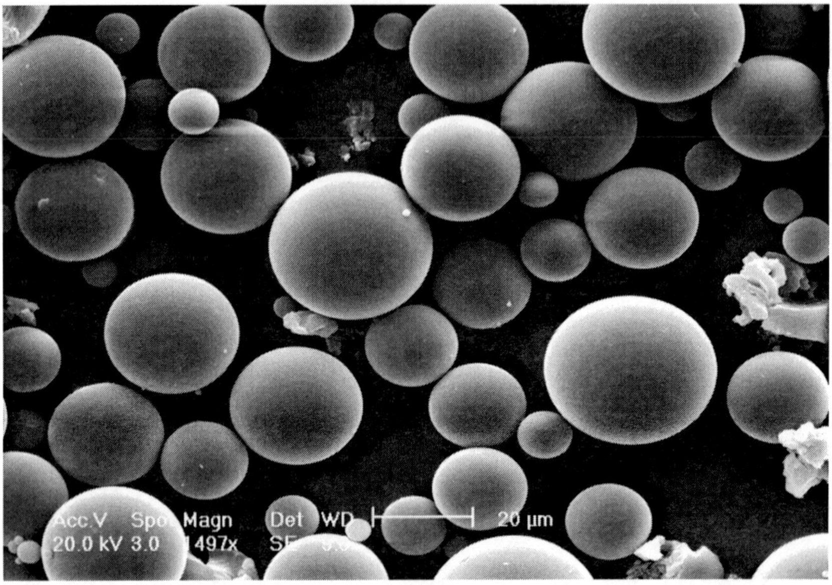

Figure 15. ESEM image of Silica-gel microcapsules loading with TMPTA monomer and Darocur® 1173 photoinitiator

Thermogravimetric Analysis - The thermogravimetric analysis of polyurethane microcapsules (without monomer) – Figure 13, presents a decreasing in weight (above 200°C) due to the deterioration of polymer. Therefore, this polymer is not a good candidate as well to encapsulate healed agent for aerospace applications.

2.3. Synthesis of Silica-Gel Microcapsules By Sol-Gel Polymerization Process

The TMPTA monomer and the Darocur® 1173 photoinitiator are encapsulated in silica gels microcapsules by sol-gel polymerization. 3-(trimethoxysilyl)propyl methacrylate (MPTS), tetraethoxysilane (TEOS), trimethylol propane triacrylate (TMPTA), Darocur® 1173, silica surfactant PEGSi (compatibility agent to increase the stability between the shell and the aqueous solution) and ethanol are mixed with a magnetic stirrer in a bottle. Then, an aqueous solution of ammoniac (16 wt. %) is added drop by drop to the mixture. After 10 minutes of continuous agitation, this solution is added to an aqueous solution of polyoxyethylene 1, 2- nonylphenyl ether

(Igepal CO-720 (NP12) at 1 wt. %. After 2 hours of reaction, the suspension is filtered under vacuum. Microparticles are collected and are dried in air at room temperature for 24-48 h. Figure 14 illustrates the schematic road to synthesis of silica-gel microcapsules loaded with TMPTA monomer [9].

Scanning Electron Microscopy - Silica microcapsules are studied by ESEM. The spherical particles have average diameters ranging from 1-30 μm, shown in - Figure 15. The shells of microparticles present a smooth surface.

Infrared Analysis - The comparison of IR spectra of empty silica-gel microparticules and silica-gel microparticles loaded with TMPTA do not enable us to prove the encapsulation of the TMPTA. In effect, the two spectra present the same absorption bands because silica-gel shell contains a methacrylate compound – Figure 16.

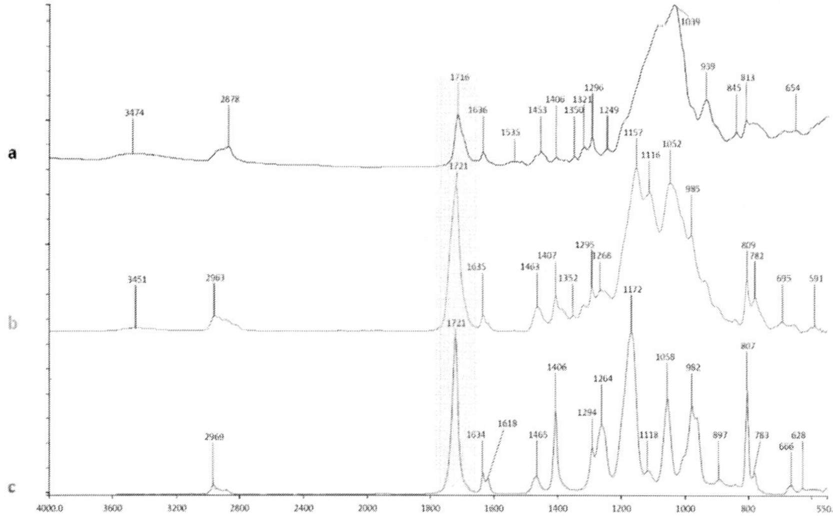

Figure 16. Infrared spectrum of (a) Empty silica-gel microcapsules (b) silica- gel microcapsules loaded with TMPTA monomer and (c) pure TMPTA

For that we have used silica-gel microcapsules (empty microcapsules and microcapsules loaded with TMPTA) with TEOS precursor alone (not methacrylate chain in this compound). The comparison between Infrared spectra of the empty silica-gel microparticles with TEOS precursor alone – Figure 17 (a) and silica-gel microparticles loaded with TMPTA – Figure 17 (c), shows the presence of new bands, in particular at 1721, 983 and 807 cm^{-1}. These bands, characteristic of the TMPTA monomer, confirm the presence of TMPTA encapsulated in silica-gel microcapsules.

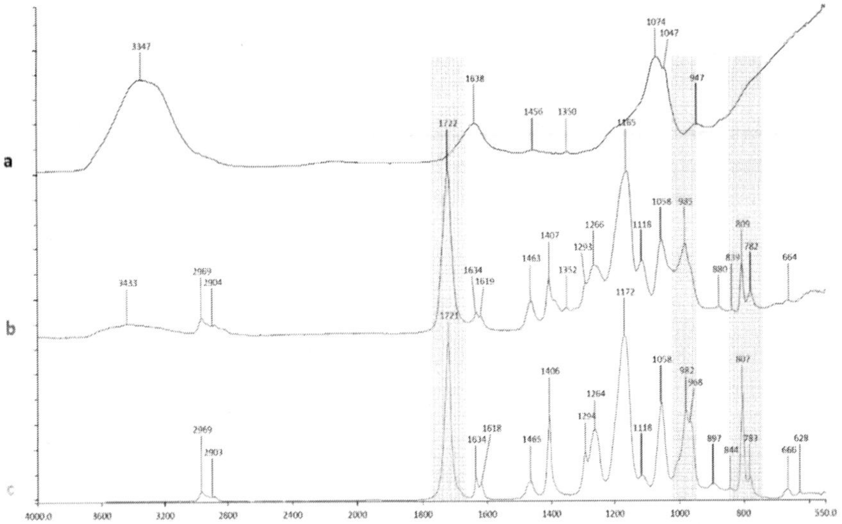

Figure 17. Infrared Spectrum of (a) empty silica-gel microcapsules with TEOS precursor alone, (b) silica-gel microcapsules with TEOS precursor alone loaded with TMPTA monomer and (c) pure TMPTA

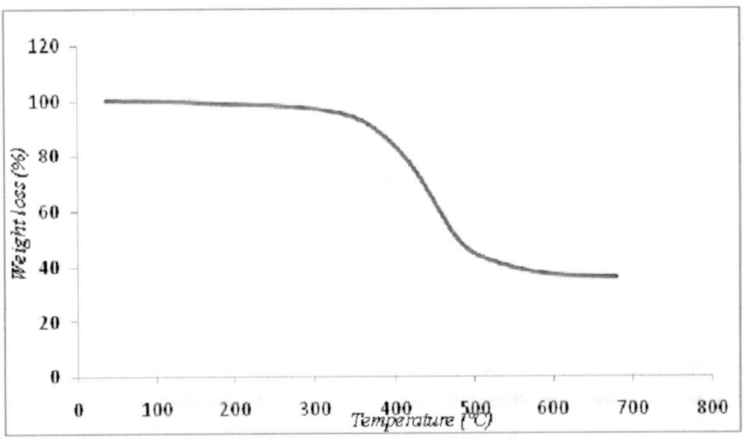

Figure 18. Thermogravimetric Analysis of Empty Silica-gel Microcapsules

Thermogravimetric Analysis - The thermogravimetric analysis of silica-gel particles (without TMPTA) indicates a good thermal resistance (above 320°C) of silica-gel membranes – Figure 18.

We have encapsulated with success TMPTA monomer by different processes of microencapsulation. Two of three techniques used cannot be applicable for aerospace application as the microcapsules are not thermal resistant above 300 °C, presenting a temperature of degradation below 200°C. However, silica membrane is much more resistant due to their degradation temperatures exceeding 300°C.

3. DISPERSION INTO THE POLYIMIDE FILM

Polyimide films are based on dianhydride of dicyclodecene tetracarbonic acid and oxydianiline ODA –Figure 19. The dianhydride was synthesized by Solar-irradiation technology of a mixture of benzene/toluene fraction with maleic anhydride to produce highly reactive monomer – benzene adduct (AB) – Figure 20.

Microcapsules loaded with TMPTA and the radical photoinitiator are mixed in 5 to 15 wt. % in solution of polyimide in DMF, cast and dry, and then subjected to 100 °C for 4 hours, followed by 4 hours at 180 °C.

Environmental scanning electronic microscopy coupled with RX (ESEM/EXD) – Figure 21(a) and21(b), show a good dispersion of the microcapsules along the polyimide film and confirm their presence.

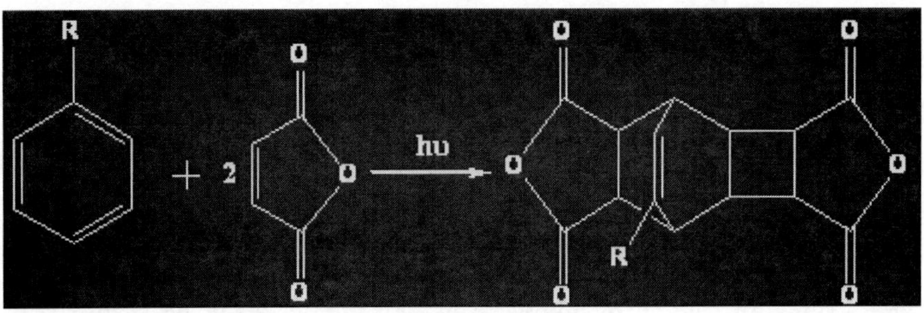

Figure 19. Photosynthesis of the dianhydride of dicyclodecene tetracarboxylic acis

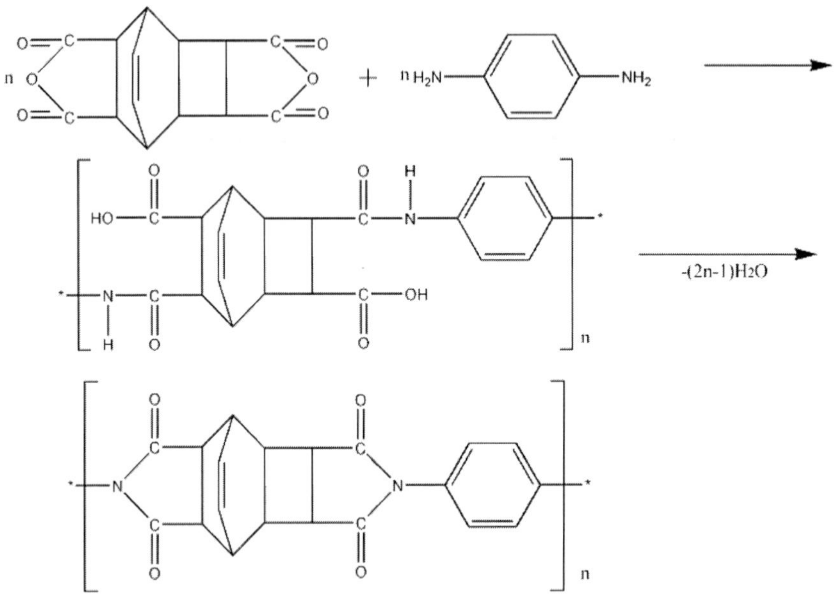

Figure 20. Polyimide based on dianhydride of dicyclodecene tetracarbonic acid and ODA

EDX of polyimide film does not show any silicium peak - Figure 21 (a) but this peak is present in the analysis of the polyimide with silica-gel microcapsules.

4. AUTO-REPARATION OF POLYIMIDE FILM

Different formulation of Polyimide films have been loaded with variable concentration of microcapsules from 5 wt. % to 15 wt. %. Then the films casted are exposed to UV light (365 nm) once cracks were produced – Figures 22 & 23.

Self-healing process was followed by morphological analysis shown by environmental scanning electron microscopy ESEM.

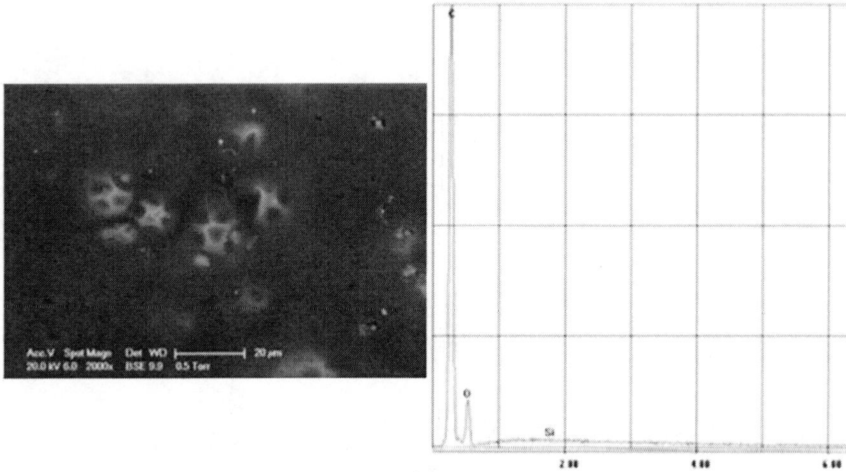

(a)

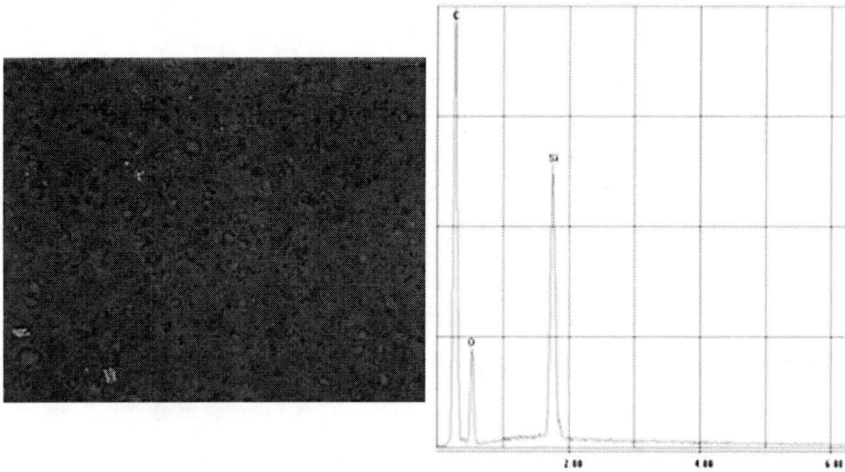

(b)

Figure 21. (a). Surface aspect of the polyimide film without microcapsules & EDX(b). Surface aspect of the polyimide film with microcapsules & EDX

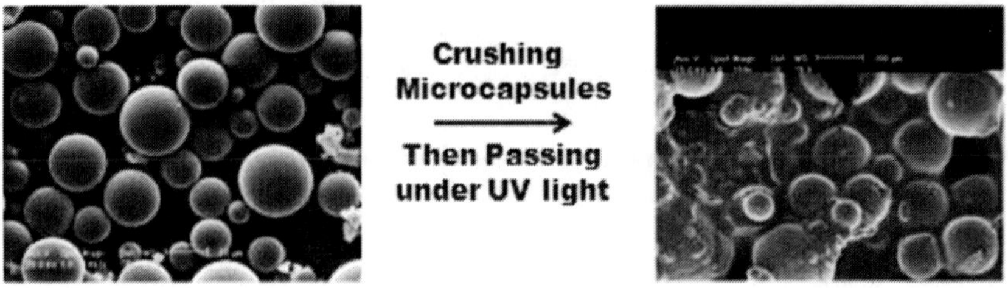

Figure 22. Auto-repairing after cracks

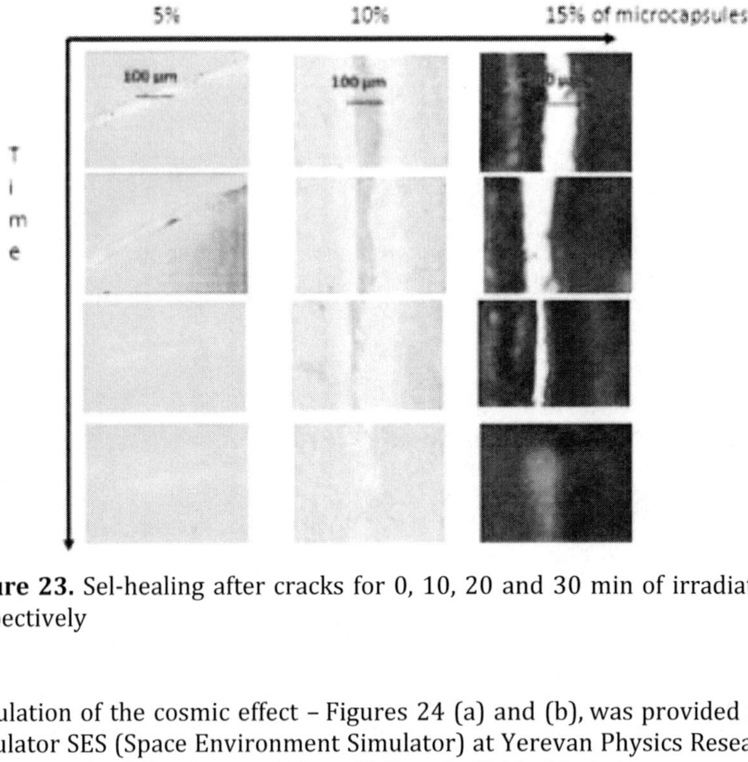

Figure 23. Sel-healing after cracks for 0, 10, 20 and 30 min of irradiation respectively

Simulation of the cosmic effect – Figures 24 (a) and (b), was provided in a simulator SES (Space Environment Simulator) at Yerevan Physics Research Institute at the following conditions (260 nm) –Table 1 below.

PI films loaded with 15 wt. % of Si-microcapsules loaded with TMPTA monomer exposed 30 min are equivalent to 6 month exposure at the real space condition over geostationary orbit of the earth.

Table 1. Simulator parameters

Experimental Conditions	Units
Vacuum	$1x10^{-5}$ Torr
Electrons Bombardment	5.0 MeV
Temperature	−50 °C
UV Solar	2.0 kW / sq.m.
Proton Bombardment pfu	> 10 MeV

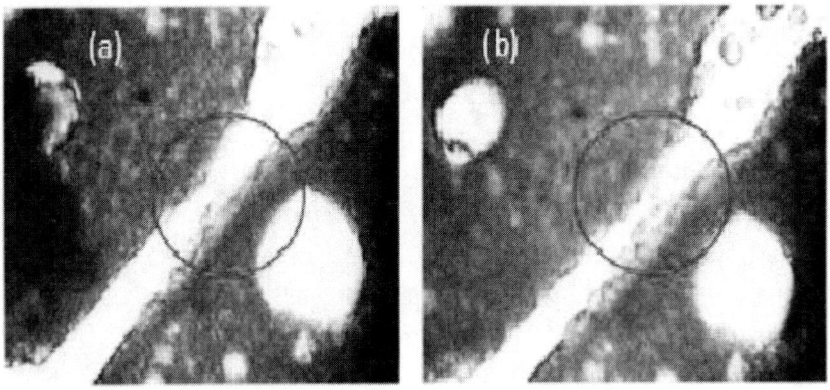

Figure 24. Cracks at time zero (a) – Cracks at exposure time 5 min under 260 nm (b)

5. MECHANICAL PERFORMANCE OF THE COMPOSITE FILM

The capsules were the defects of the structure in which cracks mainly grew from. The increases in capsules concentration led to a more brittle fracture as shown in Figure 25.

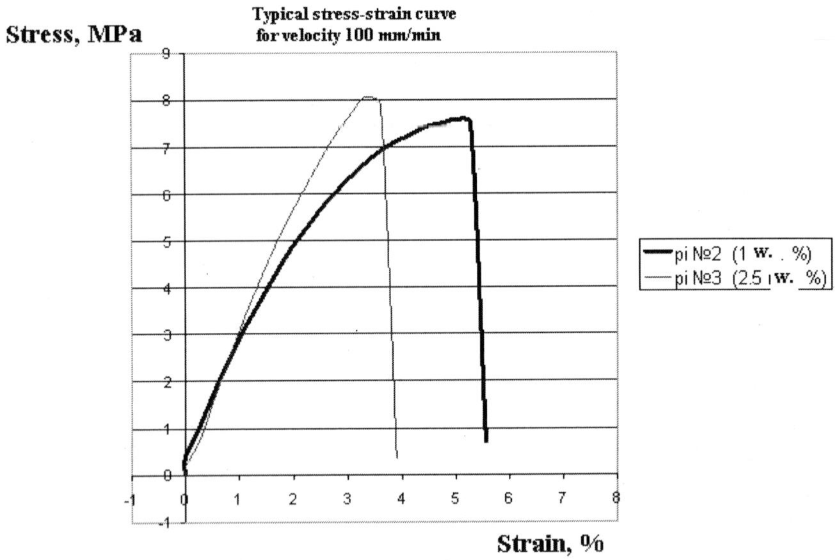

Figure 25. Effect of the wt. % loading on the strain-stress curve

Tensile curves were measured using a universal "Shimadzu" Autograph AGS machine. The test samples were strips with working size of 5x30 mm. The thickness of samples was measured by a magnetic "PosiTector 6000" instrument.

Then, a circular hole (diameter = 0.3 mm) was made in each specimen with the least (0.5 wt. %) and greatest capsules concentration (5 wt. %): Specimens pi-1 and pi-4 respectively. The holes were made by special drill press.

According to the theory of elasticity, the stress concentration coefficient near circular hole is equal to 3; in fact the real stress near the hole is approximately one. So, the circular hole might be considered as the standard stress concentrator.

Stress – strain curves for Specimens pi N°1 (0.5 wt. %) and pi N°4 (1.5 wt. %) with hole are represented in Figure 26.

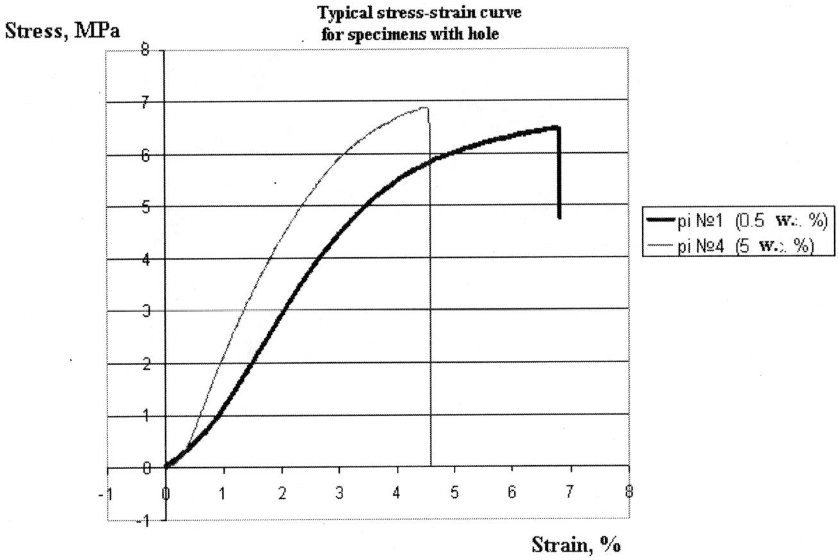

Figure 26. Effect of the wt. % loading on the strain-stress curve of the composite film with hole

The experiment showed that for Specimens pi N°1, the crack passed through the hole. So stress concentration near capsules was less than 3. But for Specimens pi N°4 the crack went through the hole. With the help of this method we can separate films with capsules into two classes: with effective stress concentration less and more than 3.

Using the polarizing microscope POLAR 312, we have detected stress fields near the capsules, even before the loading. It must be taken into account that heat input might be needed for stress relaxation to occur. On Figure 27, stress fields are represented by bright spot.

Around some capsules, internal strain remained even after rupture of the specimen. In the experiment the film was elongated manually under the microscope by special micromachine.

Stress fields were detected near the hole too as depicted in Figure 28.

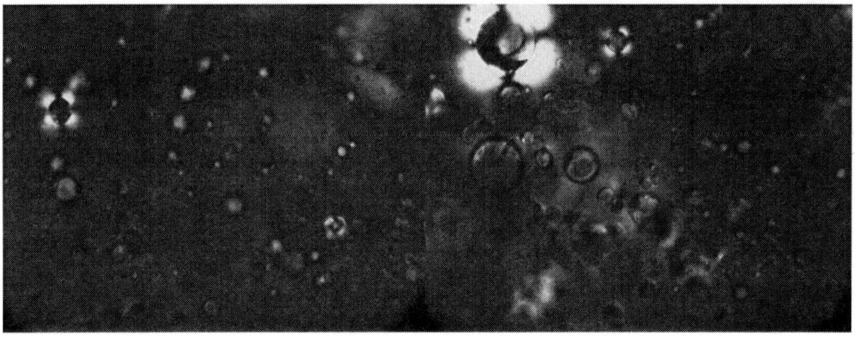

Figure 27. Stress field near hole

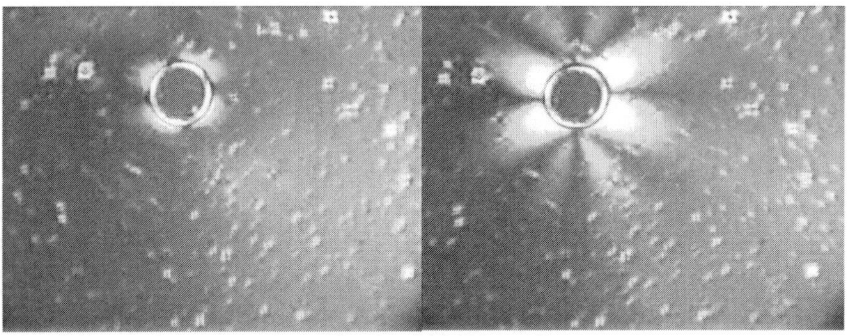

Figure 28. Stress field near hole

The increasing strain near the hole was illustrated in the photos placed in order. The cracks grew through each hole visually. For low concentration of capsules, cracks led to rupture. For high concentration, big crack arose from capsules. It meant that local concentration of stress depended on overlapping of stress fields near capsules too. Theoretically, such interaction appeared when distance between capsules was less than 5 capsule diameters – Figure 29.

These observations allowed us to conclude that stress fields near capsules are the predominant effect; capsules concentration influenced the fracture when stress fields overlapped near capsules.

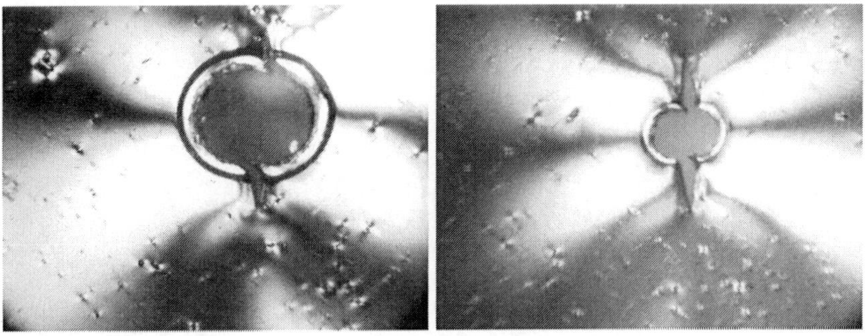

Figure 29. Stress field near hole after elongation

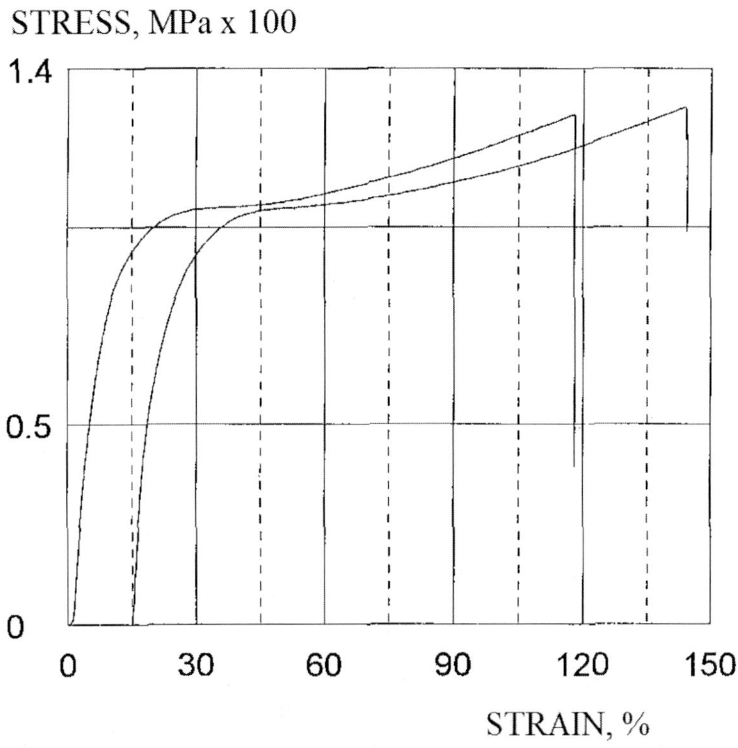

Figure 30. Typical stress-strain curve for pure PI film (without spherical particles)

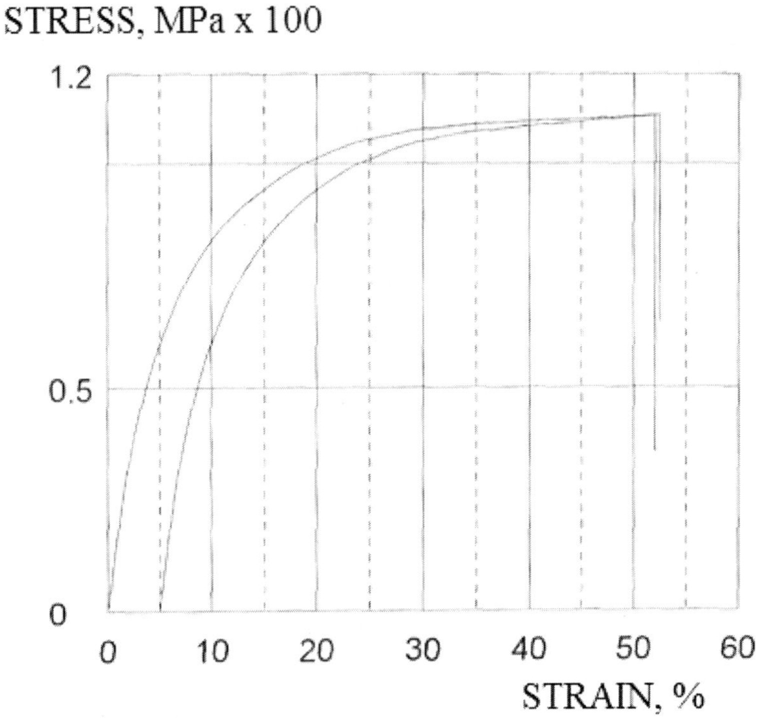

Figure 31. Typical stress-strain curves for PI containing 10 wt. % of spherical particles with healing resin

Test pieces were cut from polyimide films. Straight samples, 10 mm in width and 20 mm in length, were tested in tension at 2 mm/min speed with "Shimadzu" "Autograph" machine. The thickness of samples varied from sample to sample and was approximately 0.12 mm. After testing, failed samples were studied under an optical LOMO Mikmed-2 microscope. Hitachi S-520 scanning electronic microscope was used to get a high-magnification image of samples' top and side surfaces, and fracture surfaces of samples were studied with microscopes.

Figure 30 shows the typical stress-strain curves for pure PI film (without particles containing healing resin). The diagram is typical for plastic materials. Initially, the material deformed elastically, and yielding at strain more than 5 % was observed. The Young modulus of fracture deformation of samples was very high, around 130 %.

Figure 31 shows typical stress-strain curves for PI containing 10 wt. % of particles with healing resin. The Young modulus of composite polymer was calculated from the slope of the stress-strain curves to be 2.6 GPa. The curves were similar to that of pure PI. However, the fracture strain was lower at approximately 50 %, due to the stress concentration near particles.

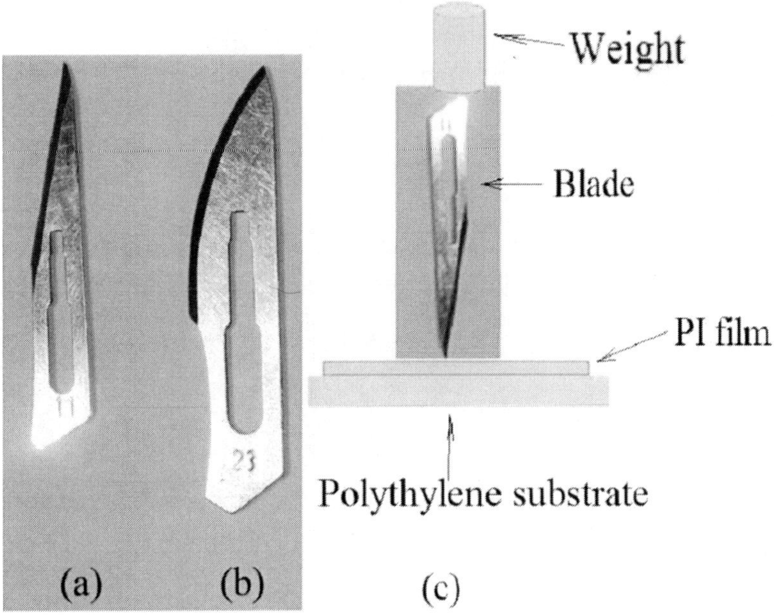

Figure 32. Changeable scalpel blades (a) and (b) used to make a crack in PI film samples. (c) – Method of making through-the-thickness crack

In each sample, two types of cracks were made. First type of cracks was through-the-thickness. These cracks were made by applying 100 g force on a changeable scalpel blade as shown in Figure 32a-c.

The resulting crack is shown in Figure 33 below where spherical particles of different size are observed.

Figure 34a shows a typical stress-strain curve for composite PI sample with a crack similar to that inFigure 33. The curve is similar to that of composite PI without a crack. The fracture strain was about 18 %, which was typical for ductile materials. However, fracture strain was lower than that in samples without a crack.

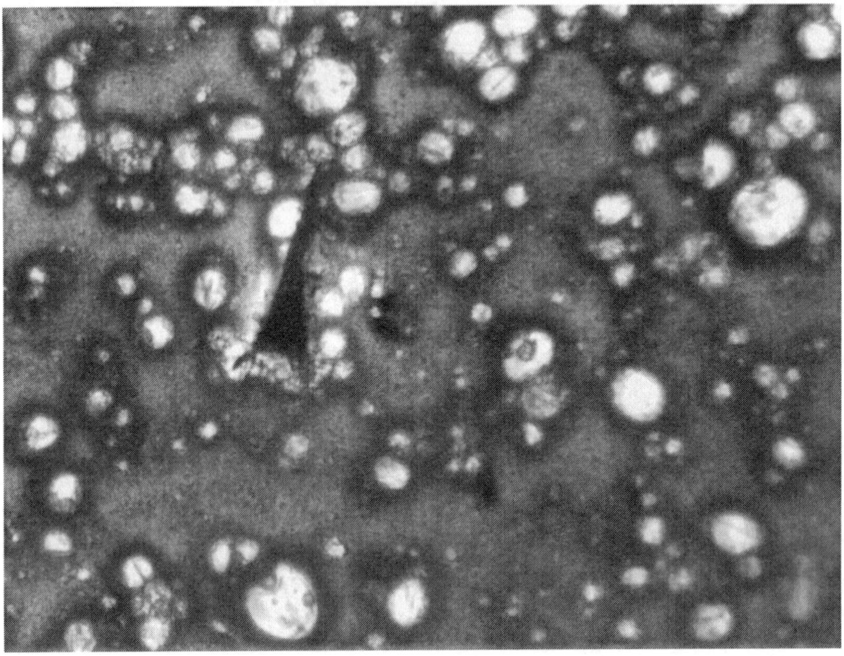

Figure 33. Throught-the-thickness triangular crack with the length of approximately 0.6 mm

Figure 34b shows a stress-strain curve of a similar sample (with a crack) after healing with 200 Watts UV lamp for 30 min. The strength and fracture strain is the same as in Figure 31. Thus, this implied that the healing did not improve mechanical properties if the crack was thick. Possibly, the amount of the liquid resin was not enough to fill the entire volume of the open crack.

The fracture toughness G_{Ic} of filled PI may be estimated from the Figure 31 with the equation below:

$$G_{Ic} = \frac{\pi \sigma^2 c}{E} \quad (1)$$

where σ is fracture stress, c is the half of the crack length and E is the Young modulus of material. Substituting σ = 80 MPa, c = 0.6 mm and E = 2.6 GPa, the fracture toughness G_{Ic} is estimated to be 2.6 kJ/m^2. This value is typical for very tough materials.

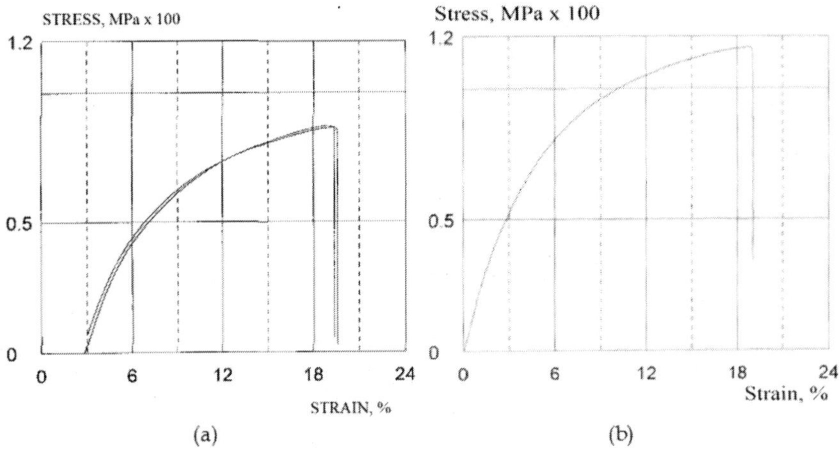

Figure 34. Typical stress-strain curve for filled PI sample with a crack. (a) – without UV irradiation; (b) – after UV irradiation

Different results were obtained from samples with crack of the second type created as shown in Figure 35.

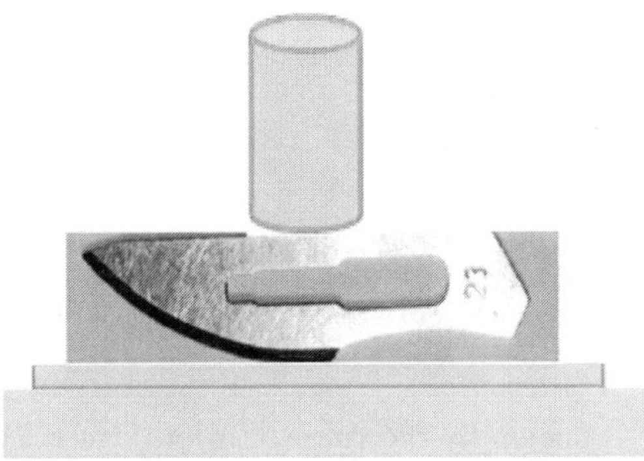

Figure 35. Method of making surface cracks

In this case the crack was on the surface and not deep. These cracks were made by putting 100 g force on a curved scalpel blade. The crack was not wide with low opening, and could not be measured with an optical microscope.

Figure 36 shows typical stress-strain curves for samples with crack subjected to UV irradiation. In this case, two different types stress-strain curves were observed.

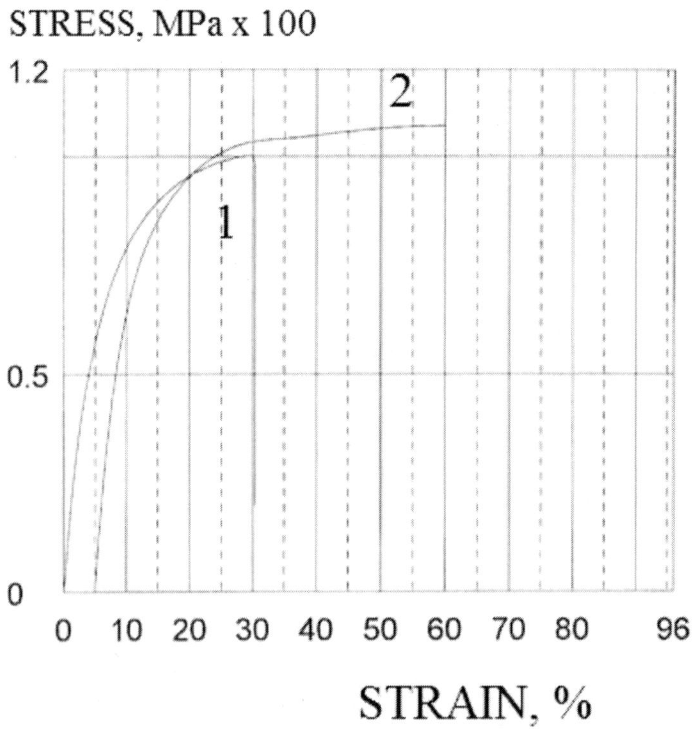

Figure 36. Two types of stress-strain curves for samples with surface cracks made as shown in Figure 35 and irradiated with UV light

The first one is shown by the curve 1 in the Figure 36. In this case the fracture strain is equal to approximately 30 %. This value is higher than that before UV irradiation (20-22 %). The stress-strain curve of the second type is shown by the curve 2 in the Figure 36. In this case the fracture strain is approximately 55 %, which was much higher than in the first case.

In the composite samples two layers were observed : the first one was pure PI and the second was PI filled with spherical particles. Samples with double-layered structure can be observed in Figure 37.

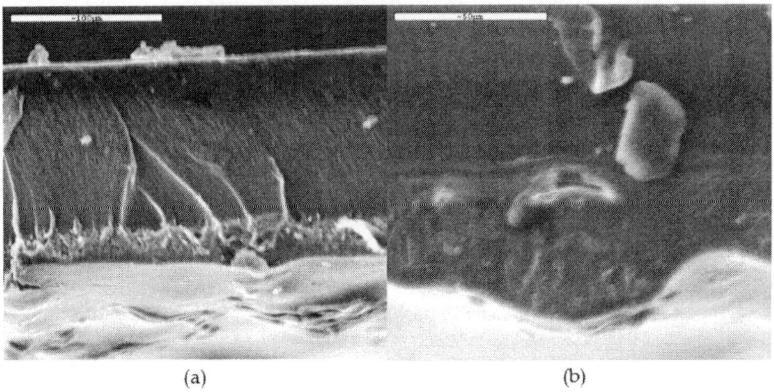

Figure 37. Samples with double-layered structure frozen in liquid nitrogen, at different magnifications. In the right photo, a round particle can be noticed

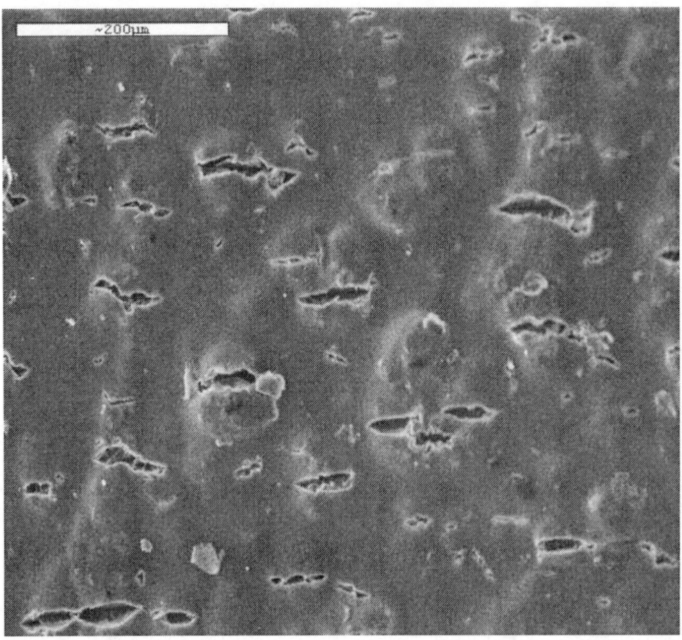

Figure 38. Local cracking of surface layer with PI filled by spherical particles

Study of the fractured samples showed that the layer of the composite PI debonded from the pure PI layer, and this explained higher fracture toughness in this case. Local cracking and debonding of composite surface layer under tensile load can be observed in Figure 38.

Schematic drawing of debonding of a surface filled layer near a crack is shown in Figure 39.

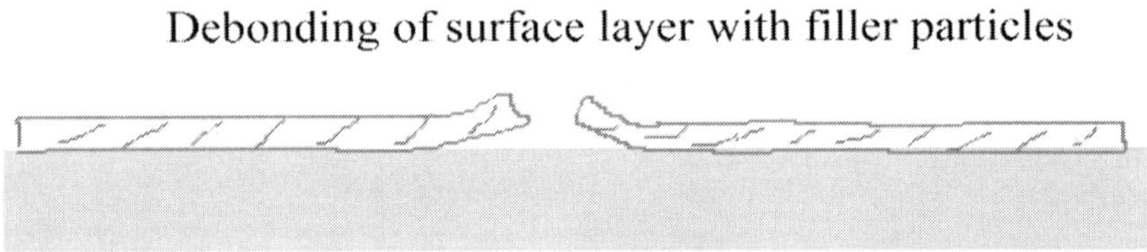

Figure 39. Schematic of debonding of composite layer at the surface near a crack

6. FORMULATION FOR LOW ORBITAL SPACEFLIGHT

In the condition of low orbital spaceflight (under 30 km), the presence of oxygen might reduce the photospeed of the crosslinking when radical process is used. To overcome this problem we have investigated the use of epoxies initiated by a cationic process of which kinetic is non sensitive to the presence of oxygen [10-13].

For that purpose we have considered a multifunctional epoxy formulations based on dicyclopentadiene and phenolic epoxy, such as solid Epiclon-HP720 – Figure 40.

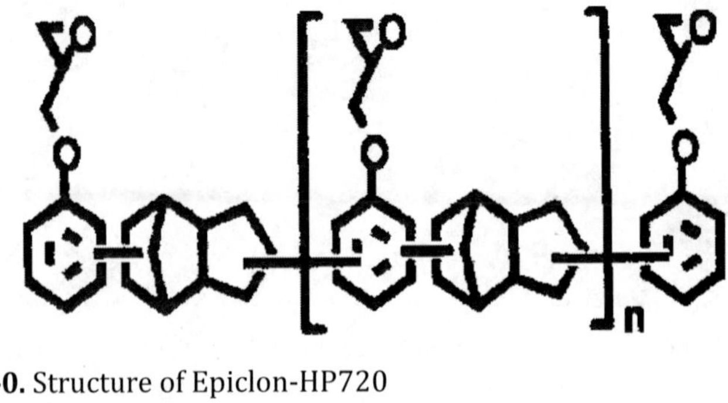

Figure 40. Structure of Epiclon-HP720

Different types of co-solvents have been studied in formulating solid epoxy resin (Epiclon-HP7200) based either:

1. on epoxy: two cycloaliphatic diepoxy monomers, both having terminal epoxy functional groups at opposite sides of the molecule, being separated by one functional carboxylate, so-called, "space" group in the case of 3,4-epoxycyclohexylmethyl-3,4-epoxycyclohexane carboxylate - Cyracure® UVR-6105 - Figure 41, and by two carboxylate "space" groups in the case of Bis-(3,4-epoxycyclohexyl) adipate - Cyracure® UVR-6128 - Figure 42, and having thus different length of chemical structure have been selected.

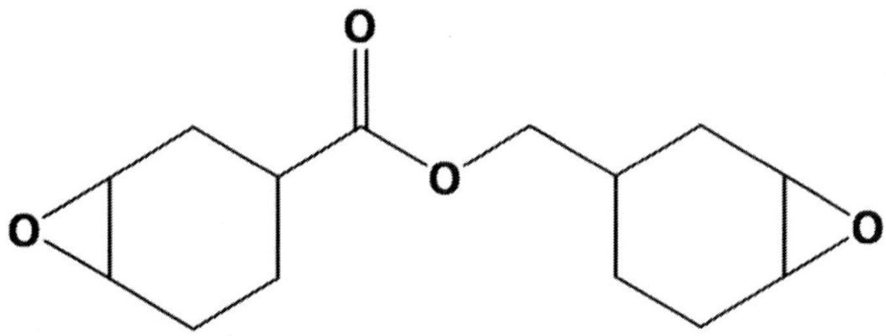

Figure 41.

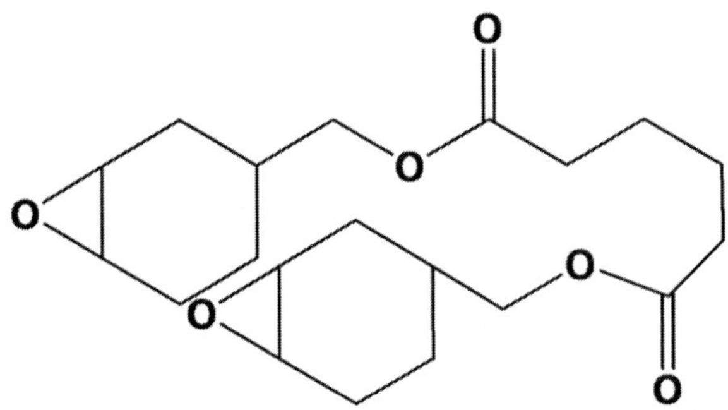

Figure 42.

2. or divinyl ether, triethylene glycoldivinyl ether - Figure 28.

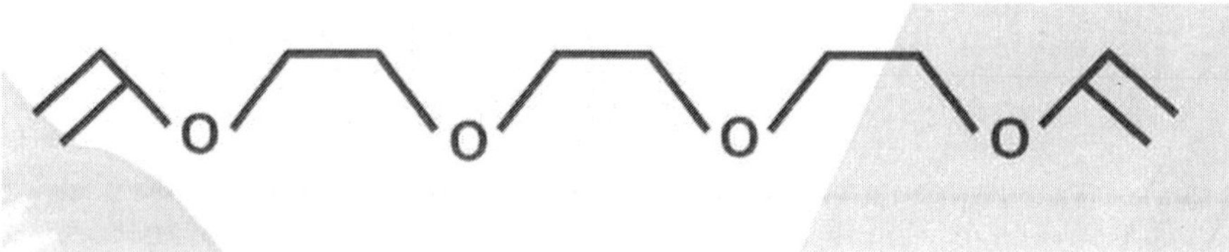

Figure 43. Chemical structure of Triethyleneglycol divinyl ether (Rapid-Cure® DVE-3)

The final objective was to formulate a system based on epoxy having the highest photoreactivity for self-healing applications.

The Photo-Differential Scanning Calorimetry (photo-DSC) [10] was used to study the cure kinetics of UV-initiated photo-polymerization of epoxy resin monomers and vinyl ether in presence of cationic photo-initiator, mainly Cyracure® UVI-6976, former Cyracure® UVI-6974, which consists of a mixture of dihexafluoroantimonate of S, S, S', S'-tetraphenylthiobis (4,1-phenyllene) disulfonium and hexafluoroantimonate of diphenyl (4 phenylthiophenyl) sulfonium (CAS no. 89452-32-9 and 71449-78-0) at 50 wt. % in propylene carbonate - Figure 44 [11].

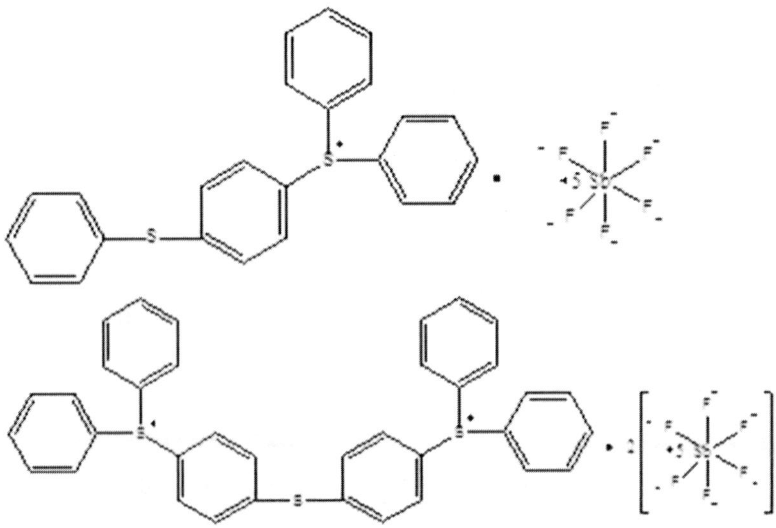

Figure 44. Cationic photoinitiator Cyracure® UVI-6976

We have studied the effect of the temperature for different solvents on the rate coefficient k according to the simplified Sestak and Berggren equation [14] and use the Arrhenius equation to calculate their activation energy Ea – Table 2.

Table 2. Activation energy for different solvents and epoxy Epiclon-HP720 solution

Solvents & Epoxy	Ea (kJ/mol)
Carboxylate (Cyracure® UVR 6105)	7.02 ±0.80
Adipate (Cyracure® UVR 6128)	4.03±0.20
Vinyl ether (Rapi-Cure® DVE-3)	3.60 ±0.42
Epiclon HP-7200/Rapi-Cure® DVE-3 (40/60 wt. %)	10.60±0.75

Photocuring kinetic of UV-initiated cationic photo-polymerization of divinyl ether shows a very low activation energy comparing to the two epoxies studied above. The Epiclon HP-7200/ Rapi-Cure DVE-3 system has been optimized by Differential Photo Calorimetry (DPC) in presence of cationic photoinitiator Cyracure® UVI-6976.

We have shown that the mixture of a multi-functional epoxy resin Epiclon HP-7200 and divinyl ether Rapid-Cure® DVE-3 for the ratio Epiclon/DVE (60 wt. % / 40 wt. %), is a valuable candidate for UV encapsulation and for further self-healing test.

7. CONCLUSIONS

Silica-oragnic microcapsules based on 50 wt. % MPTS and 50 wt. % TEOS have been selected to encapsulate the monomer trimethylol propane triacrylate TMPTA and the photoinitiator Darocur®1173. The microcapsules prepared have diameter size of 4.00 ± 0.30 μm in optimal synthesis conditions: reaction time of 2 h and stirring rate of 450 rpm.

These silica-gel macrocapsules have been chosen over urea/formaldehyde or polyurethane due to their thermal stability up to 375 °C.

We have demonstrated the feasibility of self-healing of Polyimide films as coating using a photo-radical mechanism to crosslink the TMPTA monomer.

Mechanical properties of film coatings have been investigated and found that -Stress fields near capsules are the essential factor -Large through-the-thickness and side cracks are not healed -Short surface cracks are healed and Debonding of the surface layer containing spherical particles is the most effective mechanism of healing.

For low orbital spaceflight (under 30 km) where there is the presence of small amount of oxygen, the photoreactivity of different formulations based on epoxies and vinyl ethers have been optimized by DPC, showing the feasibility of self-healing in using a photo-cationic mechanism.

REFERENCES

1. M. R Kessler- Sefl, Healing: a New Paradigm in Material Designs. Proc. IMechE 221Part G: J. Aerospace Engineering, 479-495 (2007
2. S. R White, N. R Sottos, J. S Geubelle, M. R Moore, S. R Kessler, E. N Sriram, S Brown, Viswanathan- Autonomic Healing of Polymer Composites. Nature, 409, 794-797 (2001
3. S Cho, H. M Andersson, S. R White, N. R Sottos, P Braun- Environmentally, Stable Polydimethysiloxane-Based Self-Healing of Polymers. Advanced Materials 18 (8) 997-1000 (2006
4. S Cho, H. M Andersson, S. R White, N. R Sottos, P Braun- Environmentally, Stable Polydimethysiloxane-Based Self-Healing of Polymers. Advanced Materials 18 (8) 997-1000 (2006
5. A. S Jones, J. D Rule, J. S Moore, N. R Sottos, S. R White- Life, Extension of Self-Healing Polymers with Rapidly Growing Fatigue Cracks. Special Issue of Self-Healing Polymers and Composites, J. Royal Society: Interface, 4, 395-403 (2007
6. A. P Esser-kahn, S. A Odom, N. R Sottos, S. R White, J. S Moore- Triggered, Release from Polymer Capsules. Macromolecules, 44, 5539-5553 (2011
7. C Dry- U, S Patent, 2004Multiple Function, Self-Repairing Composites with Special Adhesives. (06/30/2006)

8. A. A Périchaud, M Devassine- Self, Repairing Composition, Self-Repairing Materials, Self-Repairing Methods and Applications. Patent WO 2009A1

9. R Iskakov, A Périchaud, A Muzdubaeva, L Caserta, A Mirsakieva, B Khudaibergenov, I Razumovskaya, P Apel-Trimethylolpropanetriacrylate, Loaded Nanonpores Pc Films with Self-Healing, Potential Proceeding 3rd International Conference on Self-Healing Materials Bath, UK. 2729June 2011

10. M. J. M Abadie, V Yu, Voytekunas- New Trends in UV Curing, Eur.Chem.Tech.Journal, 6 67-77 (2004

11. M. J. M Abadie, N. K Chia, F. Y. C Boey- Cure, Kinetics for the Ultraviolet Cationic Polymerization of Cycloliphatic and Diglycidyl Ether of Bisphenol-A (DGEBA) Epoxy Systems with Sulfonium Salt Using an Auto Catalytic Model. J. Appl. Polym. Sc. 7, 86-97 (2002

12. L. L Ionescu-vasii, M. J. M Abadie- Kinetic, Model of Photoinduced Polymerisation of Phenyl Glycidyl Ether Monomer. Polym. International 1998; 47, 221-225 (1998

13. A Lebel, J Couve, M. J. M Abadie- Study, of the Photolysis of Sulfonium Salts. Cationic Photoinitiators. C.R. Acad Sci. Paris, Série II, 1, 201-207 (1998

14. J Sestak, G Berggren- Study, of the Kinetics of the Mechanism of Solid-State Reactions at Increasing Temperatures. Thermochimica Acta 19713111

10
CHAPTER

High Performance Green Barriers Based on Nanocellulose

Sandeep S Nair[1], JY Zhu[2], Yulin Deng[3] and Arthur J Ragauskas[4*]

[1]*School of Chemistry and Biochemistry, Georgia Institute of Technology, 500 10th Street, N.W, Atlanta, GA 30332, USA*

[2]*USDA Forest Service, Forest Products Laboratory, One Gifford Pinchot Drive, Madison, WI 53726, USA.*

[3]*School of Chemical and Biomolecular Engineering, Georgia Institute of Technology 500 10th Street, N.W, Atlanta, GA 30332, USA*

[4]*Department of Chemical and Biomolecular Engineering, Department of Forestry, Wildlife, and Fisheries, Center for Renewable Carbon, University of Tennessee, Knoxville, TN 37996-2200, USA*

ABSTRACT

With the increasing environmental concerns such as sustainability and end-of-life disposal challenges, materials derived from renewable resources such as nanocellulose have been strongly advocated as potential replacements for packaging materials. Nanocellulose can be extracted from various plant resources through mechanical and chemical ways. Nanocellulose with its nanoscale dimensions, high crystalline nature, and the ability to form hydrogen bonds resulting in strong network makes it very hard for the molecules to pass through, suggesting excellent barrier properties associated with films made from these material. This review paper aim to summarize the recent developments in various barrier films based on nanocellulose with special focus on oxygen and water vapor barrier properties.

Keywords: Green barriers Nanocellulose Cellulose nanofibrils Cellulose nanocrystals

INTRODUCTION

Packaging materials are widely used to prevent food and drink, healthcare, cosmetics and other consumer goods against physical, biochemical, and microbiological deterioration. They should provide sufficient barrier against oxygen, water vapor, grease, and microorganisms. Currently, the packaging materials are largely based on glass, aluminum and tin, and fossil derived synthetic plastics. These materials possess high strength and barrier properties. However, they are unsustainable, some are fragile such as glass, and their weight adds to energy costs for shipping [1-6]. The global consumer packaging demand is valued at approximately US$400b-$500b and is one of the faster-growing markets, forecasted to grow at ~4% per year until 2015 [7].

With the increased environmental concerns over sustainability and end-of-life disposal challenges, materials derived from renewable resources have been strongly advocated as potential replacements [8]. Cellulose is the most abundant polymer in nature and accounts for approximately 40% of lignocellulosic biomass. Cellulose paper-based packaging is lightweight, low-cost, and most important, sustainable. Unfortunately, common paper made from lignocelluloses does not provide sufficient barrier for water, oxygen or oil. Currently, paper based packages are made with unsustainable coatings of wax, plastics, or aluminum. Cellophane is the only cellulose based material (not modified or coated) currently used for barrier packaging due to its high gas barrier. However, the production of cellophane is via a viscose route which produces byproducts and uses reagents (CS_2 and H_2S) that are harmful to the environment [9].

The production of cellulose nanomaterial such as cellulose nanofibrils (CNFs) and cellulose nanocrystals (CNCs) have opened vast possibilities of utilizing cellulose based materials for packaging. Cellulose nanomaterial has diameter in the range of 2–50 nm with large surface area [10-12]. The ability to form hydrogen bonds resulting in strong network makes it very hard for the molecules to pass through, excellent for barrier applications [13]. This review paper aim to summarize the recent developments in various barrier films based on nanocellulose with special focus on oxygen and water vapor barrier properties.

Nanocellulose and its preparation

Cellulose nanofibrils (CNFs) or microfibrils have diameter in the range of 2–50 nm and lengths up to several micrometers depending on their origin [10-12]. CNFs have exceptional optical and mechanical properties, and therefore can be used as a building block for a variety of high-performance materials [14-16]. Intensive mechanical treatment is required to disintegrate the cellulose fiber to nanofibrils [17]. Several methods of mechanical fibrillation have been used for the production of CNFs such as homogenizers [18,19], microfluidizers [20], and grinders [21,22]. Cellulose nanocrystals (CNCs) are often prepared by treating cellulose fiber with sulfuric acid or hydrochloric acid. Strong acidic condition leads to aggressive hydrolysis to attack the noncrystalline fractions within the cellulose fiber which results in the formation of low aspect cellulose fibril aggregates known as CNCs [23-25].

Mechanical fibrillation of cellulose fibers to CNFs are very energy intensive with reported values ranging from 4500–10000 kWh/tonne [21,22]. Chemical pretreatments and enzymatic pretreatments before mechanical fibrillation have been used to reduce this energy consumption. Cellulose fibers were oxidized by a 2,2,6,6-tetramethylpiperidine-1-oxyl radical (TEMPO)-mediated system. These treated fibers were further mechanically fibrillated to CNFs while reducing the energy costs [26]. Aulin et al. [27] carboxymethylated softwood pulp and then fibers were mechanically fibrillated using a high-pressure fluidizer to produce CNFs of diameter of 5–10 nm. Enzyme pretreatment of biomass followed by mechanical homogenization has been used as an environmentally friendly alternative to chemical pretreatment for nanocellulose production [28-30]. Figure 1 shows the morphological difference between the CNCs and CNFs.

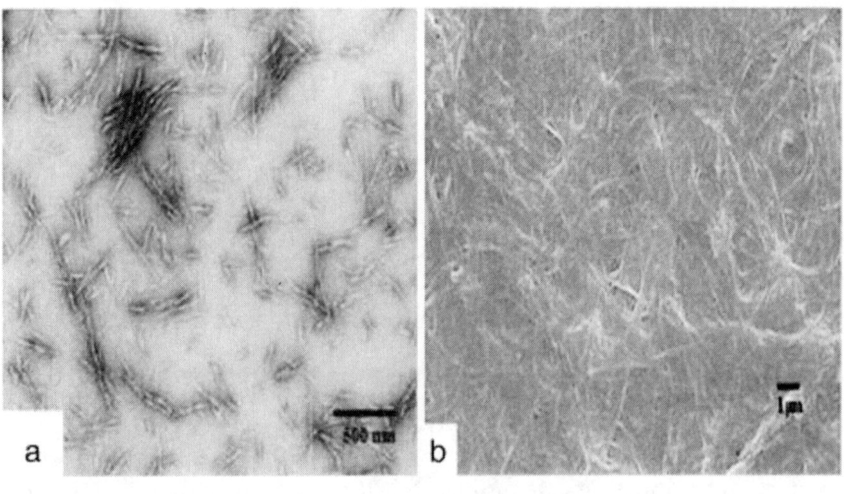

Figure 1. Images of (a) CNCs [**24**], and (b) CNFs [**18**].

Migration process of molecules through nanocellulose film

Migration of molecules between two adjacent volumes separated by a thin film of solid polymer or membrane occurs in three basic steps. In the first step, the diffusing molecule is adsorbed on to the sample surface. In the second step, the molecule then diffuses through the film or membrane. Finally, the diffusing molecule exits the sample by desorbing from the surface. The gas permeability through a nanocellulose film mainly depends on the dissolution of gas and its rate of diffusion in the film. Thus, permeability of gas molecules through nanocellulose film can be expressed as:

$$P = DS \tag{1}$$

Where P is the permeability, D is the diffusion coefficient, and S is the solubility coefficient.

The permeability coefficient P is obtained from the application of Henry's law of solubility to Fick's law of diffusion,

$$P = DS = ql/At\Delta p \tag{2}$$

Where q is the amount of material passing through the film, l is the thickness, A is the cross sectional area, t is time, and Δp is the pressure difference between the two sides of film.

The gas molecules should be first dissolved in the membrane or film before diffusing. Even though the surface of films influences the permeating gas molecules, the most dominant factor in molecular migration is bulk flow, i.e., rate of molecule diffusion in the membrane or film [31]. The good oxygen barrier properties of nanocellulose can be attributed to the dense network formed by nanofibrils with smaller and more uniform dimensions. Considering this fact, the pores within the films serve as the major path for permeating oxygen molecules. The dense nanofibrils forms more complex and smaller pores compared to pure cellulose fibers which are in micro scale. This complex dense network increases the tortuosity within the film and thereby decreases the permeability within the films [13]. Figure 2 shows a schematic representation of increased diffusion path within the nanocellulose films. Also, the high crystalline structure within the nanofibrils or whiskers contributes to the gas barrier properties [23,31]. Cellulose is composed of both crystalline and disordered regions. High crystallinity ranging from 40-90% has been reported for the nanocellulose, with CNCs showing higher crystallinity than the CNFs because strong acid hydrolyzes disordered cellulose to result in highly crystalline CNCs [21,32]. Even though the CNCs have higher crystallinity than CNFs, mechanically fibrillated CNF films were found to have much lesser oxygen permeability than CNCs. Both showed similar solubility, but the oxygen molecules penetrated more slowly though the CNF films. This is mainly due to the structural organization within the films. The CNF films have higher entanglements within the film which increased the tortuosity factor or increase the diffusion path [33]. Even though nanocellulose provides a high barrier for oxygen, the water vapor barrier properties are low. This is mainly due to high affinity between water and the nanocellulose film. Nanocellulose is much better water vapor barrier than cellulose fiber. Nanocellulose has a strong reducing effect on water vapor diffusion due to its size, and swelling constraints formed due to rigid network within the films. However, at a high relative humidity these structural organizations can be disrupted due to high swelling and can lose barrier properties for both oxygen and water vapor [34].

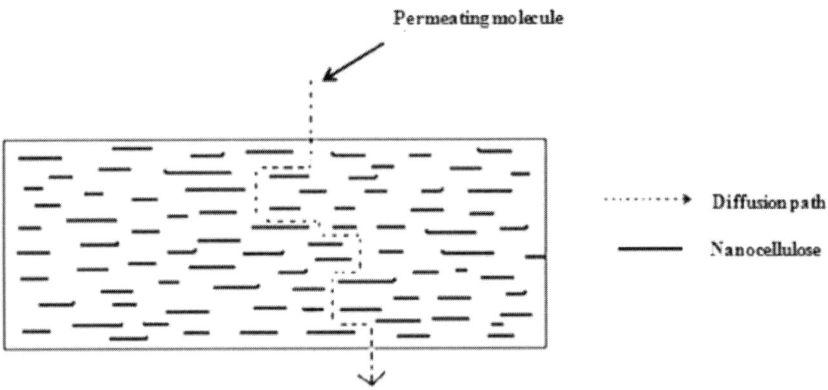

Figure 2. Shows schematic representation of increased diffusion path within the nanocellulose films.

CNFs for barrier application

CNFs is a strong gas barrier material. Compared to CNCs, CNFs consists of crystalline and disordered regions. In most of the cases, crystallinity ranging from 40-75% has been reported for the CNFs obtained from softwoods and hardwoods [11,12,21,22]. Saito and Isogai (2004) showed that the degree of crystallinity varied from 78-91% for CNFs produced from TEMPO oxidation of cotton linter [35]. Films made purely of mechanically fibrillated CNFs have very high air and oxygen barrier property. The oxygen transmission rates (OTR) of CNF films with thickness of 21 μm were as low as 17 ± 1 ml m^{-2} day^{-1}. These values are competitive with other best synthetic polymers such as ethylene vinyl alcohol (EVOH) (3–5 ml m^{-2} day-

$^{1)}$ and polyvinylidene chloride (PVdC) coated polyester films (9–15 ml m^{-2} day^{-1}) of approximately same thickness with respect to OTR [13]. Recently, Osterberg et al. [36] demonstrated a rapid method of making robust CNF films with high oxygen barrier property. The CNF solutions were first filtered followed by hot pressing at high pressure followed by air drying. At a relative humidity below 65%, the oxygen permeability of these films was below 0.6 cm^3 μm m^{-2} d^{-1} kPa^{-1}. However, oxygen permeability of CNF films increases with the increase in relative humidity. This is mainly due to the plasticizing and swelling of nanofibrils through the adsorption of water molecules at high relative humidities.

CNFs also have tunable barrier properties. Rodionova et al. [5] showed that both pure and partially acetylated CNF films can be used for modified atmosphere packaging (MAP) with permeability below 10–20 mL m^{-2} day^{-1}. Similarly, carboxymethylated CNF films with thicknesses of 2.54 and 3.19 μm have very low oxygen permeability of 0.009 and 0.0006 cm^3 μm m^{-2} day^{-1} kPa^{-1}, respectively [27]. High oxygen performance barrier films with permeability of 0.007 cm^3 μm m^{-2} day^{-1} kPa^{-1} can be made by controlled thermal treatment at different temperatures. It was demonstrated that after thermal treatment at 175°C, the oxygen permeability of the CNF films was reduced by 96% compared to the films without heat treatment [37]. Table 1 shows the oxygen permeability of films made from nanocellulose compared with those from commercially available petroleum based materials and other polymers. Also, CNFs are used to coat the surface of films to enhance the barrier properties. Fukuzumi et al. [38] prepared high oxygen barrier CNF films by TEMPO mediated oxidation. The 0.1 μm thick CNF film when coated on a polylactic acid film, reduced the oxygen permeability from 746 ml m^{-2} day^{-1} Pa^{-1} for pure PLA film to 1 ml m^{-2} day^{-1} Pa^{-1} for PLA film with CNF layer. The oxygen barrier of PET films was raised by more than 350 times through a 1 μm thick layer of the same TEMPO-oxidized CNFs [39]. Similarly, Fujisawa et al. [40] found that the use of TEMPO-oxidized CNFs with sodium carboxylate groups improved the oxygen barrier of PET films more than when coated with TEMPO-oxidized CNFs with carboxyl groups. Hult et al. [41] showed that the oxygen transmission rate of paper substrates were decreased by several logarithmic units by the deposition of CNFs and shellac coating layers. CNFs have also been used as filler material to obtain a nanocomposite material. Plackett et al. [42] showed that the addition of 15 wt% of CNFs substantially increased the oxygen barrier of amylopectin films. The use of CNFs in xylan films showed very low oxygen permeability of 0.19 cm^3 μm m^{-2} day^{-1} kPa, which is comparable to previously reported values for 100% CNF films [43]. Several studies have been conducted on making composites using CNF with nanoclay [44], and talc [45]. Oxygen barrier properties of these composites were highly competitive and much better than various synthetic polymers currently used in the packaging industry.

Table 1. Oxygen permeability of nanocellulose film compared to those made form commercially available petroleum based materials and other polymers

Material	Oxygen permeability (cc.µm/m^2.day.kPa)	References, conditions
CNF	0.6	[35], at 65% RH
CNF (carboxymethylated)	0.0006	[27], at 0% RH
CNF (carboxymethylated)	0.85	[27], at 50% RH
Cellophane	0.41	[46], at 0%RH
PVdC	0.1-3	[47], at 50% RH
Polyvinyl alcohol (PVOH)	0.20	[47], at 0% RH
EVOH	0.01-0.1	[47], at 0% RH

Even though the oxygen barrier properties of CNF films are competitive with current commercial films made from synthetic polymers, their water vapor barrier remains very low or the water vapor transfer rate remain very high. Table 2 shows the water vapor transmission rate (WVTR) of nanocellulose films than those of commercially available films from petroleum based materials and other polymers. This is mainly due to the strong hydrophilic nature of the cellulose nanofibers. However, the water vapor transmission rate (0.02 g/m^2 day/m) was significantly lower compared with films made from bleached softwood fibers (0.11 g/m^2 day/m). The CNF film structure is more compact, therefore less water can penetrate through the film compared to the films made of cellulose fibers. The water contact angle was substantially increased for the CNF film compared to films made from bleached softwood film. The hydrophilic nature is not the only important criteria for water vapor barrier property. The films with high lignin content created more hydrophobic surface with higher water contact angles but showed the least water vapor barrier enhancement. However, the CNF films without lignin produced a more compact structure. The presence of lignin hindered the hydrogen bonding and created more hydrophobic pores which aided in more water vapor transmission [34]. Recently, researchers have modified the properties of CNFs using various pretreatments to enhance the water vapor barrier property. Sharma et al. [37] showed that the water vapor permeability was reduced by 50% upon heat treatment at 175°C for 3 hours as compared to the untreated CNF films. They attributed that the enhancement in barrier property was mainly due to increase in hydrophobicity as evidenced by the increase in water contact angle and reduced porosity by heat treatment. Rodionova et al. [5] showed that the WVTR of pure CNF films ($234\ g\ m^{-2}\ day^{-1}$) was reduced to $167\ g\ m^{-2}\ day^{-1}$ by 30 mins of acetylation treatment. The acetylation created a much higher hydrophobic film surface indicates by higher water contact angles. Minelli et al. [48] showed that the use of carboxymethylated pretreated CNFs produced a more homogenous and less porous film than enzyme pretreated CNFs and showed a much enhancement in water vapor barrier property below 80% relative humidity. Recently, the water vapor permeability of the paper substrates was substantially decreased with a multilayer coating of nanocellulose and alkyd resins. The WVTR reached very low and reached values considered as high barriers in packaging applications [49].

Table 2. WVTR of nanocellulose compared to commercially available petroleum based materials and other polymers

Material	Water vapor transmission rate (WVTR) (g/m^2 day)	Average thickness of the film (μm)	References, conditions
CNF	234	42	[5], 50% RH
CNF (acetylated for 0.5 h)	167	46	[5], 50% RH
PVdC	3.07	12.7	[50], 100% RH
Polyethylene (PE)	16.8	18.3	[50], 100% RH
Plasticized polyvinyl chloride (PVC)	118.56	12.7	[50], 100% RH

CNCs for barrier application

Contrary to CNFs, very few studies have been directed toward study of 100% pure CNC film or treated CNC films. Belbekhouche et al. [33] compared the gas barrier properties between CNF and CNC films. They found that the films made of CNCs were more permeable to oxygen than those made of CNFs. The oxygen molecules penetrated much more slowly within CNF film due to the higher fibril entanglements within the film which increased the tortuosity factor. CNCs, which have crystallinity greater than 60% combined with their ability to form a dense hydrogen bonded network can increase gas barrier property. Bacterial cellulose nanocrystals (BNCs) films present excellent oxygen barrier at low relative humidity, but their high moisture sensitivity results in dramatically decreased barrier when the relative humidity is higher than 70%. The oxygen permeability of $6.99 \times 10^{-22}\ m^3\ m/m^2\ s\ Pa$ at 0% humidity increased to

5.97×10^{-18} m^3 m/m^2 s Pa at 80% humidity. However, this permeability was reduced by 97% and 74% when BNC films were coated with annealed PLA electro spun nanostructured fibers and 3-aminopropyl) trimethoxysilane (APTS), respectively [51]. Herrara et al. [52] studied thin spin coated films made from CNCs prepared with sulfuric acid and hydrochloric acid. The hydrochloric acid made CNCs resulted in films with low permeability for oxygen, while the sulfuric acid made CNCS resulted in films with higher permeability.

CNCs have been studied as filler for various natural polymers for enhancing the barrier properties. Saxena et al. [23] produced nanocomposite film with low oxygen permeability by casting an aqueous solution containing xylan, sorbitol and nanocrystalline cellulose. Oxygen permeability of films prepared from xylan, sorbitol and 50% by weight of sulfonated CNC exhibited a significantly reduced oxygen permeability of 0.1799 $cm^3.\mu m/m^2$.d.kPa compared with films prepared solely from xylan and sorbitol with an oxygen permeability of 189.1665 $cm^3.\mu m/m^2$.d.kPa. Poly lactic acid (PLA) nano-biocomposites containing 5 wt% of nanocrystals exhibited the highest oxygen barrier. The OTR for PLA nanocomposites with 5% w/w of unmodified CNCs was 17.4 ± 1.4 cm^3 mm m^{-2} day^{-1}, while that for modified CNCs with an acid phosphate ester of ethoxylated nonylphenol in a 1/4 (wt/wt) ratio was 15.8 ± 0.6 cm^3 mm m^{-2} day^{-1} [53]. Addition of 1 wt% of silver nanoparticles to these modified CNC- PLA composites further decreased the OTR to 12.6 ± 0.1 cm^3 mm m^{-2} day^{-1} [54]. The OTR values of ternary systems consisting of PLA, PHB (poly hydroxybutyrate) and 5 wt% unmodified CNCs was 15.3 cm^3 mm m^{-2} day^{-1}, while that for modified CNCs with an acid phosphate ester of ethoxylated nonylphenol in a 1/1 (wt/wt) ratio was 13 cm^3 mm m^{-2} day^{-1}. Water contact angle measurements showed that the ternary system had high hydrophobicity and the presence of sulphate groups with low polarity on the surface of CNCs increased the surface hydrophobicity of the final composite material [55].

CNCs were used as fillers in polyvinyl alcohol (PVOH) matrix. The addition of 5 wt% CNCs decreased the WVP of pure PVOH films from 0.61 ± 0.04 g.mm/kPa.h.m^2 to 0.44 ± 0.01 g.mm/kPa.h.m^2 [56]. The reinforcement of natural biopolymers with CNCs was found to reduce WVTR of the resulting nanocomposites. The films prepared using xylan as reinforcement polymer with 10% sulfonated CNCs exhibited a 74% reduction in specific water transmission properties compared with the film without CNCs and a 362% improvement compared with xylan films reinforced by 10% softwood kraft fibers. The xylan/sulphonated CNC nanocomposites showed a WVTR of 174 g/hm^2 [57]. They also compared xylan films reinforced with CNC made from hydrochloric acid with those reinforced with sulphonated CNC. Even though, films showed a significant reduction in water transmission, the reduction was not as significant as those using sulfonated CNCs. The xylan/ hydrochloric acid made CNC films showed a WVTR of 281 g/hm^2 [58]. Khan et al. [59] showed that the values of water vapor permeability (WVP) decreased sharply as the content of CNCs increased in the methyl cellulose based films. The WVP of control films (without CNCs) was 6.3 g.mm/m^2.day.kPa, while those films in cooperated with 1 wt% CNC showed a permeability of 4.7 g.mm/m^2.day.kPa.

CONCLUSIONS

Nanocellulose such as CNFs and CNCs have opened vast possibilities of utilizing cellulose based materials. The use of CNFs in films, composites, and coatings has found to substantially reduce the oxygen permeability within these materials. The oxygen barrier efficiency of pure CNF films is highly competitive and even be comparable with commercial synthetic polymers. The improvement of oxygen barrier properties by CNFs can be attributed to the dense network formed by nanofibrils with smaller and more uniform dimensions. Even though CNCs have higher crystallinity than CNFs, mechanically fibrillated CNF films were found have much lesser oxygen permeability than CNCs. The CNF films have higher entanglements within the film which increases the diffusion path for gas molecules. Also, nanocellulose has a strong reducing effect on water vapor diffusion due to its size, and swelling constraints formed due to rigid network within the films compared to cellulose fibers. The use of CNFs and CNCs in various natural polymer based composites has found to substantially reduce the gas permeability within these composites.

ACKNOWLEDGEMENTS

This work was partially supported by the USDA Forest Service R & D special funding on Cellulose Nano-Materials (2012).

AUTHORS' CONTRIBUTIONS

All the authors have contributed to the literature review and manuscript writing. All authors read and approved the final manuscript.

REFERENCES

1. Bayer IS, Fragouli D, Attanasio A, Sorce B, Bertoni G, Brescia R, Di Corato R, Pellegrino T, Kalyva M, Sabella S, Pompa PP, Cingolani R, Athanassiou A: **Water-repellent cellulose fiber networks with multifunctional properties.** *ACS Appl Mater Interfaces* 2011, **3:**4024–4031.

2. Hansen NML, Plackett D: **Sustainable films and coatings from hemicelluloses: a review.***Biomacromolecules* 2008, **9:**1493–1505.

3. Priolo MA, Gamboa D, Holder KM, Grunlan JC: **Super gas barrier of transparent polymer – clay multilayer ultrathin films.** *Nano Lett* 2010, **10:**4970–4974.

4. Reis AB, Yoshida CMP, Reis APC, Franco TT: **Application of chitosan emulsion as a coating on Kraft paper.** *Polym Int* 2011, **60:**963–969.

5. Rodionova G, Lenes M, Eriksen O, Gregersen O: **Surface chemical modification of microfibrillated cellulose: improvement of barrier properties for packaging applications.** *Cellulose* 2011, **18:**127–134.

6. Spence KL, Venditti RA, Rojas OJ, Pawlak JJ, Hubbe MA: **Water vapor barrier properties of coated and filled microfibrillated cellulose composite films.** *BioResources* 2011, **6:**4370–4388.

7. Neil-Boss N, Brooks K: **Unwrapping the packaging industry: seven factors for success.** In ;2013. http://www.ey.com/Publication/vwLUAssets/Unwrapping_the_packaging_industry_%E2%80%93_seven_factors_for_ success/$FILE/EY_Unwrapping_the_packaging_industry_-_seven_success_factors.pdf.

8. Nair SS, Wang SQ, Hurley DC: **Nanoscale characterization of natural fibers and their composites using contact-resonance force microscopy.** *Compos Part A* 2010, **41:**624–631.

9. Hyden WL: **Manufacture and properties of regenerated cellulose films.** *Ind Eng Chem* 1929, **21:**405–410.

10. Stelte W, Sanadi AR: **Preparation and characterization of cellulose nanofibers from two commercial hardwood and softwood pulps.** *Ind Eng Chem Res* 2009, **48:**11211–11219.

11. Nair SS, Zhu JY, Deng Y, Ragauskas AJ: **Hydrogels prepared from cross-linked nanofibrillated cellulose.***ACS Sustainable Chem Eng* 2014, **2:**772–780.

12. Hoeger IC, Nair SS, Ragauskas AJ, Deng Y, Rojas OJ, Zhu JY: **Mechanical deconstruction of lignocellulose cell walls and their enzymatic saccharification.** *Cellulose* 2013, **20:**807–818.

13. Syverud K, Stenius P: **Strength and barrier properties of MFC films.** *Cellulose* 2009, **16:**75–85.

14. Abraham E, Deepa B, Pothan LA, Jacob M, Thomas S, Cvelbar U, Anandjiwala R: **Extraction of nanocellulose fibrils from lignocellulosic fibers: a novel approach.** *Carbohydr Polym* 2011, **86:**1468–1475.

15. Kaushik A, Singh M: **Isolation and characterization of cellulose nanofibrils from wheat straw using steam explosion coupled with high shear homogenization.** *Carbohydr Res* 2011, **346:**76–85.

16. He L, Li X, Li W, Yuan J, Zhou H: **A method for determining reactive hydroxyl groups in natural fibers:application to ramie fiber and its modification.** *Carbohydr Res* 2012, **348:**95–98.

17. Uetani K, Yano H: **Nanofibrillation of wood pulp using a high-speed blender.** *Biomacromolecules* 2011,**12:**348–353.

18. Chinga-Carrasco G, Syverud K: **On the structure and oxygen transmission rate of biodegradable cellulose nanobarriers.** *Nanoscale Res Lett* 2012, **7:**192.

19. Chinga-Carrasco G, Kuznetsova N, Garaeva M, Leirset I, Galiullina G, Kostochko A, Syverud K: **Bleached and unbleached MFC nanobarriers: properties and hydrophobisation with hexamethyldisilazane.** *J Nanopart Res* 2012, **14:**1280.

20. Henriksson M, Berglund LA, Isaksson P, Lindstrom T, Nishino T: **Cellulose nanopaper structures of high toughness.** *Biomacromolecules* 2008, **9:**1579–1585.

21. Nair SS, Zhu JY, Deng Y, Ragauskas AJ: **Charaterization of cellulose nanofibrillation by micro grinding.** *J Nanopart Res* 2014, **16:**2349.

22. Wang QQ, Zhu JY, Gleisner R, Kuster TA, Baxa U, McNeil SE: **Morphological development of cellulose fibrils of a bleached eucalyptus pulp by mechanical fibrillation.** *Cellulose* 2012, **19:**1631–1643.

23. Saxena A, Elder TJ, Kenvin J, Ragauskas AJ: **High oxygen nanocomposite barrier films based on xylan and nanocrystalline cellulose.** *Nano-Micro Lett* 2010, **2:**235–241.

24. Wang QQ, Zhu JY, Reiner RS, Verril SP, Baxa U, Mc Neil SE: **Approaching zero cellulose loss in cellulose nanocrystal (CNC) production: recovery and characterization of cellulosic solid residues (CSR) and CNC.** *Cellulose* 2012, **19:**2033–2047.

25. Beck-Candanedo S, Roman M, Gray DG: **Effect of reaction conditions on the properties and behavior of wood cellulose nanocrystal suspensions.** *Biomacromolecules* 2006, **6:**1048–1054.

26. Saito T, Kimura S, Nishiyama Y, Isogai A: **Cellulose nanofibers prepared by TEMPO-mediated oxidation of native cellulose.** *Biomacromolecules* 2007, **8:**2485–2491.

27. Aulin C, Gallstedt M, Lindstrom T: **Oxygen and oil barrier properties of microfibrillated cellulose films and coatings.** *Cellulose* 2010, **17:**559–574.

28. Henriksson M, Henriksson G, Berglund LA, Lindström T: **An environmentally friendly method for enzyme-assisted preparation of microfibrillated cellulose (MFC) nanofibers.** *Eur Polym J* 2007, **43:**3434–3441.

29. Hayashi N, Kondo T, Ishihara M: **Enzymatically produced nano-ordered short elements containing cellulose I-beta crystalline domains.** *Carbohydr Polym* 2005, **61:**191–197.

30. Zhu JY, Sabo R, Luo XL: **Integrated production of nano-fibrillated cellulose and cellulosic biofuel (ethanol) by enzymatic fractionation of wood fibers.** *Green Chem* 2011, **13:**1339–1344.

31. Lagaron JM, Catala R, Gavara R: **Structural characteristics defining high barrier properties in polymeric materials.** *Mater Sci Technol* 2004, **20:**1–7.

32. Guo J, Catchmark JM: **Surface area and porosity of acid hydrolyzed cellulose nanowhiskers and cellulose produced by Gluconacetobacter xylinus.** *Carbohydr Polym* 2012, **87:**1026–1037.

33. Belbekhouche S, Bras J, Siqueira G, Chappey C, Lebrun L, Khelifi B, Marais S, Dufresne A: **Water sorption behaviour and gas barrier properties of cellulose whiskers and microfibrils films.** *Carbohydr Polym* 2011,**83:**1740–1748.

34. Spence KL, Venditti RA, Rojas OJ, Habibi Y, Pawlak JJ: **The effect of chemical composition on microfibrillar cellulose films from wood pulps: water interactions and physical properties for packaging applications.***Cellulose* 2010, **17:**835–848.

35. Saito T, Isogai A: **TEMPO-mediated oxidation of native cellulose. The effect of oxidation conditions on chemical and crystal structures of the water- insoluble fractions.** *Biomacromolecules* 2004, **5:**1983–1989.

36. Osterberg M, Vartiainen J, Lucenius J, Hippi U, Seppala J, Serimaa R, Laine J: **A fast method to produce strong NFC films as a platform for barrier and functional materials.** *ACS Appl Mater Interfaces* 2013,**5:**4640–4647.

37. Sharma S, Zhang X, Nair SS, Ragauskas AJ, Zhu JY, Deng Y: **Thermally enhanced high performance cellulose nano fibril barrier membranes.** *RSC Adv* 2014, **4:**45136–45142.

38. Fukuzumi H, Saito T, Wata T, Kumamoto Y, Isogai A: **Transparent and high gas barrier films of cellulose nanofibers prepared by TEMPO-mediated oxidation.** *Biomacromolecules* 2009, **10:**162–165.

39. Rodionova G, Saito T, Lenes M, Eriksen O, Gregersen O, Fukuzumi H, Isogai A: **Mechanical and oxygen barrier properties of films prepared from fibrillated dispersions of TEMPO-oxidized Norway spruce and Eucalyptus pulps.** *Cellulose* 2012, **19:**705–711.

40. Fujisawa S, Okita Y, Fukuzumi H, Saito T, Isogai A: **Preparation and characterization of TEMPO-oxidized cellulose nanofibrils films with free carboxyl groups.** *Carbohydr Polym* 2011, **84:**579–583.

41. Hult EL, Lotti M, Lenes M: **Efficient approach to high barrier packaging using microfibrillar cellulose and shellac.** *Cellulose* 2010, **17:**575–586.

42. Plackett D, Anturi H, Hedenqvist M, Ankerfors M, Gallstedt M, Lindstrom T, Siro I: **Physical properties and morphology of films prepared from microfibrillated cellulose and microfibrillated cellulose in combination with amylopectin.** *J Appl Polym Sci* 2010, **117:**3601–3609.

43. Hansen NML, Blomfeldt TOJ, Hedenqvist MS, Plackett DV: **Properties of plasticized composite films prepared from nanofibrillated cellulose and birch wood xylan.** *Cellulose* 2012, **19:**2015–2031.

44. Aulin C, Salazar-Alvarez G, Lindstrom T: **High strength, flexible and transparent nanofibrillated cellulose – nanoclay biohybrid films with tunable oxygen and water vapor permeability.** *Nanoscale* 2012, **4:**6622–6628.

45. Liimatainen H, Ezekiel N, Sliz R, Ohenoja K, Sirvio JA, Berglund L, Hormi O, Niinimaki J: **High-strength nanocellulose – talc hybrid barrier films.** *ACS Appl Mater Interfaces* 2013, **5:**13412–13418.

46. Wu J, Yuan Q: **Gas permeability of a novel cellulose membrane.** *J Membr Sci* 2002, **204:**185–194.

47. Lange J, Wyser Y: **Recent innovations in barrier technologies for plastic packaging—a review.** *Packag Technol Sci* 2003, **16:**149–158.

48. Minelli M, Baschetti MG, Doghieri F, Ankerfors M, Lindstrom T, Siro I, Plackett D: **Investigation of mass transport properties of microfibrillated cellulose (MFC) films.** *J Membr Sci* 2010, **358:**67–75.

49. Aulin C, Strom G: **Multilayered alkyd resin/nanocellulose coatings for use in renewable packaging solutions with a high level of moisture resistance.** *Ind Eng Chem Res* 2013, **52:**2582–2589.

50. Steven MD, Hotchkiss JH: **Comparison of flat film to total package water vapor transmission rates for several commercial food wraps.** *Packag Technol Sci* 2002, **15:**17–27.

51. Martinez-Sanz M, Lopez-Rubio A, Lagaron JM: **High-barrier coated bacterial cellulose nanowhiskers with reduced moisture sensitivity.** *Carbohydr Polym* 2013, **98:**1072–1082.

52. Herrera MA, Mathew AP, Oksman K: **Gas permeability and selectivity of cellulose nanocrystals films (layers) deposited by spin coating.** *Carbohydr Polym* 2014, **112:**494–501.

53. Fortunati E, Peltzer M, Armentano I, Torre L, Jimenez A, Kenny JM: **Effects of modified cellulose nanocrystals on the barrier and migration of PLA nano-composites.** *Carbohydr Polym* 2012, **90:**948–956.

54. Fortunati E, Peltzer M, Armentano I, Jimenez A, Kenny JM: **Combined effects of cellulose nanocrystals and silver nanoparticles on the barrier and migration properties of PLA nano-biocomposites.** *J Food Eng*2013, **118:**117–124.

55. Arrieta MP, Fortunati E, Dominici F, Rayon E, Lopez J, Kenny JM: **PLA-PHB/cellulose based films: mechanical, barrier and disintegration properties.** *Polym Degrad Stab* 2014, **107:**139–149.

56. Pereira ALS, do Nascimento DM, Souza Filho MM, Morais JPS, Vasconcelos NF, Feitosa JPA, Brigida AIS, Rosa MF: **Improvement of polyvinyl alcohol properties by adding nanocrystalline cellulose isolated from banana pseudostems.** *Carbohydr Polym* 2014, **112:**165–172.

57. Saxena A, Ragauskas AJ: **Water transmission barrier properties of biodegradable films based on cellulosic whiskers and xylan.** *Carbohydr Polym* 2009, **78:**357–360.

58. Saxena A, Elder TJ, Ragauskas AJ: **Moisture barrier properties of xylan composite films.** *Carbohydr Polym*2011, **84:**1371–1377.

59. Khan RA, Salmieri S, Dussault D, Uribe-Calderon J, Kamal MR, Safrany A, Lacroix M: **Production and properties of nanocellulose-reinforced methycellulose-based biodegradable films.** *J Agric Food Chem*2010, **58:**7878–7885.

11
CHAPTER

High Level Expression and Facile Purification of Recombinant Silk-Elastin-Like Polymers in Auto Induction Shake Flask Cultures

Raul Machado[1*], João Azevedo-Silva[1], Cristina Correia[1], Tony Collins[1], Francisco Javier Arias[2,3], Jose Carlos Rodríguez-Cabello[2,3] and Margarida Casal[1]

[1] *CBMA (Centre of Molecular and Environmental Biology), Department of Biology, University of Minho, Campus de Gualtar, 4710-057, Braga, Portugal.*

[2] *Bioforge (Group for Advanced Materials and Nanobiotechnology), Centro I+D, Universidad de Valladolid, Valladolid, Spain.*

[3] *Networking Research Centre on Bioengineering, Biomaterials and Nanomedicine (CIBER-BBN), 47011, Valladolid, Spain.*

ABSTRACT

Silk-elastin-like polymers (SELPs) are protein-based polymers composed of repetitive amino acid sequence motifs found in silk fibroin (GAGAGS) and mammalian elastin (VPGVG). These polymers are of much interest, both from a fundamental and applied point of view, finding potential application in biomedicine, nanotechnology and as materials. The successful employment of such polymers in such diverse fields, however, requires the ready availability of a variety of different forms with novel enhanced properties and which can be simply prepared in large quantities on an industrial scale. In an attempt to create new polymer designs with improved properties and applicability, we have developed four novel SELPs wherein the elastomer forming sequence poly(VPGVG) is replaced with a plastic-like forming sequence, poly(VPAVG), and combined in varying proportions with the silk motif. Furthermore, we optimised a simplified production procedure for these, making use of an autoinduction medium to reduce process intervention and with the production level obtained being 6-fold higher than previously reported for other SELPs, with volumetric productivities above 150 mg/L. Finally, we took advantage of the known enhanced stability of these polymers in developing an abridged, non-chromatographic downstream processing and purification protocol. A simple acid treatment allowed for cell disruption and the obtention of relative pure SELP in one-step, with ammonium sulphate precipitation being subsequently used to enable improved purity. These simplified production and purification procedures improve process efficiency and reduce costs in the preparation of these novel polymers and enhances their potential for application.

Keywords: Silk-elastin-like polymers Recombinant protein expression Auto induction Acid-based cell lysis Ammonium sulphate purification Non-chromatographic purification

INTRODUCTION

Nature has refined structural proteins to perform defined functions through the association of specific amino acid sequences in various combinations, creating diverse materials with remarkable properties. These have served as a source of inspiration for a new class of bio-engineered materials, the protein-based polymers (PBPs), founded on conservative amino acid motifs found in nature. These bioinspired materials exhibit the properties of natural proteins but frequently also display functions that are not found in nature (Rodriguez-Cabello et al.2009; van Hest and Tirrell 2001; Rabotyagova et al. 2011). Furthermore, with advances in recombinant DNA technology it is possible to design and produce tailored synthetic genes, allowing for the creation of multifunctional complex PBPs with total control over composition, molecular weight and structure.

Silk-elastin-like polymers (SELPs) are PBPs composed of multiple repeats of both silk- and elastin-like motifs that combine the outstanding mechanical and biological properties of both these proteins. Most commonly, they are based on repetitions of the silk fibroin amino acid motif GAGAGS and the mammalian elastin conserved motif VPGVG, with the proportions, number and sequence of these repeated motifs governing the properties of the polymer. The silk-like blocks spontaneously form hydrogen-bonded β-sheet crystals and impart thermal and chemical stability (Megeed et al. 2002) while the periodic inclusion of the elastomeric sequence reduces the overall crystallinity of the system and increases its flexibility and aqueous solubility (Cappello et al. 1990). These copolymers have already demonstrated their potential importance in several biomedical applications (Megeed et al. 2002; Gustafson and

Ghandehari 2010) and more recently, in the fabrication of micro- and nanostructures (Nagarajan et al. 2007; Ner et al. 2009; Qiu et al. 2009). Nevertheless, to fully develop the potential of SELPs, new polymer designs with altered and improved properties are required, allowing for a better understanding of the design of these but also leading to an expanded range of novel polymers with potentially enhanced applicability.

In the present study we have designed and synthesised four novel SELPs in which the most commonly used elastin pentamer sequence VPGVG is altered to VPAVG. This simple substitution has been shown to significantly alter the mechanical properties of poly(VPGVG), changing its mechanical response from elastic to plastic deformation (Luan and Urry 1999; Nagapudi et al. 2005). Indeed, this material has unique properties among elastin-like polymers; its Young's modulus is two orders of magnitude higher than that for poly(VPGVG) (Luan and Urry 1999) and while it does display a reversible inverse transition on heating (as do all other elastin-like polymers), this is found to be characterised by an acute hysteresis behaviour (Reguera et al. 2003; Machado et al. 2009). Hence, the preparation of SELPs based on combinations of this unique repetitive motif (VPAVG) with that of silk (GAGAGS) should give rise to a new set of polymers with expanded and potentially enhanced properties and applicability.

One of the principle limitations to the successful commercialisation of any recombinant protein or polymer is associated with the production and purification on a commercially viable level. SELPs are currently mainly produced in *Escherichia coli* using the *lacI* regulated T7 promoter-driven system by batch production in rich media, with volumetric productivities on the low miligram/L scale (i.e. approx. 20 mg/L) being reported (Nagarsekar et al. 2002; Haider et al. 2005; Dandu et al. 2009; Xia et al. 2011). Most commonly, the Sambrook protocol (Sambrook and Russell 2001) is used, with induction of protein production by the synthetic lactose analogue isopropyl-β-D-thiogalactopyranoside (IPTG) addition at the middle of the exponential growth phase. On the other hand, the use of auto-induction media whereby lactose added during the initial media preparation acts to automatically induce protein production (Studier 2005) and thereby circumvent the need for monitoring cell growth and addition of inducer, should allow for a more automated and efficient production method. Indeed, for high-throughput approaches this provides major advantages, avoiding intermediate steps during fermentation and minimising culture handling. In the present study, we examined the production of novel SELP copolymers with an auto-induction approach.

Purification of SELPs is most commonly carried out by immobilized metal affinity chromatography (IMAC) with affinity tags, in particular histidine-tags (6xHis) (Nagarsekar et al. 2002; Haider et al. 2005; Dandu et al. 2009; Xia et al. 2011). This approach is somewhat cumbersome however, requiring pretreatment steps such as cell disruption (e.g. by sonication) and separation of soluble cellular content (e.g. by centrifugation), and requiring the use of specialised and somewhat expensive matrices and equipment (Chow et al. 2008). In contrast, the use of non-chromatographic approaches can allow for a more economical, simplified and higher throughput purification process facilitating scale up to an industrial level (Chow et al. 2008). Indeed, the unique characteristics and known extreme chemical and thermal stability of fibrous proteins as well as of bio-engineered polymers based on these have been exploited in the development of simplified purification protocols for these. Elastin like polymers are frequently purified by temperature cycling, making use of their reversible inverse transition from soluble to insoluble form on heating above the inverse transition temperature (Meyer and Chilkoti 1999). High temperature treatment has been documented for the purification of recombinant spider silk protein (Scheller et al. 2001) and resilin-like polypeptides (Lyons et al. 2007). While these approaches are obviously advantageous as compared to affinity chromatography, they are nonetheless multistep processes requiring downstream processing and pretreatment. Moreover, they cannot be applied to our novel SELPs as temperature accelerates the irreversible gelation process (Haider et al. 2004). In contrast, the use of extremes of pH, and in particular acidic pHs at which fibrous proteins are known to be stable, may lead to an unfolding and precipitation of the host proteins and allow for both cell disruption and protein purification in one simple step. Here we examined the use of acidic pH combined with ammonium sulphate precipitation for a simplified downstream processing and purification protocol, thereby avoiding cumbersome cell disruption and chromatographic approaches and reducing process time and costs.

In this work, the genetic construction, development and optimisation of the autoinduction production protocol as well as a simplified non-chromatographic purification protocol for four novel SELPs (SELP-1020-A, SELP-520-A, SELP-59-A, SELP-109-A) is described and discussed.

Materials and methods

MATERIALS

The cloning steps were performed in *Escherichia coli* strain XL1-Blue (Stratagene) and protein expression was achieved with *E. coli* strain BL21(DE3) (Novagen). Unless otherwise stated, chemicals were obtained from Panreac. Isopropyl β-D-1-thiogalactopyranoside (IPTG) was obtained from Formedium, tryptone from BD Biosciences and yeast extract from Cultimed, α-lactose was from BDH Chemicals and tris base from Fisher Scientific. Ethylenediamine tetraacetic acid disodium salt dihydrate (EDTA) and ampicillin were obtained from Sigma-Aldrich. Dodecyl sulphate sodium salt (SDS) and beta-mercapto-ethanol were obtained from Merck. HisTrap™ HP columns were from GE Healthcare. Restriction enzymes, T4 DNA Ligase, GeneRuler™ 1 kb DNA ladder and Taq Polymerase were obtained from Fermentas. Accuzyme Mix was obtained from Bioline and Antarctic phosphatase from NEB (New England Biolab). Plasmid extraction was performed with GenElute™ Plasmid Miniprep Kit (Sigma-Aldrich) and DNA purification from agarose gel by Nucleospin® Extract II (Macherey-Nagel). Cloning vector pDrive was obtained from Qiagen and modified as previously described (Rodriguez-Cabello et al. 2012). Expression vector pET-25b(+) was obtained from Novagen.

Construction of SELP copolymers

In this work we designed four different SELP copolymers with formulation (*SxEy*)*n*, where S represents the silk block with the sequence GAGAGS, *x*: the number of silk blocks; *E* represents the elastin-like block with the sequence VPAVG, *y*: the number of elastin-like blocks and *n* is the number of repetitions of the overall copolymer. All the hybrid copolymers were designed to have a similar molecular weight of 55 kDa but different silk to elastin block ratios or different number of silk blocks (Figure 1): SELP-1020-A with composition (S10E20)4; SELP-520-A with composition (S5E20)5; SELP-59-A with composition (S5E9)9 and SELP-109A with composition (S10E9)7. For nucleotide and amino acid sequences see [Additional file 1].

Genetic constructions were obtained by recursive directional ligation (Meyer and Chilkoti 2002; McDaniel et al. 2010) using previously described standard molecular biology procedures (Machado et al. 2009; Rodriguez-Cabello et al. 2012 codon usage while minimizing sequence repetition. All constructions and concatenations were performed in a modified pDrive cloning vector (Rodriguez-Cabello et al. 2012). The gene sequences were selected to emphasize *Escherichia coli*), transformed into *E. coli* strain XL1 Blue by the SEM method (Inoue et al.1990) and confirmed by automated DNA sequencing. Vector dephosphorylation with antarctic phosphatase was performed prior to ligation with T4 DNA ligase. Synthetic DNA duplexes flanked by *Eam* 1104I recognition sites and coding for 5 repetitions of the monomer GAGAGS (S5) were generated by polymerase chain reaction amplification with *Taq* polymerase. The elastin block was obtained by PCR amplification of two synthetic primers, $VPAVG_{FW}$ with 173 nucleotides and

VPAVG$_{REV}$ with 171 nucleotides, flanked by *Eam* 1104I recognition sites; this allowed obtaining the amplification of 9 and 11 repetitions of VPAVG (E9 and E11, respectively). Each amplicon was subsequently cloned into the modified pDrive cloning vector and selected by PCR colony screening. After cloning, each monomer was obtained by restriction digestion with *Eam* 1104I and isolated and purified with a DNA gel extraction kit. To obtain the dimer S10, the monomer was subjected to self-ligation, while the block E20 was obtained by combination of E9 and E11. The hybrid polymers *SxEy* were constructed by ligation of the basic blocks and then subjected to concatenation ligation reactions to get the required *n* tandem repeats corresponding to the predicted molecular weight of about 55 kDa. The DNA sequence encoding each final SELP construct was digested with *Eam* 1104I, blunt-ended and subcloned into pCM13 (a modified pET-25(+) expression vector). Blunt-ends were produced by filling the sticky-ends of DNA fragments with Accuzyme mix at 72°C for 30 minutes.

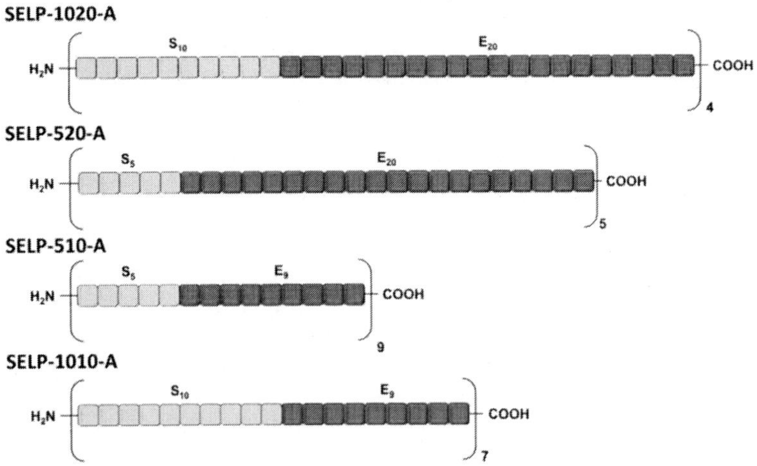

Figure 1. Schematic representation of the block units for the four SELP copolymers, where Sx represents the number of silk-like blocks and Ey the number of elastin-like blocks. Each hybrid block (SxEy) was self-ligated through a series of ligation reactions to achieve a number of tandem repeats corresponding to a molecular weight of approximately 55 kDa.

Construction of expression vector pCM13

To allow for a correct insertion of all SELP encoding genes, the original pET-25b(+) expression vector was mutated by site-directed mutagenesis according to the method previously described by Ansaldi, M. *et al* (Ansaldi et al. 1996). The point mutations were introduced by PCR using the oligonucleotides described in Table 1.

Table 1. Primers pairs for introduction of selected mutations in pET-25b(+) expression vector

Pair #1	Reverse *Eco72* I	5′ GAGCAGCAGACCAGCAGCAGCGGTCGGCAGCA**C**GT**G**TT**C**CATATGTATATCTCCTT 3′
	Forward *Eco72* I	5′ CGCTGCTGCTGGTCTGCTGCTCCTCGCTGCCCAGCCGGC 3′
Pair #2	Reverse *Eco147* I	5′ GGGGTCTTCCGGGGCGAGTTCTGGCTGGCTAG**G**CC**T**TTTGATCTCGAGTGCGG 3′
	Forward *Eco147* I	5′ AGAACTCGCCCCGGAAGACCCCGAGGATGTCGAGCACCA 3′

Site-directed mutagenesis was performed by PCR with the Accuzyme Mix (high fidelity polymerase) (Queiros et al. 2001); initially by using pair #1 of primers and secondly by using pair #2 of primers (Table 1). Parental DNA was digested with *Dpn* I and PCR products were transformed to *E. coli* XL1 Blue. Point-mutations were confirmed by restriction digestion of plasmid DNA with *Eco* 72 I for mutations introduced by pair #1, and *Eco* 147 I for mutations introduced by pair #2, as well as by sequencing.

The aforementioned modified pET-25b(+), titled pCM13, was prepared for cloning by restriction digestion with the blunt-end producing enzymes, *Eco* 72I and *Eco* 147I, gel purified and dephosphorylated. This was then ligated with the end-filled SELP-1020-A, SELP-520-A, SELP-59-A and SELP-109-A DNA constructs. Insertion and correct orientation was evaluated by digestion with *Ppu* 21I and automated DNA sequencing. The expression vectors pCM13(SELP-1020-A), pCM13(SELP-520-A), pCM13(SELP-59-A) and pCM13(SELP-109-A) were then transformed to *E. coli* BL21(DE3) for protein expression.

Production in shake flasks

Overnight pre-cultures of 10 ml Lysogeny Broth (LB: 10 g tryptone, 5 g yeast extract, 5 g NaCl, per litre) were used to inoculate fresh cultures in shake flasks to an optical density of 0.01 (OD_{600}) and allowed to grow for 24 hours. All the media used were supplemented with 150 μg/ml of ampicillin. Protein expression was analysed in LB with IPTG induction (0.5 and 1 mM); LB supplemented with 2 g/L of α-lactose (LB+Lac) to allow auto-induction; Terrific Broth (TB: 12 g tryptone, 24 g yeast extract, 0.17 M KH_2PO_4, 0.72 M K_2HPO_4, 5 g glycerol, per litre) with IPTG induction (0.5 and 1 mM) and TB supplemented with 2 g/L of α-lactose (TB+Lac) to allow auto-induction. Throughout this work, the term lactose will refer α-lactose. Other parameters considered include, temperature (30 and 37°C) and culture volume to flask volume ratio (1:2, 1:3, 1:4 and 1:5, corresponding to

500 ml, 333 ml, 250 ml and 200 ml of medium in 1 L Erlenmeyer flask). In cases where IPTG induction was used, cell cultures were allowed to grow to an OD_{600} of 1.0 and protein expression was induced by adding IPTG to a final concentration of 0.5 or 1 mM for 4 hours. In media supplemented with lactose, cells were allowed to grow for 24 hours. All the experiments were conducted in standard flat-bottomed glass Erlenmeyer 1 L flasks with cotton plugs and a standard 200 rpm agitation. Data represents mean values of at least 3 independent experiments.

Analysis of protein production levels from cell crude extracts

For evaluation of protein production levels, samples of cell crude extracts were analysed by sodium dodecyl sulphate polyacrylamide gel electrophoresis (SDS-PAGE) and stained with copper chloride 0.3 M (Lee et al. 1987). Evaluation of protein expression levels in the different bacterial cell cultures was carried out by direct comparison between samples normalized for the same cell density. For normalisation, 1 ml of cell culture was collected, centrifuged and resuspended in 100 µl of TE buffer (50 mM Tris, 1 mM EDTA at pH 8.0) and supplemented with an additional 25 µl of sample loading buffer (10% w/v SDS, 10 mM beta-mercapto-ethanol, 20% v/v glycerol, 0.2 M Tris–HCl pH 6.8, 0.05% w/v bromophenol blue). Samples were kept at 4°C for 30 minutes and centrifuged at the same temperature for 20 minutes at 14,000×g. For electrophoresis, the volume of the total soluble cell crude extract (supernatant) to be applied in each lane was calculated according to the following formula: $ODi/ODf = Vi/Vf$, where ODi is the OD_{600} of the cell culture, ODf is 0.05 (this value was optimized for copper staining), Vi is 125 (100 µl of TE + 25 µl of sample loading buffer) and Vf is the volume of supernatant (µl) to apply. In the case of samples applied in the same gel, protein expression was visually estimated, supported by area calculation with the image processing software *ImageJ* (Rasband1997–2011) and expressed as a relative percentage of the highest value. Whenever there was the need to compare expression levels in different gels, analysis was carried out by visually comparing band intensity/thickness. The same volume of 3 µl of molecular weight marker was used in every gel. No modifications were made to the images other than cutting, pasting and resizing.

Protein production and purification

For all SELP copolymers, bacterial cell cultures were grown in TB+lac containing ampicillin, with a liquid to flask volume ratio of 1:5 (200 ml of medium in 1 L flask) at 37°C. After fermentation time, cells were harvested by centrifugation and frozen until use. For cell lysis, cells were thawed, resuspended in the adequate buffer and lysed by ultrasonic disruption (sonication) using a Vibra cell™ 75043 (Bioblock Scientific, 750 W max power) with a solid probe of 25 mm diameter. Ultrasonic treatment was applied with an amplitude of 65% for 3 seconds, followed by a 9 second pulse-off delay with a total sonication time of 30 minutes. Samples were kept on ice throughout the process. Following disruption, cellular debris was removed by centrifugation at 11,500×g for 20 minutes at 4°C.

Recombinant SELP-1020-A was purified by immobilized metal affinity chromatography (IMAC) and compared with the combined acid pH treatment and ammonium sulphate precipitation. For all the remaining copolymers, purification was obtained by the combined purification method. Protein purity, molecular weight and recovery rates were assessed by SDS-PAGE with copper chloride (0.3 M) staining at each protein purification stage.

Purification by immobilized metal affinity chromatography (IMAC)

Recombinant protein from SELP-1020-A culture was purified by IMAC with a HisTrap HP 5-ml column already prepacked with a Ni Sepharose™ High Performance matrix. The cells pellet was thawed, resuspended in binding buffer (20 mM sodium phosphate, 0.5 M NaCl, 20 mM imidazole, pH 7.4) and lysed by ultrasonic disruption (sonication) as described above. The insoluble debris was removed by centrifugation and the clear supernatant filtered through a 0.45 µm filter to avoid column clogging. Elution was performed with buffers containing 0.5 M NaCl, 20 mM phosphate buffer, pH 7.4 and increasing concentrations of imidazole (20 mM, 40 mM, 80 mM, 150 mM, 200 mM, 250 mM and 500 mM), by following the manufacturer's instructions. The fractions were analysed for SELP content by SDS-PAGE with copper chloride staining. Final polymer purification from the soluble fraction was performed with buffer containing 80 mM imidazole, 0.5 M NaCl and 20 mM phosphate buffer, pH 7.4.

Acid pH treatment

Cultures with pCM13(SELP-1020-A)-BL21(DE3) were grown in the previously optimised conditions for 22 hours. Cells were collected, resuspended in TE buffer (50 mM Tris, 1 mM EDTA, pH 8) and lysed by sonication as described above. The insoluble debris was removed by centrifugation and the supernatant adjusted to pH 4, 3.5 or 3 with 1 M hydrochloric acid (HCl), while maintaining on ice with constant agitation. After an incubation period of 30 minutes at the corresponding pH, the precipitated supernatant was centrifuged at 11,500×g for 20 minutes at 4°C. The insoluble debris was removed and the clear supernatant analysed by SDS-PAGE.

Ammonium sulphate precipitation

Cell pellets were thawed and resuspended in TE buffer and lysed by sonication as described above. The insoluble debris was removed by centrifugation and the clear supernatant was adjusted to pH 3.5 with HCl 1 M, leading to precipitation of most of the *E. coli* proteins. The precipitated proteins were removed by centrifugation at 11,500×g for 20 minutes at 4°C. To assess the optimal concentration of ammonium sulphate for purification of recombinant SELP copolymers, the supernatant of each recombinant protein (SELP-1020-A, SELP-520-A, SELP-59-A and SELP-109-A) was precipitated with ammonium sulphate at final concentrations of 10, 15, 20, 25, 30, 35 and 40% saturation. Ammonium sulphate was added slowly to the supernatant while maintaining on ice with agitation and the resulting mixture was kept on ice for 30 minutes prior to centrifugation at 11,500×g for 20 minutes at 4°C. After centrifugation, the precipitated polymer fraction was resolubilized in deionized water for 2–3 hours at 4°C and centrifuged or filtered to remove the remaining insoluble contents. The clear polymer solution was then dialysed against water at 4°C, filtered with a 0.45 μm filter and lyophilized.

Evaluation of cell disruption by acidic pH treatment

For evaluation of the efficiency of the acidic pH treatment for cell disruption, SELP-59-A producing cultures were used. Cells were collected, resuspended in TE buffer at final wet cell weights of 0.4 g/ml and 0.8 g/ml and subjected to acid treatment or sonication as described above. For the acid treatment, suspensions were adjusted to pH 3.5 with 1 M HCl and incubated overnight at 4°C with mild agitation (acid-based cell lysis). Percentage of cell survival was compared in cells submitted to 1) acid-based cell lysis, 2) sonication, 3) acid-based cell lysis followed by sonication and 4) untreated cell crude extract. After each treatment, samples were taken, plated on LB agar without antibiotic at appropriate dilutions (10^{-7} for the untreated sample and 10^{-6} for the remaining) and incubated overnight at 37°C. Cells were counted and the results expressed as percentage of cell survival compared to the untreated sample. All experiments were done in triplicate. Evaluation of protein release was conducted in cell extracts submitted to acid-based cell lysis and acid-based cell lysis followed by sonication. After each treatment the cell extract was centrifuged for 20 minutes at 11,500×g and the supernatant was collected for further analysis by SDS-PAGE. The untreated cell extract was prepared according to the method previously described in Protein Analysis in Material and Methods. The same volume of cell extract/sample was used for all experiments as well as for SDS-PAGE.

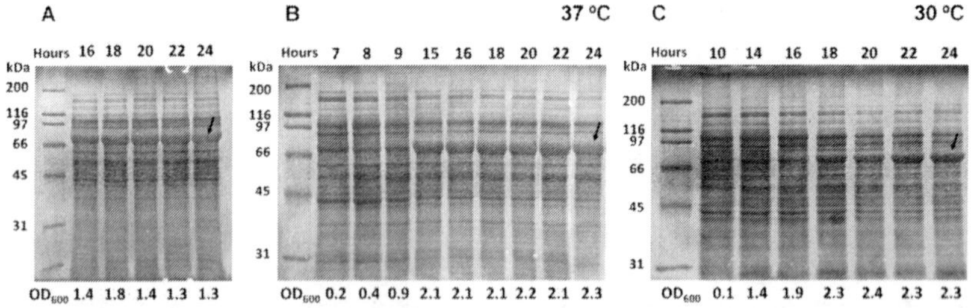

Figure 2. Electrophoretic patterns of cell crude extracts during growth of auto-induced pCM13(SELP-1020-A)-BL21(DE3) cultures in A) LB+lac at 37°C, B) TB+lac at 37°C and C) TB+lac at 30°C. To facilitate direct comparison of protein expression, all samples were normalised for the same cell density before loading on gel. The sampling time is given above each lane whereas the OD600 at that time point is indicated below. Target recombinant protein is indicated by arrows and the molecular weight marker is in the left of each gel. The abnormal gel mobility of the recombinant protein was previously observed by other authors (Teng et al. 2009; Lyons et al.2007; McPherson et al. 1992).

RESULTS

Optimization of culture conditions

E. coli BL21 (DE3) transformed with pCM13(SELP-1020-A) was chosen as the representative sample for production optimization studies. Initial studies aiming at protein production were performed in LB and TB media supplemented with lactose (LB+lac and TB+lac), at 37°C with a liquid to flask volume ratio of 1:5 and analysed through a 24 hour fermentation period. Figure 2 represents the cell crude extract of protein samples taken at different time periods during the 24 hour shake flask fermentation, showing that higher ODs (1.3 for LB+lac in Figure 2A and2.3 for TB+lac in Figure 2B) as well as higher production levels could be obtained with TB supplemented with lactose at each time period.

Visually, SELP-1020-A expression levels seem to reach a steady state after 18 hours of fermentation in LB+lac (Figure 2A) which was further corroborated by calculating the area of the protein bands (Table 2). Regarding SELP-1020-A production at 37°C in TB+lac, protein expression starts in the period between 9 and 15 hours of fermentation and reaches a steady state after 18 hours (Table 2) with little variation in protein band area or cell density. When the temperature of the bacterial cell culture was reduced to 30°C, while maintaining all the other conditions (TB+lac, agitation and flask volume ratio), protein expression was delayed, starting only after 16 hours of fermentation and showing lower protein expression (Figure 2C). In this case, protein expression increased gradually over time as determined by calculating the area of the expressed SELP-1020-A band (Table 2. For SDS-PAGE analysis, samples were reverse stained with copper chloride according to the method previously described by Lee *et al.* (Lee et al. 1987). This staining method was preferred over conventional Coomassie blue staining not only due to the greater sensitivity (≈10 ng protein/lane) but also due to the poor Coomassie staining of SELP-1020-A bands that appear extremely diffuse and make band analysis difficult. Although copper staining can be a coarse method for comparing samples in different gels, by applying normalised samples for the same cell density, evaluation of protein expression levels can be directly compared on the basis of the protein band thickness.

In shake flasks, the oxygen transfer rate (OTR) is reversely proportional to the volume of culture (Maier and Büchs 2001). Higher OTR is obtained in flasks with less culture volume or by increasing the agitation rate, however one should take in account that very low amounts can only be used for short-term fermentations otherwise the medium will evaporate. To study the effect of OTR in SELP-1020-A production, a preliminary study of protein expression with different liquid to flask volume ratios was carried out with the previously optimised conditions (TB+lac, 22 hours fermentation at 37°C). Considering SELP-1020-A, high protein expression levels with comparable band intensities were obtained for all the ratios investigated (Figure 3). The relative percentage of protein production is summarized in Table 2demonstrating that for all the volume ratios tested there was little or no variation.

Table 2. Relative productivities of SELP-1020-A as determined by area calculation of protein bands

	Relative percentage (%)						
	LB +lac 37°C	TB +lac 37°C	TB +lac 30°C	Flask volume ratios (TB+lac, 37°C) 1:2	1:3	1:4	1:5
15 hours	n.d.	89	n.d.	n.d.	n.d.	n.d.	n.d.
16 hours	86	87	37	n.d.	n.d.	n.d.	n.d.
18 hours	100	100	50	n.d.	n.d.	n.d.	n.d.
20 hours	99	96	67	n.d.	n.d.	n.d.	n.d.
22 hours	93	99	91	100	99	100	98
24 hours	95	97	100	n.d.	n.d.	n.d.	n.d.

Bacterial cells were cultured in LB+lac at 37°C, TB+lac at 37°C, TB+lac at 30°C and TB+lac at 37°C with different liquid to flask volume ratios (1:2, 1:3, 1:4 and 1:5). Band areas were calculated with ImageJ and expressed as relative percentage. n.d. – not determined.

Regarding bacterial cell growth, lower volume of medium in the flasks resulted in higher cell density as determined by the OD_{600}. For all the cases, the OD increased with increasing volume ratios with values of 1.1, 1.4, 1.8 and 2.2 for volume ratios of 1:2, 1:3, 1:4 and 1:5, respectively. For the subsequent production of all SELP copolymers the standard parameters were set as a 22 hour fermentation period in TB+lac, with a flask volume ratio of 1:5 (200 ml of medium in 1 L flask) at 37°C.

SELP-1020-A production with IPTG induction

As the reference model of induction, we have chosen the widely used protocol for recombinant protein expression described in both, the Sambrook laboratory (Sambrook and Russell 2001) and the pET system (Novagen 2006) manuals. In cell cultures in LB medium, for the two concentrations of IPTG tested (0.5 mM and 1 mM), the best expression levels were obtained after 4 hours of induction, with similar band intensities for both concentrations (Figure 4). Comparable results with similar expression levels were obtained with TB (data not shown) however, and as expected, the final optical density reached higher values (approximately 2.4 for TB and 1.6 for LB for both concentrations of IPTG).

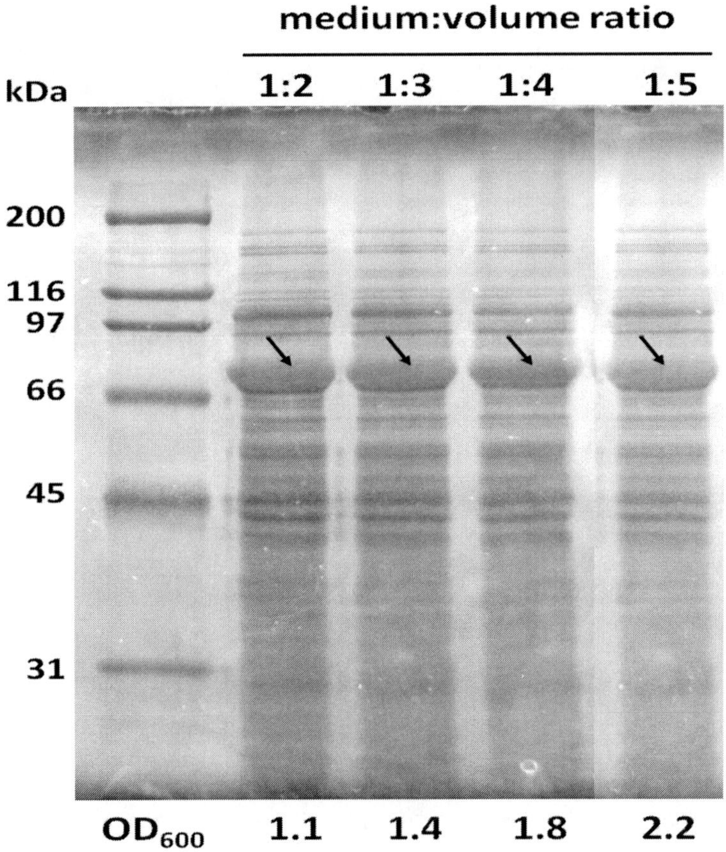

Figure 3. Cell growth and SELP-1020-A expression using auto-induction medium with different flask volume ratios after 22 hours of fermentation. All samples were normalised for the same cell density prior to loading on gel. This figure was assembled from pictures of different gels. No modifications were made to the images other than cutting, pasting and resizing. Cell density is indicated below each gel with target recombinant protein indicated by arrows. Molecular weight marker is represented on the right and left lanes.

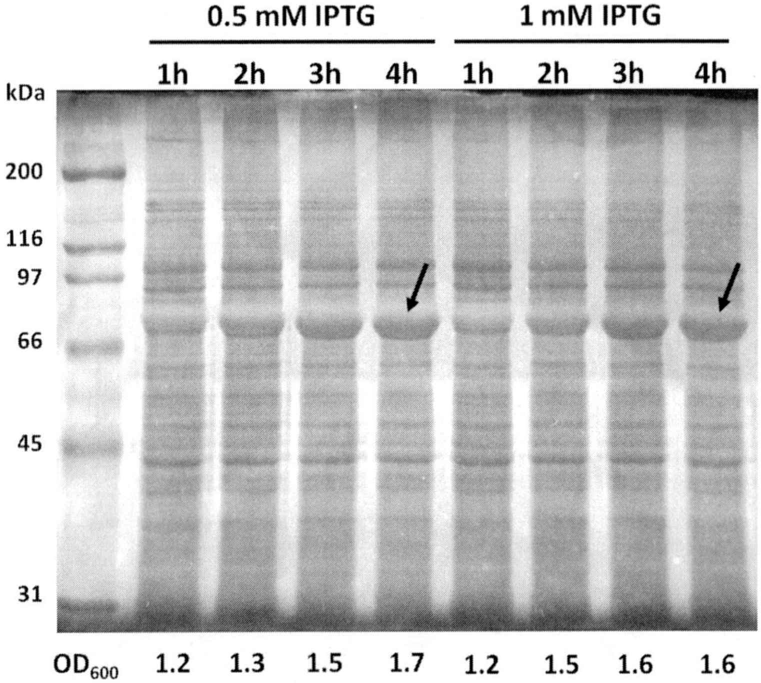

Figure 4. Expression levels with respect to the concentration of inducer and time course of induction. Cultures of *E. coli* harboring pCM13(SELP-1020-A) in LB were induced with 0.5 mM or 1 mM of IPTG when the OD_{600} reached 1. Samplings were taken at the indicated intervals. Cell density is indicated below each lane with target recombinant protein indicated by arrows.

Copolymers purification

During the optimisation of culture conditions for protein production, we realized that SELP-1020-A was expressed in the soluble form and allowing for purification by IMAC. SELP-1020-A was expressed in TB+lac, in the conditions here reported and cells were treated as described in the Materials and Methods section. The recombinant polymer was purified from the clear supernatant with the HisTrap HP nickel column using sequential elutions with buffers containing imidazole. A highly purified, as indicated by SDS-PAGE analysis, SELP-1020-A polymer fraction was obtained by eluting with buffer containing 80 mM of imidazole (Figure 5). Little or no SELP-1020-A losses were observed during the process and no recombinant protein was present in further elutions with higher concentrations of imidazole (150 mM, 200 mM, 250 mM and 500 mM), showing that SELP-1020-A was purified with a high recovery rate. However, an apparent molecular weight of approximately 66 kDa, determined from the SDS-PAGE analysis, was much higher than the theoretical calculation of 54117 Da for the same protein. The unexpected higher apparent molecular weight, as determined by SDS-PAGE, was previously observed by other authors (Teng et al. 2009; Lyons et al.2007; McPherson et al. 1992) and attributed to the hydrophobic nature of these protein polymers. MALDI-TOF MS was therefore used to compare the molecular weight of the chromatographically purified copolymer and showed a molecular ion peak at 54125 Da, which is in agreement within experimental error, with the theoretically calculated value of 54117 Da (Figure 5B).

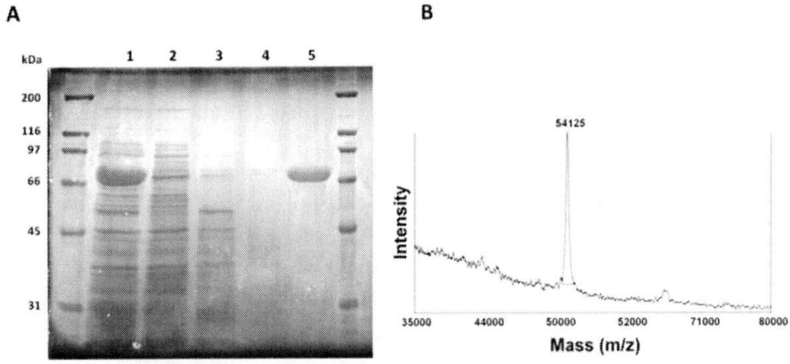

Figure 5. Purification of SELP-1020-A by A) affinity chromatography and B) molecular weight determination by mass spectrometry. A)SDS-PAGE of SELP-1020-A: lane 1 – soluble lysate before column loading; lane 2 – flow through; lane 3 – fraction eluted with 20 mM imidazole; lane 4 – fraction eluted with 40 mM imidazole; lane 5 – fraction eluted with 80 mM imidazole. Fully purified polymer fraction was obtained by elution with buffer containing 80 mM of imidazole with high recovery rate. **B)** Image represents the MALDI-TOF mass spectrum of SELP-1020-A. The molecular ion peak at 54125 confirms the theoretical calculated value (54117 Da).

Although being a popular and generally used method for purification of recombinant proteins, immobilised metal-affinity chromatography has serious limitations, namely the difficulties associated with scale-up, high cost and column fouling. The use of non-chromatographic methods would overcome these drawbacks and hence we investigated acid pH treatment and ammonium sulphate precipitation as methods for SELP purification.

Indeed the latter approach has already been applied successfully as part of a purification protocol for recombinant spider silk protein (Scheller et al. 2001) and resilin-like proteins (Lyons et al. 2007; Kim et al. 2007). These alternative purification methods are less labour intensive and less expensive and should allow for high recovery rates of recombinant proteins known to be stable under the conditions used. The low pH would be expected to precipitate the endogenous *E. coli* proteins while the more stable SELPs would stay in solution.

Bacterial cells harbouring pCM13(SELP-1020-A) were allowed to grow in the previously optimised conditions, collected and lysed. After cell disruption, the insoluble debris was removed by centrifugation and the supernatant was adjusted to pH 4, 3.5 or 3. While adjusting the pH, the precipitation of proteins was clearly visible with the solution turning opaque with a whitish colour. After removal of precipitated proteins, the clear supernatant was analysed by SDS-PAGE, showing that adjusting the pH of the crude lysate to 3.0 or 3.5 was very effective in the removal of *E. coli* proteins (Figure 6). Treatment at pH 3.0 demonstrated to be more effective in the removal of contaminant proteins. However, the lower intensity of the band suggests that there is a loss of the recombinant copolymer during the process, which is probably due to a partial precipitation. Therefore, treatment at pH 3.5 was apparently more efficient for purification, without detectable SELP losses being observed during the process. In the case of SELP-1020-A, adjusting the pH of the crude lysate to 3.0 was almost enough for purification of the copolymer. However, purification with only the acidic treatment was not very reliable as in some random rare occasions the supernatant presented some contaminant proteins, although in very low amounts. In the case of SELP-59-A and SELP-520-A some contaminants do remain after acidic pH treatment and therefore ammonium sulphate precipitation was examined as a further non-chromatographic step for completion of purification.

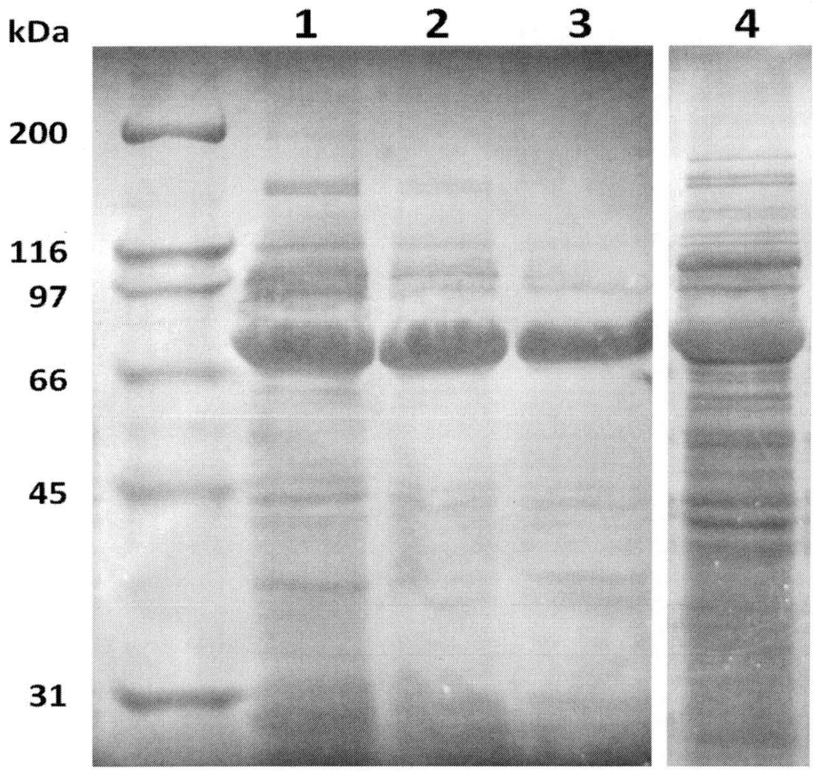

Figure 6. Effect of the acidic treatment of the soluble lysate. The clear supernatant of the lysate from SELP-1020-A (lane 4) was adjusted to pH 4 (lane 1), 3.5 (lane 2) and 3 (lane 3). After pH adjustment, the precipitated pellet was removed and the clear supernatant was analysed by SDS-PAGE. Nearly pure SELP-1020-A was obtained after the acidic treatment at pH 3.

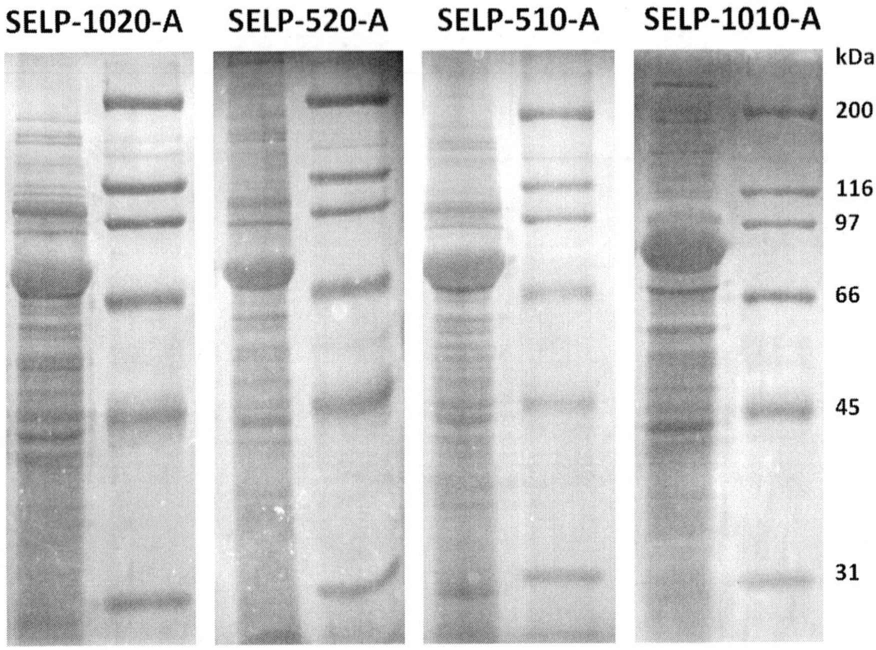

Figure 7. Cell crude extract obtained from bacterial cell cultures producing the different copolymers. This figure was assembled from pictures of different gels. No modifications were made to the images other than cutting, pasting and resizing.

Due to the hydrophobic nature of SELPs, all the remaining copolymers (SELP-1020-A, SELP-520-A and SELP-59-A) precipitated from the acid-treated lysate at relatively low concentrations of ammonium sulphate. For all the copolymers, 20% saturation was optimal for precipitation from the soluble lysate with most of the *E. coli* endogenous proteins being retained in the supernatant (Figure 8).

All the recombinant SELPs were produced in the optimised conditions. SDS-PAGE analysis of the cell crude extracts revealed that high levels of protein expression were obtained for all the recombinant copolymers (Figure 7). Although showing a good protein expression as mentioned above, no further attempts were made for the purification of SELP-109-A as its precipitation occurred even during cell lysis (data not shown).

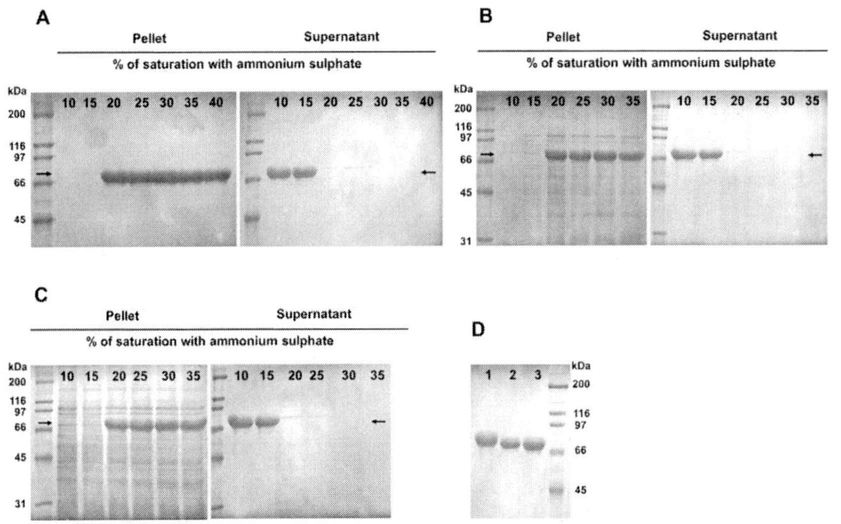

Figure 8. Purification of the recombinant copolymers by ammonium sulphate precipitation. The soluble acid-treated lysates of SELP-1020-A **(A)**, SELP-520-A **(B)** and SELP-59-A **(C)** were saturated with increasing concentrations of ammonium sulphate (indicated above each gel). Depending on the concentration used, the copolymers either precipitated or stayed in the supernatant. **(D)** (Increased purity was obtained by resuspending the precipitated copolymer (with 20% ammonium sulphate) in water and letting at 4°C with agitation followed by centrifugation or filtration (lane 1 – SELP-1020-A, lane 2 – SELP-520-A, lane 3 – SELP-59-A). The abnormal gel mobility of the recombinant protein was previously observed by other authors (Teng et al. 2009; Lyons et al. 2007; McPherson et al. 1992) and attributed to the hydrophobic nature of the proteins.

Figure 8A shows the fractions (pellets and supernatants) resulting from the precipitation of SELP-1020-A acid-treated extracts with increasing concentrations of ammonium sulphate. Due to the treatment with pH, the polymers, and in particular SELP-1020-A, were already highly pure. Precipitation with 20% ammonium sulphate allowed for both purification and concentration of the samples and in this way avoided high volumes for dialysis and subsequent lyophilisation. Furthermore, no protein losses were detected by SDS-PAGE, indicating high recovery rates of recombinant protein.

Finally, increased purity of the three copolymers was obtained by resuspending the precipitates in deionized water for 2–3 hours at 4°C with agitation. The recombinant copolymers were readily dissolved in the aqueous solution while the *E. coli* contaminants remained in the insoluble state. Removal of the insoluble contaminant debris was achieved by centrifugation and further filtration allowing for obtaining of highly pure polymer fraction (Figure 8D).

After lyophilisation, the volumetric productivity (milligrams of recombinant protein per litre of culture) of the recombinant copolymers produced in the optimised conditions and purified by ammonium sulphate precipitation were 185, 151 and 198 mg/L for SELP-1020-A, SELP-520-A and SELP-59-A, respectively.

Acid-based cell lysis and protein release

To further simplify the purification protocol, we have investigated if an acid-based cell lysis could be employed as a single step method for both cell disruption and removal of contaminant *E. coli* proteins. To assess if bacterial cells were disrupted by the acidic treatment, therefore releasing the SELP copolymer, we measured the content of protein release by SDS-PAGE. Cell death was analysed by counting the number of colony-forming units (CFUs), and compared with the untreated cell crude extract.

As can be seen in Figures 9A and 9B, only about 4% cell survival was observed after sonication, represented by a low number of colony-forming units (CFUs). However, the acid-based cell lysis proved to be even more efficient as no CFUs were detected in solid medium, thus representing 0% cell survival. As expected, no CFUs were formed in the sample submitted to both acidification and sonication. In terms of protein release, the acid-based cell lysis proved to be very efficient at the two concentrations tested, showing that up to 0.8 g/ml of wet cell weight can be successfully disrupted by this method (Figure 9C). In fact, the supernatants of the treated samples exhibit similar levels of recombinant copolymer as that present in the cell crude extract while also allowing for the removal of most of the *E. coli* endogenous proteins. The additional sonication treatment did not influence the amount of protein in the supernatant of the acid-lysate, thus suggesting that a sole acidification step is enough to promote cell lysis and partial protein purification. Increased purity can then be achieved by further purification with ammonium sulphate as described above.

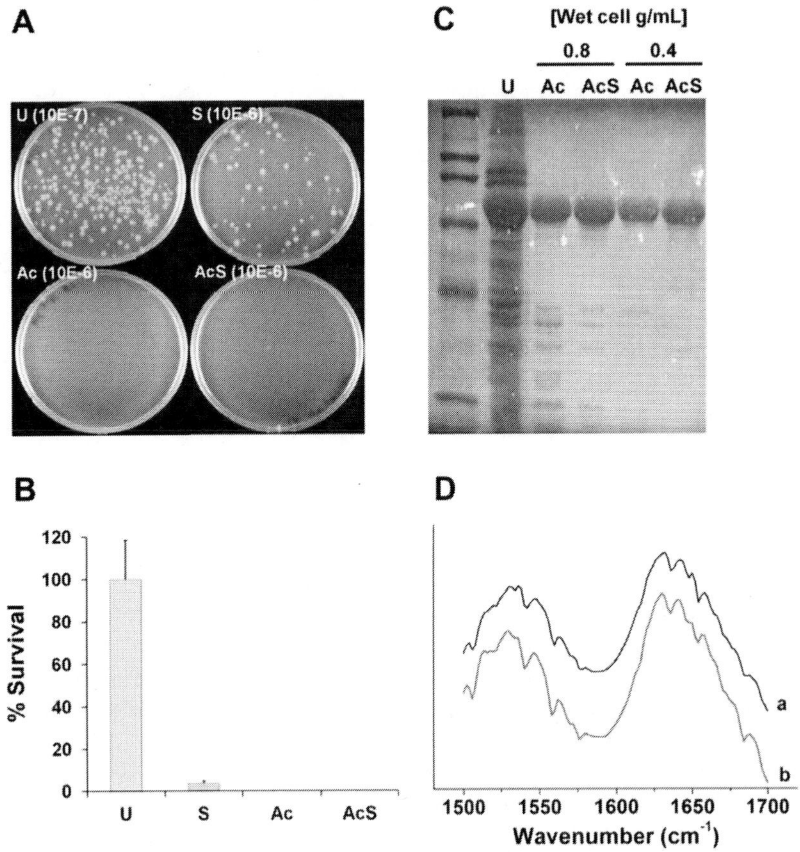

Figure 9. Evaluation of cell death and protein release in bacterial cultures of SELP-59-A. (A) – number of CFUs in untreated cell crude extracts (U) and in cell extracts submitted to sonication (S), acid-based cell lysis (Ac) and acid-based cell lysis followed by sonication (AcS). Cell dilutions are represented in parenthesis. **(B)** – Graphic representation of the percentage of cell survival for the different treatments with results expressed as percentage of survival as compared to the untreated sample. **(C)** – Evaluation of protein release in samples at different concentrations of wet cell weight after acid-based cell lysis (Ac) and acid-based cell lysis followed by sonication (AcS). The same volume was applied in each lane. Samples legend: U – untreated cell crude extract; Ac – acid-based cell lysis; AcS – acid-based cell lysis followed by sonication. **(D)** – FTIR spectra displaying the amide I (1600 – 1700 cm-1) and amide II (1500 – 1600 cm-1) band regions of pure lyophilized SELP-59-A samples without **(a)** and with **(b)** the acidic treatment.

In order to evaluate if the acidic treatment would induce any conformational changes in the molecular structure of the copolymers, Fourier transform infrared spectroscopy (FTIR) was employed to assess possible structural changes between samples purified with and without the acidic treatment. Figure 9D shows the FTIR spectra in the amide I (1600 – 1700 cm^{-1}) and amide II (1500 – 1600 cm^{-1}) regions of pure lyophilized SELP-59-A samples, demonstrating that the acidic treatment did not induce significant structural modifications.

DISCUSSION

In this study we successfully synthesised and overexpressed four novel SELPs of similar size (55 kDa) but with variable content and ratios of silk to elastin i.e. 1:1 (SELP-109-A), 1:2 (SELP-1020-A, SELP-59-A) and 1:4 (SELP-520-A). Due to the novel elastin block (VPAVG) and the various silk to elastin ratios used processing of these polymers should give rise to a novel family of SELPs with unique biological and mechanical properties and thereby potentially extend the functionality of SELPs in general. Indeed, those polymers with a higher proportion of the elastin-like motif may be expected to be characterised by a reduced tensile strength, higher resilience, lower propensity for hydrogel formation, increased solubility and reduced crystallinity as compared to those with a higher proportion of silk (Dandu et al. 2009; Gustafson and Ghandehari 2010; Haider et al. 2005). The four polymers prepared in this study have been tailored to cover a large range of many of these properties and studies are currently underway to characterise these.

We examined the expression of these polymers with the currently most used method (Sambrook and Russell 2001) and found all proteins to be expressed intracellularly in a soluble form. In an attempt to simplify the production protocol we investigated the use of an auto-induction approach in which the inducing agent (in this case lactose) was added directly during media formulation. The best expression levels were obtained with TB+lac at 37°C when compared with LB+lac at the same temperature. However, despite the high expression levels, low cell densities ($OD_{600} \approx 2$) were reached. This can be further optimised by adding low concentrations of glucose as previously demonstrated by Studier (Studier 2005), where the addition of glucose at low amounts (0.05% w/v) allowed cells to have an initial burst in growth. However, it should be noted that addition of glucose causes saturated cultures to become very acidic, limiting saturation density (Studier 2005). In this study we have chosen not to add glucose to medium composition, as our intention was to perform a preliminary study in the evaluation of protein expression levels where expression is induced in the early stages of growth. It is well known that the presence of glucose exerts catabolite repression, preventing the uptake and utilization of lactose, whilst glycerol does not interfere with lactose induction or the ability to metabolize this sugar for energy. In fact, addition of glycerol in auto-induction media was demonstrated to double the yield of target recombinant protein when compared to medium relying on lactose as the primary energy source. We have found that expression of the recombinant copolymers only occurs in the period between 9 to 15 hours of growth with no recombinant protein being produced earlier. This delay in the induction of protein expression is most likely attributed to the yeast extract used in the complex media formulation. The yeast extract supplies a variety of metabolites including complex carbohydrates that allow for fast initial growth rates. However, the yeast extract is also rich in amino acids that may prevent induction of target proteins by lactose in the early stages of growth. Indeed, in the absence of glucose, the presence of amino acids appears to inhibit induction until growth slows upon approach to saturation (Studier 2005).

One of the major factors affecting both bacterial cell growth and recombinant protein expression is the concentration of dissolved oxygen in the medium. It is commonly accepted, that high oxygen transfer rates lead to higher cell densities in bacterial cell cultures (Losen et al. 2004, Collins et al. *unpublished*). In shake flask fermentations, the OTR is influenced by several factors namely shaking frequency and shaking diameter, filling volume, flask shape and size, surface properties of the flask material and the physical-chemical properties of the liquid (Maier and Büchs 2001). Increasing the shaking frequency or lowering the filling volume leads to higher OTR due to liquid mixing and a high area of contact that allows efficient oxygen diffusion (Maier and Büchs 2001; McDaniel et al. 1965; McDaniel and Bailey 1969). In this work, we have performed a preliminary study on the effect of oxygen on the recombinant protein expression by evaluating the expression levels in shake flask cultures with different liquid to flask volume ratios. Cell density was shown to be inversely proportional to the filling volume without any deficiencies in protein expression (Figure 3) and thus suggesting that an optimization of oxygen levels could play a major factor in improving cell densities.

Auto-induction is more convenient than IPTG induction as it does not require any intermediate steps. The expression host is simply inoculated in the medium and let to grow without the need to monitor cell culture and add inducer at the appropriate time. In our study, induction of protein expression by lactose showed similar or even higher expression levels when compared with the conventional IPTG induction method. This work provides an insight into small/laboratory scale fermentations where auto-induction can provide major advantages as no extra culture handling is required. All copolymers were found to be successfully overexpressed in this auto-induction media, demonstrating its high potential as a facile method for production of recombinant SELPs in small/laboratory scale fermentations, but also for the production of several other proteins. In fact, in our laboratory, several other recombinant proteins were produced by this auto-induction method and purified for biotechnological purposes, namely, structural protein-based polymers, human growth factors, serine proteases, hydrolases, esterases, lipocalines, antimicrobial peptides and plant transcription factors (*personal communication, Casal, M.*). In all the studies performed above we used purified lactose but other sources of lactose can be used. Waste pollutants derived from the dairy industry, like cheese whey, were previously used for the production of recombinant proteins with expression levels similar to those obtained with IPTG (Viitanen et al. 2003; De León-Rodríguez et al. 2006).

All four newly developed copolymers were expressed at high levels using the optimised fermentation conditions. Treating the soluble crude lysate with an acidification treatment allowed removal of most of the *E. coli* endogenous proteins. This methodology was previously employed for the purification of a recombinant major ampullate spidroin I (MaSp1) protein from the spider *Nephila clavipes* (Xia et al. 2010). The crude lysate of MaSp1 producing cells was adjusted to pH 4 and led to precipitation of host cell proteins whereas the recombinant protein remained in the supernatant after centrifugation. However, the supernatant from the acidic treatment was still enriched with host cell proteins. In our study, we demonstrated that acidification at pH 3.0-3.5 is much more efficient for removal of *E. coli* endogenous proteins than pH 4. In fact, upon acidification of the SELP-1020-A crude lysate, a highly pure polymer fraction was obtained in just one step. Minor contaminants remained for SELP-520-A and SELP-59-A and these were removed by a simple ammonium sulphate precipitation step, followed by resuspension in cold water and dialysis against water.

SELPs are structural proteins lacking complex folding mechanisms and spontaneously self-assemble into β-sheets by hydrogen-bonding mediated processes. The stability of SELPs at low pHs is well known with many SELP copolymers being extensively processed into materials in formic acid solutions without any reported negative effects (Nagarajan et al. 2007; Qiu et al. 2009,2010). Furthermore, it was previously demonstrated that pH had no effect on the equilibrium swelling ratio of SELP hydrogels between pH 2.0 and pH 12.0, indicating an absence of pH sensitivity (Dinerman et al. 2002; Haider et al. 2005; Dandu et al. 2009). However, the substitution of an amino acid in the elastin-like block by a glutamic acid generated a pH stimuli-sensitive polymer (Nagarsekar et al. 2003, 2002). Here, we took advantage of this known acidic pH stability to separate the novel SELP polymers from the unstable contaminants. The acidic pH stability was further supported by comparing the FTIR spectra between pure lyophilized copolymer samples with and without the acidic treatment.

Concerning SELP-109-A, purification was not carried out as the polymer precipitated during cell lysis, possibly due to the high content of silk blocks that promote gelation (Megeed et al. 2002; Cappello et al. 1998). With the appropriate composition, SELPs undergo an irreversible sol–gel transition through crystallization of the silk-like blocks. This spontaneous transition is influenced by the number of silk-like blocks as well as by environmental conditions like temperature (Haider et al. 2004). As the formation of hydrogen-bonds is the major driving force behind gelation, copolymers with larger silk-like blocks will provide more contact points for potential inter-chain junctions (Haider et al. 2005; Cresce et al. 2008). On the other hand, the inclusion of larger elastin-like blocks will increase the spacing between the silk-like blocks, increasing its flexibility and aqueous solubility (Haider et al. 2005). In fact, when analysed by solution viscometry and under the same conditions, the rate of gelation was found to reflect the increase in silk-like block content (Cappello et al. 1998). In the case of SELP-109-A and due to the large silk-like and small interrupting elastin-like blocks, gelation occurs rapidly. A protocol for purification of SELP-109-A will be further developed, possibly involving the use of chaotropic agents in the lysis buffer in order to inhibit hydrogen bonding. However it is beyond the purpose of this work where our main focus is to report the genetic construction and especially, a facile method for recombinant SELP expression and purification. For the remaining copolymers, precipitation with ammonium sulphate revealed to be an easy method for completion of SELP purification, allowing for the obtaining of highly pure polymer fractions with relatively low concentrations of ammonium sulphate. The volumetric productivities after lyophilisation were 185, 151 and 198 mg/L for SELP-1020-A, SELP-520-A and SELP-59-A, respectively, in comparison to the previously described 25–30 mg/L purified with IMAC (Dandu et al. 2009; Haider et al. 2005; Nagarsekar et al. 2002).

In an attempt to improve and simplify the purification process further we investigated whether the pH 3.5 treatment may also allow for cell disruption and hence enable both protein release and purification in one-simple step. Our analysis indicated that the treatment was indeed just as efficient at cell disruption, as monitored by SDS-PAGE, as was sonication. Up to 0.8 g of cell pellets (wet weight) per ml of resuspension buffer, which was the maximum cell concentration that could be easily resuspended, were found to be efficiently disrupted by incubation at pH 3.5 with, in addition, the removal of the majority of contaminating proteins being also achieved. Indeed a comparative investigation of cell viability following treatment further supported the suitability of the approach for cell disruption as it was found to be even more effective in reducing cell viability than sonication. Further purification and protein concentration can be obtained by precipitation with 20% ammonium sulphate with subsequent resuspension and dialysis against water followed by lyophilisation allowing for the obtention of highly pure, stable polymers. Hence we have developed a simple non-chromatographic method that allows for both cell disruption and the obtention of highly pure protein with reduced cost. This facile method is suited for proteins expressed in the soluble form and stable at low pH as in the case of the structural proteins hereby exemplified.

SELP copolymers purified by the method developed in this study did not show any cytotoxicity in mouse myoblast cell line C2C12 and even presented a slight increase in cell viability [see Additional file 2]. This suggests that these novel SELPs can be processed into materials and explored for biomedical applications; which is currently under study at our group. Although this work is focused on lab-scale fermentations, the purification method has already been validated in a high-density fermentation experiment with a 500 L fermenter, showing full recovery of the recombinant copolymer with high purity.

ACKNOWLEDGEMENTS

This work was financially supported by the European Comission via the 7th Framework Programme project EcoPlast (FP7-NMP-2009-SME-3, collaborative project number 246176), by Portuguese funding from FEDER through POFC – COMPETE and PEst project C/BIA/UI4050/2011 (Portugal), by MICINN (MAT 2009-14195-C03-03, IT2009-0089, ACI2009-0890, MAT2010-15310 and MAT2010-15982), the JCyL (VA034A09, VA030A08 and VA049A11-2) and CIBER-BBN. E.C. (Spain). Raul Machado was the receiver of a fellowship from Fundação para a Ciência e Tecnologia (SFRH/BD/36754/2007), co-funding by FSE through QREN/POPH (Portugal). Tony Collins is thankful to the Fundação para a Ciência e Tecnologia for its support through Programa Ciência 2008.

REFERENCES

1. Ansaldi M, Lepelletier M, Mejean V: **Site-specific mutagenesis by using an accurate recombinant polymerase chain reaction method.** *Anal Biochem* 1996,**234**(1):110–111.

2. Cappello J, Crissman J, Dorman M, Mikolajczak M, Textor G, Marquet M, Ferrari F: **Genetic engineering of structural protein polymers.** *Biotechnol Prog* 1990,**6**(3):198–202.

3. Cappello J, Crissman JW, Crissman M, Ferrari FA, Textor G, Wallis O, Whitledge JR, Zhou X, Burman D, Aukerman L, Stedronsky ER: **In-situ self-assembling protein polymer gel systems for administration, delivery, and release of drugs.** *J Control Release* 1998,**53**(1–3):105–117.

4. Chow D, Nunalee ML, Lim DW, Simnick AJ, Chilkoti A: **Peptide-based Biopolymers in Biomedicine and Biotechnology.** *Materials science & engineering R, Reports: a review journal* 2008,**62**(4):125–155.

5. Cresce AV, Dandu R, Burger A, Cappello J, Ghandehari H: **Characterization and real-time imaging of gene expression of adenovirus embedded silk-elastinlike protein polymer hydrogels.** *Mol Pharm* 2008,**5**(5):891–897.

6. Dandu R, Von Cresce A, Briber R, Dowell P, Cappello J, Ghandehari H: **Silk-elastinlike protein polymer hydrogels: Influence of monomer sequence on physicochemical properties.** *Polymer* 2009,**50**(2):366–374.

7. De León-Rodríguez A, Rivera-Pastrana D, Medina-Rivero E, Flores-Flores JL, Estrada-Baltazar A, Ordóñez-Acevedo LG, de la Rosa APB:**Production of penicillin acylase by a recombinant Escherichia coli using cheese whey as substrate and inducer.** *Biomol Eng*2006,**23**(6):299–305.

8. Dinerman AA, Cappello J, Ghandehari H, Hoag SW: **Swelling behavior of a genetically engineered silk-elastinlike protein polymer hydrogel.** *Biomaterials* 2002,**23**(21):4203–4210.

9. Gustafson JA, Ghandehari H: **Silk-elastinlike protein polymers for matrix-mediated cancer gene therapy.** *Adv Drug Deliv Rev*2010,**62**(15):1509–1523.

10. Haider M, Megeed Z, Ghandehari H: **Genetically engineered polymers: status and prospects for controlled release.** *J Control Release* 2004,**95**(1):1–26.

11. Haider M, Leung V, Ferrari F, Crissman J, Powell J, Cappello J, Ghandehari H: **Molecular engineering of silk-elastinlike polymers for matrix-mediated gene delivery: biosynthesis and characterization.** *Mol Pharm* 2005,**2**(2):139–150.

12. Inoue H, Nojima H, Okayama H: **High efficiency transformation of Escherichia coli with plasmids.** *Gene* 1990,**96**(1):23–28.

13. Kim M, Elvin C, Brownlee A, Lyons R: **High yield expression of recombinant pro-resilin: lactose-induced fermentation in E. coli and facile purification.** *Protein Expr Purif* 2007,**52**(1):230–236.

14. Lee C, Levin A, Branton D: **Copper staining: a five-minute protein stain for sodium dodecyl sulfate-polyacrylamide gels.** *Anal Biochem* 1987,**166**(2):308–312.

15. Losen M, Frölich B, Pohl M, Büchs J: **Effect of oxygen limitation and medium composition on escherichia coli fermentation in shake-flask cultures.** *Biotechnol Prog* 2004,**20**(4):1062–1068.

16. Luan C-H, Urry DW: **Elastic, plastic, and hydrogel-forming protein-based polymers.** In *Polymer Data Handbook*. Edited by: Mark JE. Oxford University Press, New York, NY; 1999:78–89.

17. Lyons RE, Lesieur E, Kim M, Wong DCC, Huson MG, Nairn KM, Brownlee AG, Pearson RD, Elvin CM: **Design and facile production of recombinant resilin-like polypeptides: gene construction and a rapid protein purification method.** *Protein Eng Des Sel* 2007,**20**(1):25–32.

18. Machado R, Ribeiro AJ, Padrao J, Silva D, Nobre A, Teixeira JA, Arias FJ, Cunha AM, Rodriguez-Cabello JC, Casal M: **Exploiting the sequence of naturally occurring elastin: construction, production and characterization of a recombinant thermoplastic protein-based polymer.** *J Nano Res* 2009, **6:** 133–145. 10.4028/www.scientific.net/JNanoR.6.133

19. Maier U, Büchs J: **Characterisation of the gas–liquid mass transfer in shaking bioreactors.** *Biochem Eng J* 2001,**7**(2):99–106.

20. McDaniel LE, Bailey EG: **Effect of shaking speed and type of closure on shake flask cultures.** *Appl Microbiol* 1969,**17**(2):286–290.

21. McDaniel LE, Bailey EG, Zimmerli A: **Effect of oxygen supply rates on growth of Escherichia coli.** *Appl Microbiol* 1965, **13:** 109–114.

22. McDaniel JR, Mackay JA, Quiroz FG, Chilkoti A: **Recursive directional ligation by plasmid reconstruction allows rapid and seamless cloning of oligomeric genes.** *Biomacromolecules* 2010,**11**(4):944–952.

23. McPherson DT, Morrow C, Minehan DS, Wu J, Hunter E, Urry DW: **Production and purification of a recombinant elastomeric polypeptide, G-(VPGVG)19-VPGV, from Escherichia coli.** *Biotechnol Prog* 1992,**8**(4):347–352.

24. Megeed Z, Cappello J, Ghandehari H: **Genetically engineered silk-elastinlike protein polymers for controlled drug delivery.** *Adv Drug Deliv Rev* 2002,**54**(8):1075–1091.

25. Meyer DE, Chilkoti A: **Purification of recombinant proteins by fusion with thermally-responsive polypeptides.** *Nat Biotechnol*1999,**17**(11):1112–1115.

26. Meyer DE, Chilkoti A: **Genetically encoded synthesis of protein-based polymers with precisely specified molecular weight and sequence by recursive directional ligation: examples from the elastin-like polypeptide system.** *Biomacromolecules* 2002,**3**(2):357–367.

27. Nagapudi K, Brinkman WT, Thomas BS, Park JO, Srinivasarao M, Wright E, Conticello VP, Chaikof EL: **Viscoelastic and mechanical behavior of recombinant protein elastomers.** *Biomaterials* 2005,**26**(23):4695–4706.

28. Nagarajan R, Drew C, Mello CM: **Polymer–micelle complex as an aid to electrospinning nanofibers from aqueous solutions.** *J Phys Chem C* 2007,**111**(44):16105–16108.

29. Nagarsekar A, Crissman J, Crissman M, Ferrari F, Cappello J, Ghandehari H: **Genetic synthesis and characterization of pH- and temperature-sensitive silk-elastinlike protein block copolymers.** *J Biomed Mater Res* 2002,**62**(2):195–203.

30. Nagarsekar A, Crissman J, Crissman M, Ferrari F, Cappello J, Ghandehari H: **Genetic engineering of stimuli-sensitive silkelastin-like protein block copolymers.** *Biomacromolecules* 2003,**4**(3):602–607.

31. Ner Y, Stuart JA, Whited G, Sotzing GA: **Electrospinning nanoribbons of a bioengineered silk-elastin-like protein (SELP) from water.** *Polymer* 2009,**50**(24):5828–5836.

32. Novagen (EMD Biosciences): *pET System Manual.* 11th edition. Novagen (EMD Biosciences), Gibbstown, NJ; 2006.

33. Qiu W, Teng W, Cappello J, Wu X: **Wet-spinning of recombinant silk-elastin-like protein polymer fibers with high tensile strength and high deformability.** *Biomacromolecules* 2009,**10**(3):602–608.

34. Qiu W, Huang Y, Teng W, Cohn CM, Cappello J, Wu X: **Complete recombinant silk-elastinlike protein-based tissue Scaffold.***Biomacromolecules* 2010,**11**(12):3219–3227.

35. Queiros O, Casal M, Moradas-Ferreira P, Leao C: **PCR-based method for the construction of gene disruption cassettes in yeast.***Biotechniques* 2001,**31**(2):264–266. 268

36. Rabotyagova OS, Cebe P, Kaplan DL: **Protein-based block copolymers.** *Biomacromolecules* 2011,**12**(2):269–289.

37. Rasband WS: *ImageJ*. U. S. National Institutes of Health, Bethesda, Maryland, USA; 1997–2012. http://imagej.nih.gov/ij/

38. Reguera J, Lagarón JM, Alonso M, Reboto V, Calvo B, Rodríguez-Cabello JC: **Thermal behavior and kinetic analysis of the chain unfolding and refolding and of the concomitant nonpolar solvation and desolvation of two elastin-like polymers.***Macromolecules* 2003,**36**(22):8470–8476.

39. Rodriguez-Cabello JC, Martin L, Alonso M, Arias FJ, Testera AM: **"Recombinamers" as advanced materials for the post-oil age.***Polymer* 2009,**50**(22):5159–5169.

40. Rodriguez-Cabello JC, Girotti A, Ribeiro A, Arias FJ: **Synthesis of Genetically Engineered Protein Polymers (Recombinamers) as an Example of Advanced Self-Assembled Smart Materials.** In *Nanotechnology in Regenerative Medicine. Methods in Molecular Biology, vol 811*. Edited by: Navarro M, Planell JA. Humana Press; 2012:17–38.

41. Sambrook J, Russell DW: *Molecular cloning: a laboratory manual.* 3rd edition. Cold Spring Harbor Laboratory Press, New York; 2001.

42. Scheller J, Guhrs KH, Grosse F, Conrad U: **Production of spider silk proteins in tobacco and potato.** *Nat Biotechnol* 2001,**19**(6):573–577.

43. Studier FW: **Protein production by auto-induction in high-density shaking cultures.** *Protein Expr Purif* 2005,**41**(1):207–234.

44. Teng W, Cappello J, Wu X: **Recombinant silk-elastinlike protein polymer displays elasticity comparable to elastin.***Biomacromolecules* 2009,**10**(11):3028–3036.

45. van Hest JCM, Tirrell DA: **Protein-based materials, toward a new level of structural control.** *Chem Commun* 2001, **19:** 1897–1904.

46. Viitanen M, Vasala A, Neubauer P, Alatossava T: **Cheese whey-induced high-cell-density production of recombinant proteins in Escherichia coli.** *Microb Cell Fact* 2003,**2**(1):2.

47. Xia XX, Qian ZG, Ki CS, Park YH, Kaplan DL, Lee SY: **Native-sized recombinant spider silk protein produced in metabolically engineered Escherichia coli results in a strong fiber.** *Proc Natl Acad Sci U S A* 2010,**107**(32):14059–14063.

48. Xia X-X, Xu Q, Hu X, Qin G, Kaplan DL: **Tunable self-assembly of genetically engineered silk–elastin-like protein polymers.***Biomacromolecules* 2011,**12**(11):3844–3850.

Index

Index

45. van Hest JCM, Tirrell DA: **Protein-based materials, toward a new level of structural control.** *Chem Commun* 2001, **19:** 1897–1904.

46. Viitanen M, Vasala A, Neubauer P, Alatossava T: **Cheese whey-induced high-cell-density production of recombinant proteins in Escherichia coli.** *Microb Cell Fact* 2003,**2**(1):2.

47. Xia XX, Qian ZG, Ki CS, Park YH, Kaplan DL, Lee SY: **Native-sized recombinant spider silk protein produced in metabolically engineered Escherichia coli results in a strong fiber.** *Proc Natl Acad Sci U S A* 2010,**107**(32):14059–14063.

48. Xia X-X, Xu Q, Hu X, Qin G, Kaplan DL: **Tunable self-assembly of genetically engineered silk–elastin-like protein polymers.***Biomacromolecules* 2011,**12**(11):3844–3850.